Nuclear Science and Technology

This interdisciplinary book series publishes monographs and edited volumes in all areas of nuclear science and technology, covering developments in theoretical research, experiments, technology and applications. The series attempts to bring together experts and professionals from all over the world, to provide readers with a library on a wide range of topics in physics, engineering, medicine and environmental sciences. Topics include but are not limited to:

Nuclear physics and interdisciplinary research;
Nuclear energy science and engineering;
Nuclear fusion and high temperature plasmas;
Synchrotron radiation applications, beamline technology;
Accelerator, ray technology and applications;
Nuclear chemistry, radiochemistry, radiopharmaceuticals, nuclear medicine; and
Nuclear electronics and instrumentation.

The books are intended for researchers, professionals, and students in the field, as well as those interested in staying current with the latest advances.

Lei Yu · Jianli Hao · Wenzhen Chen

Nuclear Technology and Safety

中国原子能出版社

Springer

Lei Yu
College of Nuclear Science and Technology
Naval University of Engineering
Wuhan, Hubei, China

Jianli Hao
College of Nuclear Science and Technology
Naval University of Engineering
Wuhan, Hubei, China

Wenzhen Chen
College of Nuclear Science and Technology
Naval University of Engineering
Wuhan, Hubei, China

ISSN 2948-1856 ISSN 2948-1864 (electronic)
Nuclear Science and Technology
ISBN 978-981-95-2855-4 ISBN 978-981-95-2856-1 (eBook)
https://doi.org/10.1007/978-981-95-2856-1

Translation from the Chinese Simplified language edition: "Nuclear Technology and Nuclear Safety" by Lei Yu et al., © China Atomic Energy Publishing & Media Company Ltd. 2021. Published by China Atomic Energy Publishing & Media Company Ltd. All Rights Reserved.

The original submitted manuscript has been translated into English. The translation was done using artificial intelligence. A subsequent revision was performed by the author(s) to further refine the work and to ensure that the translation is appropriate concerning content and scientific correctness. It may, however, read stylistically different from a conventional translation.

This Springer imprint is published by the registered company Springer Nature Singapore Pte Ltd.
The registered company address is: 152 Beach Road, #21-01/04 Gateway East, Singapore 189721, Singapore

Preface

Since the beginning of the new century, with the significant advancement in nuclear power, construction of nuclear facilities, and application of nuclear energy in relevant countries worldwide, nuclear safety and nuclear technology have received unprecedented attention and application. After the Fukushima nuclear accident, many people, especially those working in nuclear-related institutions, have an urgent desire to acquire knowledge in this discipline.

In terms of nuclear technology, it can be said that it is an important component of modern science and technology, a major milestone in human civilization of the 20th century, and one of the most important high-tech fields in contemporary times. In recent years, although there have been some monographs or textbooks available for teaching in colleges and universities and for students' learning, these books are basically limited to specific fields of nuclear technology involved. There is a lack of learning materials that provide a comprehensive and systematic introduction to its connotations, applications, and development, and that are suitable for personnel in relevant nuclear-related units.

In the aspect of nuclear safety, the safe utilization of nuclear energy is currently one of the most important factors restricting the development of nuclear power. Nuclear safety issues are not only related to a country's nuclear power development and the building of nuclear capabilities, but also, in the event of a nuclear accident, will endanger the public and the environment, directly affecting a country's politics, economy, and diplomacy. The effective guarantee of nuclear safety is inseparable from high-quality nuclear-related talents and long-term education and awareness of nuclear safety. In response to the above issues, this textbook is intended to be compiled.

This book provides a relatively comprehensive introduction to the main knowledge content and application fields of nuclear technology and nuclear safety. In addition to taking the basics of nuclear physics as public foundational knowledge, it focuses on explaining the basic concepts, theoretical methods, and applications in nuclear technology and nuclear safety, aiming to meet the basic needs of relevant personnel in understanding nuclear technology and nuclear safety.

The book is divided into nine chapters, covering contents such as the application and development history of nuclear technology, basic concepts of nuclear technology, development history of nuclear safety, basic concepts and characteristics of nuclear safety, goals and basic principles of nuclear safety, fundamentals of nuclear physics, applications of nuclear technology, nuclear power technology and its applications, safety issues and safety design of nuclear reactors, safety assessment of nuclear reactors, nuclear safety regulation, radiation safety and protection, and nuclear safety culture.

In terms of compilation ideas, chapter arrangement, and content composition, this book adopts an open approach, emphasizing the scientificity and completeness of knowledge, as well as the systematicness and fundamentality of learning.

Wuhan, China

Lei Yu
Jianli Hao
Wenzhen Chen

Acknowledgements During the compilation of this book, relevant contents from the listed references have been incorporated. Doctors and masters such as Xi Chu, Mingrui Li, Songyang Ma, Junjie Ma, and Lei Ye have also done a great deal of proofreading and editing work. We would like to express our sincere gratitude to them all.

Contents

About the Authors

Lei Yu currently a professor at the College of Nuclear Science and Technology, Naval University of Engineering. Graduated with a Ph.D. from Naval University of Engineering in 2008. Primarily engaged in teaching and research in the field of nuclear reactor safety analysis. He has won a second-class National Science and Technology Progress Award. Published four textbooks, and received the Second Prize in the Inaugural National Excellent Textbook Award for one of his publications.

Jianli Hao currently an associate professor at the College of Nuclear Science and Technology, Naval University of Engineering. Graduated with a Ph.D. from Naval University of Engineering in 2013. Primarily engaged in teaching and research in the field of nuclear reactor safety analysis. Published one monograph and three textbooks, and received the Second Prize in the Inaugural National Excellent Textbook Award for one of his publications.

Wenzhen Chen currently a professor at the College of Nuclear Science and Technology, Naval University of Engineering. Graduated with a Ph.D. from Huazhong University of Science and Technology in 1994, and completed his postdoctoral studies at Shanghai Jiao Tong University in 1998. Primarily engaged in teaching and research in the field of nuclear reactor safety analysis. He has won two second-class Natural Science Awards from Hubei Province, published two monographs and four textbooks, and received the Second Prize in the Inaugural National Excellent Textbook Award for one of his publications.

Chapter 1
Introduction

1.1 The Application and Development History of Nuclear Technology

In 1896, French physicist Henri Becquerel discovered the natural radioactivity of uranium (U), laying a crucial scientific foundation for the birth of nuclear technology and promoting its rapid development. The exploration and utilization of nuclear energy and radiation that began in the late 1930s enhanced our understanding of atomic nuclei and the interactions between various types of radiation emitted by atomic nuclei and matter. This knowledge was applied to modern scientific research—exploring new fields, explaining new phenomena, and confirming new substances, gradually forming a new scientific technology rich in content and integrated with multiple disciplines, namely nuclear science and technology, or nuclear technology for short. Therefore, nuclear technology is the science that studies how to practically apply the laws of atomic nucleus changes revealed in nuclear science research, as well as their inherent and accompanying physical phenomena.

Nuclear technology is an important part of modern science and technology, a significant milestone in the history of human civilization in the twentieth century, one of the high technologies, and also an important symbol of social modernization. Nuclear technology is now widely used in various fields such as defense, scientific research, industry, agriculture, medicine, communication, transportation, environmental protection, and resource development, forming a relatively independent and complete research and application system. With its knowledge-intensive, cross-penetrating, irreplaceable, high-efficiency, and wide adaptability characteristics, it has been included in the world's high-tech competition arena. Currently, more than 150 countries worldwide are conducting research, development, and application of nuclear technology.

Its rapid development has become a driving force for continuous innovation and development of new technologies, new materials, new processes, and new methods in national defense and the national economy.

L. Yu et al., *Nuclear Technology and Safety*, Nuclear Science and Technology,
https://doi.org/10.1007/978-981-95-2856-1_1

Nuclear technology has become an indispensable important field in current national defense construction and national economic development, a strategic high ground contested by world powers, and an important direction for the priority development of powerful countries. In addition to nuclear power, nuclear technology plays an irreplaceable important role in industrial applications. Its level of application development not only affects the competitiveness of a country's industry but also becomes an important factor determining a country's comprehensive strength. Therefore, developed countries such as Europe, the United States, and Japan attach great importance to it, investing heavily in research, development, and mastery of its core technology. The development of nuclear technology application disciplines has also played a pivotal role in China's scientific and technological progress, economic construction, public health, national security, and other aspects.

Nuclear technology is generally divided into isotope and radiation technology (also known as non-power nuclear technology) and nuclear power technology. The nuclear technology mentioned in this book also includes the peaceful use of nuclear energy and other content.

From the discovery of natural radioactivity in 1896 to the present, the development of nuclear energy and nuclear technology has a history of more than 120 years. This development process can be roughly divided into three stages. The first stage, from 1896 to 1942, was the stage of basic nuclear research. During this stage, through the efforts of many scientists around the world, a series of major discoveries and breakthroughs in basic research were made, such as the discovery of radioactivity, the establishment of atomic nucleus structure models, the proposal of the isotope concept, the discovery of neutrons and fission reactions, etc. The research results of this stage laid the theoretical foundation for the later development of nuclear technology. After the end of World War II, nuclear technology began to be massively transferred to civilian economic applications. In terms of peaceful applications, since the Soviet Union built the first nuclear power plant in 1954, the application of nuclear technology has seen unprecedented development. The launch of nuclear-powered ships, the maiden voyage of atomic icebreakers, radiation breeding, the eradication of pests and diseases by radiation sterilization technology, the successful development of medical gamma cameras, the establishment of radioactive immunoassay technology, etc., have sprung up like bamboo shoots after a rain, and turned to solve urgent social problems, technical problems, and problems in biomedicine. For example, environmental pollution and protection, seawater desalination, energy development, bioengineering, cancer diagnosis and treatment, etc. By the 1980s and 1990s, nuclear technology had formed emerging industries in many areas. In developed Western countries, the application of nuclear energy and technology has penetrated into all areas of the national economy, the technology is becoming mature, and new progress is constantly being made. Entering the twenty-first century, with the wide application of computers, electronics, and other new materials and technologies in the field of nuclear technology, it has entered a new stage of development, its application fields are more extensive, and its role is getting bigger and bigger. The economic and social benefits obtained from the application of nuclear technology have a greater impact. The application of nuclear technology in the field of modern medicine alone has saved the lives of tens of thousands of people. The following gives a rough history of the development and application of nuclear energy and technology (Li et al. 1994).

(1) In 1896, the existence of natural radioactivity was discovered by mankind. In 1898, French scientists Pierre and Marie Curie discovered Polonium (Po) and Radium (Ra), and in 1902, they extracted 0.1 g of radium salt and a few milligrams of polonium. The discovery of radium led to significant transformations in science and philosophy, opening the door for humans to explore the mysteries of the atomic world. In the same year, Rutherford used a strong magnet to deflect uranium rays, discovering that the rays divided into two types of oppositely charged particles. One type, which was easily absorbed, was called alpha rays; the other, which had a certain penetrating power, was called beta rays. In 1900, the French scientist Villard observed that in addition to the two types of rays mentioned above, radium also emitted a third type of ray. This ray was not affected by magnetic fields, was very similar to X-rays, and had strong penetrating power, so it was called gamma rays. In 1905, Einstein creatively proposed the relationship between mass and energy conversion, i.e., $E = mc^2$, laying the theoretical foundation for the application of nuclear energy. In 1910, scientists proposed the concept of isotopes and for the first time carried out non-destructive testing using X-ray photography, marking the beginning of the application of nuclear technology.

(2) In 1923, researchers first used radioactive tracers for biological research. In 1930, German physicists Bothe and Becker bombarded the nuclei of beryllium, lithium, and boron with alpha particles, producing a mysterious ray that could cause a Geiger counter to discharge and had strong penetrating power. In 1932, British physicist James Chadwick suggested that this "ray" was a stream of neutral particles without charge, with a mass number and proton number equivalent, and named it "neutron". The discovery of the neutron became an important tool for scientists to conduct nuclear science research and greatly promoted the rapid development of nuclear science and technology. In 1936, activation analysis technology was first applied in practice; in 1947, people determined the age of archaeological excavations by measuring the content of the radioactive isotope Carbon-14.

(3) Around 1930, British physicists Cockcroft and Walton built the first particle accelerator. In 1931, American physicist Van de Graaff built a static accelerator, and American physicists Lawrence and Livingston designed and manufactured the first cyclotron for accelerating ions. In 1947, a traveling wave electron linear accelerator and a standing wave proton accelerator were built.

(4) In 1934, Joliot-Curie bombarded aluminum foil with alpha rays from polonium, found that the aluminum foil was radioactive, and separated Phosphorus-30 from the aluminum foil using chemical methods, marking the first discovery of artificial radioactivity. They also discovered in their experiments that uranium fission produces multiple neutrons and releases a large amount of energy, predicting that a chain reaction could be achieved.

(5) In 1938, German scientist Hahn achieved uranium fission by bombarding uranium nuclei with neutrons and found that the fission process released energy. In the same year, American physicist Bethe, who was born in Germany, and German astronomer Weizsäcker independently discovered fusion reactions, or "thermonuclear reactions".

(6) In 1939, Austrian physicist Meitner and her nephew Frisch discovered the phenomenon of uranium fission and measured a fission energy of around 200 MeV. In March 1941, people irradiated uranyl sulfate with accelerated neutrons and for the first time produced 0.005 g of another fissionable material, Plutonium-239.

(7) In January 1942, in a small courtyard under the west grandstand of an abandoned football field at the University of Chicago, Italian-American physicist Fermi designed and built the world's first uranium fission reactor, CP-1 (Chicago Pile-1). On December 2, 1942, humans completed the first self-sustaining fission chain reaction experiment, and the first natural uranium-graphite reactor was successfully started.

(8) On December 20, 1951, the United States used the residual heat from its first plutonium-producing breeder reactor to generate electricity, with an electric power of 100 kW. Although the United States began researching and building various types of reactors for power generation from the early 1950s.

(9) On June 27, 1954, the Soviet Union built the world's first nuclear power reactor using nuclear fuel to generate electricity, with an electric power of 5000 kW. In December 1957, the United States built the Shippingport nuclear power plant, which is similar to today's nuclear power plants.

(10) In 1957, the American Raychem Corporation first used an accelerator to irradiate and produce heat-shrinkable materials, marking the beginning of the radiation chemical industry. In the same year, the American Firestone Company first used an accelerator to irradiate and produce car tires using vulcanization technology.

(11) In 1961, Allis Chalmers and William Myers put the first commercial Anger camera into use at Ohio University. On July 21, 1969, the American Apollo 11 spacecraft landed on the moon. The two devices that provided heat for the spacecraft were Plutonium-238 isotope heaters. In 1970, the American Ford Company first used an electron accelerator for curing paint on car parts.

(12) In 1969, British engineer Hounsfield successfully designed the Computer Tomography (CT) scanner, which was introduced in 1972, improving the detection rate and diagnostic accuracy of lesions. In 1974, the first Positron Emission Tomography (PET) scanner was introduced and began to be used clinically in the late 1990s, greatly promoting the development of nuclear medicine. In 1976, John Keyes developed the first multi-purpose Single Photon Emission Computed Tomography (SPECT) scanner.

(13) In 1981, Mach published the first radiopharmaceutical monoclonal antibody for tumor imaging. In 1983, the first industrial CT scanner was introduced. In 1991, Cytogen Corporation was the first to introduce a radiopharmaceutical monoclonal antibody for tumor imaging approved by the FDA. In the

same year, the Qinshan Nuclear Power Plant in China successfully generated electricity.

(14) In 1992, Charpak won the Nobel Prize in Physics for inventing the multi-wire proportional chamber and drift chamber.

(15) In 1994, Brockhouse and Shull won the Nobel Prize in Physics for their development of neutron scattering techniques.

(16) In 1997, a seminar by the International Atomic Energy Agency concluded that it is technically feasible to use nuclear energy for seawater desalination, and an international nuclear energy seawater desalination advisory group was established.

(17) In 2001, the U.S. Department of Energy invested in the development of a smaller isotope microbattery that could power micro-mechanical devices of a few cubic millimeters.

(18) In 2002, Russia completed the preliminary research and development work for a floating nuclear reactor at sea and began to study specific construction issues.

(19) In 2003, American scientist Paul Lauterbur and British scientist Peter Mansfield were awarded the Nobel Prize in Physiology or Medicine for their groundbreaking achievements in the field of nuclear magnetic resonance imaging technology.

(20) In 2004, Japan was the first in the world to successfully develop a miniature X-ray source with a diameter of 5 mm and a length of about 2 cm. This device, equipped with an endoscope, can be used for medical irradiation and detection.

(21) In 2005, fusion energy made significant progress. All parties involved in the negotiations for the International Thermonuclear Experimental Reactor signed a joint statement and agreed to start construction in Cadarache, France. This decision marked an important stage in the development of fusion energy—the scientific and engineering demonstration of fusion technology under conditions related to the operation of a fusion power plant.

(22) In 2007, nuclear technology and isotope technology continued to make substantial contributions in the fields of agriculture, human health, marine and terrestrial environments, and water resource management. In the field of food and agriculture, plant mutation breeding is supporting the development of new crop varieties, greatly reducing the demand for fertilizers.

(23) In 2008, the European Organization for Nuclear Research conducted beam tests through the world's largest linear particle accelerator, creating a "miniature" state of the universe immediately after the Big Bang through large hadron collisions, exploring the mysteries of the Higgs boson, dark matter, dark energy, and other unresolved mysteries.

(24) In 2010, Russia's first floating nuclear power plant was launched.

(25) In 2011, the United States designed a miniature nuclear power plant for space exploration. In the same year, Japan announced the successful invention of a radiation detection plastic.

(26) In 2012, the U.S. Nuclear Regulatory Commission issued the first construction and operation license for a full-size laser enrichment facility to General

Electric-Hitachi Nuclear Energy. In the same year, significant progress was made in photon radiation therapy for cancer, which can provide precise doses of radiation to tumors while minimizing damage to surrounding normal tissues. The United States published the first document setting priority tasks for NASA's Space Technology Department with a broad public foundation—the "Space Technology Roadmap and Priorities". This report placed nuclear power sources and nuclear propulsion technology at the forefront of priority development technologies.

(27) In 2013, the United States conducted non-nuclear tests of nuclear rocket engines. Nuclear rocket engines use nuclear reactors to heat hydrogen to extremely high temperatures, and the expansion of hydrogen through the nozzle generates thrust. Compared with traditional chemical rocket engines, its thrust and efficiency are more than doubled.

(28) In 2015, the Max Planck Institute for Plasma Physics in Germany announced that the world's largest stellarator for studying nuclear fusion reactions, "Wendelstein 7-X", had begun operation and produced helium plasma for the first time.

(29) In 2016, the Oak Ridge National Laboratory in the United States significantly reduced the cost of uranium extraction from seawater through the development of amidoxime-based polyethylene fiber material, creating conditions for industrial application. This has become another possible way to produce natural uranium, and is expected to provide sufficient fuel security for the sustainable development of nuclear energy in the future.

(30) In 2017, the United States began thermal testing of the latest prototype of a space nuclear reactor power source, initiating a three-year technical demonstration project. This indicates that the United States has resolved the reliability issues of space reactor power sources and mastered some key technologies. In the same year, Russia overcame the bottleneck in the mass production technology of Nickel-63 isotope batteries. Japan's radiation-resistant robots successfully completed their first inspection inside the damaged reactor at the Fukushima nuclear accident site.

(31) In 2018, Russian scientists developed a new coating for nuclear reactor fuel elements. This coating can slow down the oxidation reaction between steam and zirconium at 1200 °C, making the production of accident-resistant fuel possible.

(32) In 2019, Russia's first floating nuclear power station, the Akademik Lomonosov, was put into operation. In October 2020, China's "Hualong One" achieved criticality for the first time. In the same month, Russia's most powerful nuclear-powered icebreaker, the "Arktika," was commissioned.

(33) On September 21, 2020, the proton beam energy of the superconducting cyclotron independently developed by China reached 231 MeV for the first time, marking an important milestone in the proton therapy system independently developed by China National Nuclear Corporation. It has gradually realized the localization of cyclotron equipment for the diagnosis and treatment of

major diseases such as malignant tumors, cardiovascular and cerebrovascular diseases in China.

(34) On January 30, 2021, Unit 5 of the Fuqing Nuclear Power Plant, the world's first Hualong One unit, was successfully put into commercial operation. On April 18, 2022, the Karachi K-2/K-3 project, the first overseas demonstration project of Hualong One in Pakistan, was fully completed and put into production. On August 19, 2024, the executive meeting of the State Council of the People's Republic of China decided to approve five nuclear power projects including the first phase project of Xuwei in Jiangsu Province. Up to now, the total number of Hualong One units in operation and under construction at home and abroad has reached 33, making it the third-generation nuclear power technology with the largest number of units in operation and under construction in the world.

(35) On August 25, 2023, the new-generation "artificial sun"—China's HL-2M Tokamak—achieved high-confinement mode operation under a plasma current of 1 million amperes for the first time, once again breaking the operation record of China's magnetic confinement fusion devices. The demonstration project of the Huaneng Shidao Bay High-Temperature Gas-Cooled Reactor Nuclear Power Plant in China was officially put into commercial operation in December 2023.

From the above, it can be seen that the application of nuclear energy and technology has ushered humanity into the nuclear age, bringing about fundamental changes. Human life and production are now inseparable from nuclear energy and technology. We have reason to predict that the future era of nuclear energy and technology will be the most brilliant and best era in human history. At that time, people will no longer be perplexed by energy resources. The impact of nuclear energy and technology on the future development of humanity is immeasurable.

1.2 Basic Concepts in Nuclear Technology

Nuclear technology is a comprehensive modern scientific technology based on nuclear physics, radiation physics, radiochemistry, radiation chemistry, and the interaction between nuclear radiation and matter. It uses ionizing radiation from radioactive isotopes and other forms of radiation interacting with matter to observe natural phenomena, reveal natural laws, solve scientific problems, and apply them practically.

The fields and scope involved in nuclear technology and its applications are very broad (Wang 2002). From a technical perspective, it can be divided into three aspects: radiation technology, isotope technology, and supporting technology, namely:

(1) Radiation technology: radiation processing, ion beam processing, nuclear analysis technology, nuclear imaging technology, radiation detection instruments;

(2) Isotope technology: isotope preparation technology, isotope products (or labeled compounds), radioactive drugs, radiometric dating, radioactive isotope energy, isotope tracing;

(3) Supporting technology: accelerator technology, reactor technology, nuclear radiation detection and nuclear electronics, radiation source technology, radiation dosimetry and radiation protection.

From an application perspective, it can be divided into: the application of nuclear technology in industry, life sciences, agriculture, environmental science, materials science, energy field, and nuclear instruments, etc. With its unique advantages, nuclear technology has further penetrated into other basic and applied fields on the basis of decades of development, forming some interdisciplinary or marginal disciplines, such as nuclear astronomy, nuclear archaeology, nuclear geology, nuclear oceanography, etc.

1.2.1 Elements, Nuclides, and Isotopes

(1) Elements

An element refers to a group of atoms with the same number of protons. It is generally composed of several isotopes, such as ^{1}H (Protium, P), ^{2}H (Deuterium, D), and ^{3}H (Tritium, T) are collectively referred to as the hydrogen (H) element.

(2) Nuclides

A nuclide refers to a type of atom or atomic nucleus with a specific number of protons and neutrons, such as ^{1}H nuclide, ^{2}H nuclide, ^{3}H nuclide. Nuclides are divided into stable nuclides and radioactive nuclides; nuclides that can spontaneously emit radiation are called radioactive nuclides. To date, more than 110 elements in the periodic table have been discovered, with about 2800 nuclides in total. Among them, less than 300 are stable or have extremely long lifetimes, and the rest are unstable nuclides. Among the naturally occurring radioactive nuclides (i.e., primordial and cosmogenic radioactive nuclides), there are only over 60 cosmogenic radioactive nuclides, and the rest are artificial radioactive nuclides produced by reactors or accelerators. Primordial radioactive nuclides are the initial radioactive nuclides that existed when the Earth was formed. Due to their long half-lives, they have not completely decayed yet, excluding radioactive nuclides produced by the decay of ^{238}U, ^{235}U and ^{232}Th.; cosmogenic radioactive nuclides refer to radioactive nuclides produced by the interaction of atomic nuclei in the atmosphere with radiation from the universe; artificial radioactive nuclides refer to radioactive nuclides produced artificially. In stable nuclides, there are strict limitations on the relative ratio of neutron number N and proton number Z.

(3) Isotopes

Isotopes are a class of nuclides with the same atomic number but different mass numbers, such as ^{1}H, ^{2}H, and ^{3}H, which are isotopes of each other. In a mixture of isotopes of an element, the ratio of the number of atoms of a specific isotope to the total number of atoms of the element is the abundance of that specific isotope.

Natural abundance refers to the abundance of a specific isotope naturally present in an element. For example, ^{234}U, ^{235}U, and ^{238}U are three isotopes of uranium, and their atomic percentages (abundance) in natural uranium are 0.0055%, 0.72%, and 99.274%, respectively.

Isotones: A class of atoms with the same number of neutrons and different numbers of protons.

Nuclear isobars: A class of atoms with the same mass number and different numbers of protons.

Nuclear isomers: Nuclides in the same isomeric state.

1.2.2 Rays and Radiation

Radiation is the phenomenon of energy being transmitted through waves or subatomic particles. Radiation can be classified into ionizing radiation and non-ionizing radiation. Ionizing radiation refers to radiation capable of causing ionization in matter. There are many types of ionizing radiation, including electromagnetic radiation (such as γ-rays and X-rays), charged particle radiation (such as α-rays, β-rays, electron beams, proton beams, deuteron beams, heavy ion beams, meson beams, etc.), and uncharged particle radiation (such as neutrons). Non-ionizing radiation refers to radiation with insufficient energy to cause ionization in matter, such as infrared rays and microwaves.

Gamma rays: A form of extremely high-energy electromagnetic radiation (electromagnetic waves), typically emitted when an atomic nucleus transitions from a higher energy state to a lower one (gamma decay). Gamma photons are neutral with zero rest mass, have wavelengths of 0.0001–0.001 nm, and possess strong penetrating power.

X-rays: Also known as Rontgen rays, these are high-energy electromagnetic waves produced by X-ray machines. Their wavelengths are longer than gamma rays, their range is slightly shorter, and their penetrating power is weaker than gamma rays.

Alpha rays: Essentially high-speed helium nuclei, alpha rays emitted by isotopes typically have energies of 4–7 MeV. They have strong direct ionization capability but weak penetrating power. For alpha particles produced by decay, a 10 cm thick air layer can completely absorb them, and a single sheet of ordinary printer paper can block them entirely.

Beta rays: Essentially high-speed electrons (traveling at nearly 1/4 to 1 times the speed of light), usually emitted from the nucleus during radioactive decay. Their energy is typically high and continuously distributed. Their range is much longer than alpha rays, with beta rays traveling several meters to tens of meters in air.

Neutrons: Uncharged particles with a mass slightly larger than protons. The type and probability of interactions between neutrons and matter depend strongly on the neutron's energy. Neutrons have strong penetrating power and indirect ionization capability.

Laser: A new light source developed in the 1960s. Lasers are also a form of electromagnetic waves with relatively long wavelengths. Due to their excellent directionality (with a deviation of only about 0.1°), high brightness per unit area, and good monochromaticity, they can cause resonant absorption in biological cells, leading to excitation or ionization of atoms and molecules, thereby inducing internal changes in organisms.

Cherenkov radiation: Electromagnetic radiation emitted when a charged particle moves through a medium at a speed greater than the speed of light in that medium.

1.2.3 Basic Parameters and Concepts

(1) Radioactivity

Radioactivity refers to the number of radioactive decays per unit time. In the International System of Units (SI), the unit of radioactivity is decays per second (decay/s), known as the Becquerel (Bq). The non-SI unit of radioactivity, the curie (Ci), is related to the Becquerel as follows: $1 \text{ Ci} = 3.7 \times 10^{10}$ Bq.

(2) Specific Activity

Specific activity refers to the radioactivity per unit volume or unit mass of a substance, also known as radioactive concentration. The radioactivity per unit volume is sometimes simply called volume-specific activity, with the SI unit Bq/m^3 and the commonly used unit Bq/L. The radioactivity per unit mass is sometimes called mass-specific activity, with the SI unit Bq/kg.

(3) Absorbed Dose

Absorbed dose refers to the amount of energy absorbed per unit mass of matter exposed to ionizing radiation. The SI unit is joules per kilogram (J/kg), and the specialized unit for absorbed dose is the gray (Gy), where 1 Gy = 1 J/kg.

Absorbed dose rate: The absorbed dose per unit time, with the unit Gy/s. Absorbed dose rate is often used to characterize the external radiation levels of neutrons, γ-rays, etc. When people say "the nuclear radiation in a certain place is very strong," they usually mean the absorbed dose rate is high.

(4) Dose Equivalent

Dose equivalent is the product of the absorbed dose at a point in tissue and the radiation quality factor at that point, representing the biological effect of 1 kg of biological tissue or the human body absorbing 1 J of radiation energy. The SI unit of dose equivalent is the sievert (Sv), where 1 Sv = 1 J/kg = 100 rem (a non-SI unit).

(5) Effective Dose

Different tissues and organs exhibit varying consequences when exposed to the same equivalent dose. The risk posed by radiation exposure depends not only on the type and energy of the radiation but also on the specific tissue or organ exposed. To comprehensively assess the potential risk to an organism, the equivalent doses to all tissues and organs are weighted and summed, resulting in the effective dose. The SI unit of effective dose is Sv.

(6) Nuclear Transition

Nuclear transition refers to the process by which a nuclear system changes from one quantum energy state to another. Examples include the transformation of one nuclide into another via α-decay or β-decay, or changes in nuclear energy levels through the absorption (or emission) of photons, orbital electrons, or electron pairs.

(7) Doppler Effect

The Doppler effect refers to the change in the observed wavelength of radiation due to the motion of the radiation source relative to the observer.

1.2.4 Nuclear Decay and Nuclear Reactions

(1) Nuclear Decay

Nuclear decay is one of the processes by which one atomic nucleus transforms into another. This physical process occurs extensively in nuclear reactors, where delayed neutrons are emitted by certain fission fragments known as "precursor nuclei" during nuclear decay. Among the approximately 2,000 nuclides discovered to date, natural nuclides are relatively few, with most being artificially created and unstable. Unstable nuclides undergo spontaneous decay and are called radioactive nuclides. Those existing in nature are termed natural radioactive nuclides, while those produced by nuclear reactions are called artificial radioactive nuclides. Types of radioactive decay include β-decay, α-decay, γ-decay, and transmutation.

β^--decay: The parent nucleus emits a negative electron (commonly called a β^--particle), accompanied by the emission of an antineutrino. The resulting nucleus has an atomic number (Z) increased by 1, while the mass number (A) remains unchanged. In this process, a neutron in the nucleus is converted into a proton.

β^+-decay: The parent nucleus emits a positron (commonly called a β^+-particle), accompanied by the emission of a neutrino. The resulting nucleus has an atomic number (Z) decreased by 1, while the mass number (A) remains unchanged. In this process, a proton in the nucleus is converted into a neutron.

α-decay: The parent nucleus emits a helium nucleus (commonly called an α-particle). The resulting nucleus has an atomic number (Z) decreased by 2 and a mass number (A) decreased by 4.

γ-decay: When a radioactive nucleus undergoes α-decay or β-decay, the resulting daughter nucleus is often in an excited state. Within an extremely short time, the daughter nucleus transitions from a higher excited state to a lower one or the ground state, emitting γ-rays. Both γ-rays and X-rays are forms of electromagnetic waves, but γ-rays have shorter wavelengths than X-rays.

Transmutation: The decay of an excited nucleus by emitting nucleons.

Internal conversion: A nuclear transition process competing with γ-radiation, where the excitation energy of the nucleus is directly converted into the excitation energy of shell electrons through electromagnetic interaction, leading to de-excitation.

Auger effect: A de-excitation process in which an excited atom emits an orbital electron instead of an X-ray due to an outer-shell electron filling an inner-shell vacancy.

Mossbauer effect: The recoilless emission and recoilless resonant absorption of γ-radiation.

Decay constant: The probability per unit time that a nucleus of a given nuclide will decay. It is a fixed constant and depends only on the nucleus itself.

Decay energy: The energy released in a given nuclear decay.

Decay scheme: A diagram detailing the energy levels, types of radiation, half-life, and other nuclear data for the decay of a radioactive nuclide.

Half-life: The average time required for half of the atomic nuclei in a sample to decay. It statistically reflects the rate of nuclear decay.

(2) Nuclear Reactions

Nuclear reaction: A general term for phenomena in which particles (including atomic nuclei) collide with atomic nuclei, causing changes in the mass, charge, or energy state of the nucleus.

Chain nuclear reaction: A reaction in which one of the products can induce similar reactions, allowing the process to continue.

Fusion reaction: Under certain conditions, two or more light nuclei combine to form a heavier nucleus, releasing enormous energy and possibly other particles in the process.

Nuclear fission: A heavy nucleus may become unstable after absorbing a neutron, splitting directly into two or more fragments, often accompanied by the emission of neutrons, γ-rays, and light charged particles.

Absorption reaction: The incident particle is captured by the target nucleus to form a compound nucleus, which may undergo radioactive decay, radiative capture, transmutation, or fission.

Scattering reaction: A reaction in which the incident particle interacts with the target nucleus, and the outgoing particle is of the same type as the incident particle. This can be divided into two cases:

(1) Elastic scattering: Momentum and kinetic energy are conserved in the reaction, and the internal state of the target nucleus remains unchanged. Elastic scattering can further be divided into resonant elastic scattering and potential scattering.
(2) Inelastic scattering: Momentum is conserved, but kinetic energy is not, with some of the kinetic energy converted into internal energy of the target nucleus, leaving it in an excited state. The nucleus then emits γ-rays to transition to a lower energy level or the ground state.

Spallation reaction: When the energy of the incident particle exceeds 100 MeV, no compound nucleus is formed. Instead, the incident particle may break the nucleus into a heavier nucleus and many lighter nuclei, such as α-particles, protons (p), or neutrons (n).

Threshold energy: The minimum kinetic energy (in the laboratory frame) of the incident particle required to induce a specific nuclear reaction.

Microscopic cross-section*: A measure of the probability of a specific interaction between an incident particle and a target particle. The unit is the barn (b), where 1 b $= 10^{-2}$ m^2, approximately equal to the cross-sectional area of a nucleon in an atom.

Fission fragments: The various nuclides produced by fission, each with a certain kinetic energy.

Fission yield: The fraction of a specific fission product produced in fission. Fission products with mass numbers near 90 and 140 typically have high fission yields.

1.2.5 Interaction of Radiation with Matter

Based on the type of radiation, interactions with matter can be classified as follows.

(1) Interaction of alpha Rays with Matter

The energy of α-particles emitted in nuclear decay is typically 4–7 MeV. Their primary interactions with matter include ionization, excitation, scattering, and nuclear reactions.

1) Ionization and Excitation

α-particles interact with the shell electrons of surrounding atoms via Coulomb forces, transferring energy to them. If an electron gains enough energy to escape the atom, becoming a free electron, it forms an ion pair with the resulting positive ion—a phenomenon called ionization. If the kinetic energy of these free electrons is sufficient, they can ionize other atoms. The former is called primary ionization, while the latter is secondary ionization. In air, creating one ion pair requires about 35 eV of energy. Thus, a 5 MeV α-particle can produce approximately 1.5×10^5 ion pairs in air. Therefore, as α-particles pass through matter, they create many ion pairs along their path. The number of ion pairs produced per unit distance is called specific ionization or ionization density. If the energy transferred to a shell electron is insufficient for it to escape the atom but enough to excite it to a higher energy level, leaving the atom in an excited state, the process is called excitation. The excited atom subsequently emits X-rays of a specific energy to return to the ground state. This excitation energy can also be transferred to an outer-shell electron, enabling it to escape the nucleus's binding and become a free electron (an Auger electron). This process is called the Auger effect.

2) Scattering

The scattering of alpha particles can be divided into elastic scattering and inelastic scattering, with elastic scattering being the most common. Experiments and theoretical calculations show that when alpha particles are incident perpendicularly on the scattering body, most of the alpha particles scatter at small angles after scattering, and the probability of large-angle scattering is relatively small.

3) Nuclear Reactions

Generally, the probability of alpha particles causing nuclear reactions is relatively small. They interact with elements such as Be, B, F, Li, Na, O, etc., to produce neutrons, which is currently the main method for preparing neutron sources in isotope production.

(2) Interaction of Beta Rays with Matter

The energy of beta particles emitted by radioactive nuclides undergoing beta decay is generally below 4 MeV. Within this energy range, the main forms of interaction between beta rays and matter are ionization, excitation, scattering, and secondary radiation.

1) Ionization, Excitation

When beta particles interact with matter, they can cause its molecules or atoms to ionize. However, the ionization value is several orders of magnitude smaller than that of alpha particles with the same energy. For example, a 5 MeV alpha particle produces about 3000 pairs of positive and negative ions per micron in water, while a 1 MeV beta particle only produces 5 pairs per micron.

The energy required for a beta particle to produce an ion pair is almost independent of its speed, and is about 32.5 eV in air.

2) Scattering

Beta particles are more easily scattered than alpha particles. They can be scattered not only by nuclei but also by orbital electrons.

3) Secondary Radiation

When beta particles are deflected due to scattering in the Coulomb field of atomic orbital electrons or in the Coulomb field and nuclear force field of atomic nuclei, their speed will change, accompanied by the emission of electromagnetic waves. This process is called bremsstrahlung. In addition to bremsstrahlung, characteristic X-rays can also be emitted when atoms in matter interact with electrons. The mechanism of characteristic X-ray production is as follows: fast electrons knock electrons from the K, L, M… shells of atoms out of the atom, creating vacancies in these shells. When outer shell electrons transition to inner shells, the energy difference between the two shells is emitted in the form of X-rays, which have a definite energy.

Furthermore, when fast electrons interact with matter, they can excite the valence electrons of its atoms to higher energy levels. When they return to the ground state, they emit visible light and ultraviolet light. These secondary radiations are collectively referred to as fluorescence.

(3) Interaction of Gamma (γ) Rays with Matter

The wavelength range of gamma rays is between 10^{-11} and 10^{-8} cm. There are more than a dozen forms of interaction with matter, mainly depending on the energy of the gamma rays. The main forms of interaction of gamma rays emitted during radioactive nuclide decay or X-rays emitted during inner shell electron transitions with matter are the photoelectric effect, Compton effect, and pair production.

1) Photoelectric Effect

The photoelectric effect refers to the phenomenon where an atom absorbs a gamma photon and transfers all its energy to an electron, causing the electron to be emitted from the atom. This electron is known as a photoelectron.

2) Compton Effect

The Compton effect refers to the phenomenon where a gamma photon collides elastically with a free electron in the outer shell of an atom, deflects by an angle θ, and the electron gains some energy and is emitted at a specific angle relative to the incident photon.

3) Pair Production

Pair production refers to the phenomenon where a gamma photon with energy greater than or equal to 1.02 MeV annihilates near an atomic nucleus to generate a pair of positrons and electrons. Part of the photon's energy is converted into the rest mass of the positron and electron, and the rest is converted into their kinetic energy.

1.2.6 Radioactive Analysis and Labeling

Radiochemistry is a branch of chemistry that studies radioactive substances. It includes the use of chemical methods to process irradiated or naturally occurring radioactive substances to obtain radioactive isotopes and their compounds, the application of chemical technology to nuclear research, and the use of radioactive substances to study chemical problems.

(1) Radioanalytical Chemistry

Radioanalytical chemistry is a discipline that performs trace analysis by measuring radioactivity or nuclear phenomena, also known as nuclear analytical chemistry.

Radioanalytical methods: Using radioactive isotopes and radiolabeled compounds as indicators, the content of the sample to be measured is determined by measuring its radioactivity. The radioanalytical method used in volumetric analysis is called radioactive titration.

Radiochemical analysis: After separating and purifying the sample using appropriate methods, the amount of radioactive material in the sample is determined by measuring its radioactivity. For example, the method of determining potassium content by measuring the radioactivity of the natural radioactive nuclide ^{40}K.

Isotope Dilution Method: A radioactive isotope or labeled compound, with a known specific activity and identical to the substance to be measured, is mixed evenly with the sample. A portion of this mixture is then purified and its specific activity is measured. The content of the substance to be measured is calculated based on the change in specific activity before and after mixing.

Activation Analysis: This method involves using nuclear reactions to transform stable isotopes in the sample into radioactive isotopes. The content of the substance to be measured is determined by measuring the radioactive activity.

Excited X-ray Fluorescence Analysis: When α, β, γ, or X-rays act on a sample, orbital electrons absorb part of their kinetic energy due to Coulomb scattering, causing the atom to be in an excited state. Characteristic X-rays are emitted when the atom returns from the excited state to the ground state. The type and content of elements are analyzed based on the energy and intensity of these characteristic X-rays.

Muon X-ray Fluorescence Analysis: When negatively charged muons (μ–) with a certain energy are shot into the sample, they are captured by the atomic nucleus due to the Coulomb force, forming muon atoms and releasing a series of characteristic X-rays, i.e., muon X-rays. This method is used to obtain the chemical composition and state of the sample.

Mössbauer Spectroscopy: This refers to nuclear γ-ray resonance spectroscopy under no recoil conditions.

Positron Annihilation Method: This method uses the annihilation lifetime of positrons to study the microscopic structure of matter.

Nuclear Magnetic Resonance Method: This method determines the molecular structure of a sample through nuclear magnetic resonance spectroscopy characteristics such as chemical shift, coupling constant, multiplicity, the width and intensity of absorption peaks, and temperature effects.

(2) Labeling

Labeling refers to the process where one or more atoms or chemical groups in the original compound molecule are replaced by easily recognizable atoms or groups. These easily recognizable atoms or groups in the labeled compound are called tracer atoms (or groups). Radioactive labeled compounds are compounds labeled with radioactive isotopes as tracers, such as Na ^{18}F, $^{14}CH_3COOOH$, where ^{18}F, ^{14}C are referred to as tracer atoms.

Stable Isotope Labeled Compounds: These are compounds in which the abundance of a stable isotope of an element is artificially changed to an observable level, such as $NH_2{}^{13}CH_2COOH$, $^{15}NH_3$, where ^{13}C, ^{15}N are referred to as tracer atoms.

Non-isotope Labeled Compounds: Under specific conditions, tracer atoms that are not isotopically related can replace certain atoms in the compound molecule, such as using ^{75}Se to replace the sulfur atom in the cysteine molecule to produce selenium-labeled cysteine.

Multi-labeled Compounds: These are compounds in which two or more types of tracer atoms are introduced into the molecule, such as $^{14}CH_3CH(^{15}NH_2)COOH$, $^{14}CH_3$ $CH(NH_2)^{13}COOH$, etc.

Positional Labeling: The labeled atom is located at a specific position in the molecule, and the number of labeled atoms is also fixed.

Uniform Labeling: The labeled atom is uniformly distributed in the labeled molecule.

Nominal Positional Labeling: The distribution of tracer atoms at specified positions is less than 95% of the total content of similar atoms in the compound. For such compounds, N or n can be used after their names.

1.2.7 Radiation Processing

Radiation Cross-linking: Under the action of radiation, the microscopic linear long chains of polymer materials produce free radicals. The free radicals between chains react or the polymer chains rearrange, forming a network T-shaped structure. The polymer chains change from a two-dimensional structure to a three-dimensional structure. This change caused by radiation is called radiation cross-linking.

Radiation Grafting: Radiation acts on the base material to produce free radicals, which react with unsaturated monomer molecules, grafting the monomer molecules onto the base material.

Radiation Polymerization: The polymerization system prepared with unsaturated monomers is irradiated to initiate free radical or ionic polymerization.

Radiation Degradation: Under the influence of radiation, free radicals are produced in polymer chains, triggering a rearrangement of the molecular chains. This rearrangement leads to the breakage of the main polymer chain, forming smaller molecular chains and altering the properties of the polymer.

Radiation Sterilization (also known as Radiation Disinfection): High-energy radiation is used to destroy the protein molecular structure and genetic material of bacteria or viruses, effectively killing the bacteria. This process is known as radiation sterilization.

1.2.8 Radiation Protection

All applications of nuclear technology and nuclear energy development activities carry the potential for releasing radioactive substances into the environment, posing risks to workers, the public, and the environment. If exposure exceeds a certain dose, it can be harmful to humans and the environment. Therefore, implementing radiation protection measures is crucial. Radiation protection is a significant branch of nuclear science and technology, studying how to minimize or prevent harm from ionizing radiation. Its primary task is to protect the health and safety of workers involved in radioactive work, the public and their descendants, and the environment, thereby promoting the development of nuclear energy. The main areas of study in radiation protection include radiation dosimetry, radiation protection standards, radiation protection technology, radiation protection evaluation, and radiation protection management.

Radiation protection is closely related to nuclear safety, with both supporting and complementing each other. Nuclear safety focuses on maintaining the normal operation of nuclear facilities, preventing accidents, and mitigating consequences after accidents to protect workers, the public, and the environment from radiation damage. The emphasis of radiation protection is on monitoring radiation levels, evaluating radiation effects, implementing radiation protection measures, and emergency response to ensure radiation doses do not exceed specified limits and to optimize radiation protection.

The radiation protection system is an essential part of the nuclear safety management system. It ensures safety and health by implementing strict preventive measures, preventing potential hazards, ensuring that the radiation dose received by workers and the public is below the limit, and maintaining radiation exposure at a reasonable, feasible, and as low as possible level, thereby ensuring health and safety. This is also an important manifestation of the principle of "ensuring safety."

1.3 Basic Concepts of Nuclear Safety

The concept of safety discussed here is actually a state of mind, a trust that individuals and the public have in the government, society, and businesses in their measures to address hazards or risks in certain areas (such as transportation, food, housing, finance, social security, natural disasters, etc.). The more comprehensive, thoughtful, and detailed these measures are, the higher the sense of safety individuals or the public have in these areas. Moreover, safety is an issue that humans have had to face since their inception. Due to the potential harm of radioactivity and its possible social, economic, and political impacts, nuclear safety has received widespread attention. As times change and technology advances, the public's understanding of nuclear safety continues to evolve, and the concept of nuclear safety continues to develop. Nuclear safety is a relative concept; absolute safety does not exist. Therefore, it is necessary to establish widely accepted nuclear safety goals. In the process of using nuclear energy and nuclear technology, it is necessary to adhere to the basic principles of nuclear safety to achieve nuclear safety goals and minimize damage to personnel, the environment, and society.

It should be noted that although countries around the world have gradually established a relatively complete regulatory system and standards in reactor design, operation, decommissioning, and radioactive waste management, nuclear emergency response, etc., due to the safety issues in nuclear energy systems such as reactivity control, heat dissipation, and radioactive containment, especially the occurrence of the Three Mile Island nuclear accident, the Chernobyl nuclear accident, and the Fukushima nuclear accident, humans are more concerned about nuclear safety than ever before.

The concept of nuclear safety has different definitions in different contexts and covers different scopes. To better understand nuclear safety, it is necessary to first understand the concept of safety. This section introduces the concept of nuclear safety and its related concepts starting from the definition of safety.

1.3.1 Safety

Safety refers to a state free from threats, danger, harm, and loss. The "Occupational Health and Safety Management System" (GB/T 28,001) defines safety as "a state free from unacceptable risk of harm." Therefore, safety can be understood as controlling risks to an acceptable level.

1.3.2 Nuclear Safety and Nuclear Accidents

(1) Nuclear Safety

Nuclear safety refers to the necessary and sufficient supervision, protection, prevention, and mitigation measures taken for nuclear facilities and nuclear materials to ensure their safety, prevent accidents caused by any technical reasons, human errors, or natural disasters, and minimize the radioactive consequences in case of accidents, thereby protecting the public, workers, and the environment from nuclear accident hazards.

The development of nuclear energy must prioritize nuclear safety. Nuclear safety is a crucial component of national security, holding a significant position within the overall national security system. It is not only related to environmental safety but also directly impacts public interest. Therefore, it is essential to implement necessary and sufficient monitoring, protection, and various accident prevention and mitigation measures for nuclear facilities, nuclear activities, nuclear materials, and radioactive substances. This is to prevent accidents and avoid causing unacceptable radioactive harm to the public and the environment.

In 2003, the International Atomic Energy Agency (IAEA) published the "Handbook on Nuclear Law" to assist member countries in establishing legal systems for the peaceful use of nuclear energy and ionizing radiation. The handbook proposes that the basic principles of nuclear law are the "Safety Principle" and the "Security Principle". The 2011 edition of the handbook further introduced the concepts of "Nuclear Safety", "Nuclear Security", and "Nuclear Safeguard", and used these concepts as the guiding principles of the handbook. According to the IAEA's definitions, nuclear safety refers to "achieving normal operating conditions, preventing accidents or mitigating their consequences, thereby protecting workers, the public, and the environment from improper radiation hazards"; nuclear security refers to "preventing, detecting, and responding to theft, sabotage, unauthorized access, illegal transfer, or other malicious acts involving nuclear materials and other radioactive substances or related facilities"; nuclear safeguard refers to "an agreement between the IAEA and one or more member countries that contains the country's or countries' commitment not to use certain items for any military purpose and authorizes the IAEA to supervise the fulfillment of this commitment". From these definitions, it can be seen that nuclear safety encompasses some aspects of nuclear security, and nuclear security overlaps with nuclear safeguard in some areas. Nuclear safety and nuclear security have a synergistic relationship in many fields. According to the definition of nuclear safety stipulated in China's "Nuclear Safety Law" enacted in 2017, it actually covers both nuclear safety and nuclear security (National People's Congress of China, 2017).

(2) Nuclear Accidents

A nuclear accident, as defined in the appendix of China's "Nuclear Safety Law", refers to radioactive, toxic, explosive, or other hazardous accidents caused by nuclear fuel, radioactive products, radioactive waste in nuclear facilities, or nuclear materials transported into or out of nuclear facilities, or a series of such accidents.

In the event of a nuclear accident, if the dedicated safety facilities cannot function as designed, the release of radioactive substances may reach unacceptable levels. The International Nuclear Event Scale (INES) classifies safety-significant events occurring within nuclear facilities into seven levels according to the severity of the nuclear accident. During the 60 years of peaceful use of nuclear energy by mankind, three major nuclear accidents have occurred: the Three Mile Island nuclear accident in the United States in 1979, the Chernobyl nuclear accident in the Soviet Union in 1986, and the Fukushima nuclear accident in Japan in 2011. Nuclear accidents are characterized by their suddenness, complexity in handling, severity of consequences, and public sensitivity. The development of the nuclear industry is not absolutely safe, but relatively safe. However, as human understanding and corporate technology continue to improve, and national regulatory capabilities continue to increase, and lessons are learned from past nuclear accidents and events, safety improvements can be made, and nuclear safety can be effectively guaranteed.

1.3.3 Other Related Concepts

To better understand nuclear safety, the following introduces some basic concepts related to it.

(1) Risk

As mentioned earlier, the development of the nuclear industry is not absolutely safe, but relatively safe. For governments, nuclear power builders, and the public, a reasonable measure is needed to determine an acceptable safety level for nuclear facilities. To compare the safety levels of different things, a safety measure needs to be established, and risk is usually used as a safety measure.

Risk refers to the combination of the probability of a specific dangerous event (accident or unexpected event) occurring and the consequences it produces. It is a combination of two factors: the probability of the danger occurring, i.e., the danger probability; and the consequences produced after the dangerous event occurs.

As the environment in which humans live changes, people's safety judgments about a thing will also change, and the level of risk that people can accept will also change. Nuclear safety issues are a specific case of all safety issues and must be thought about in the same way.

In the field of nuclear safety, the IAEA's definition of risk includes: ① a multi-attribute quantity that represents the probability of hazards, dangers, or damages or injuries related to actual or potential exposure, involving the probability of producing specific harmful consequences and the severity and characteristics of such consequences; ② the mathematical mean (expected value) of an appropriate measure for a specific consequence (usually undesirable); ③ the probability of a specific health effect occurring in an individual or group due to radiation exposure.

(2) Nuclear Facilities

In China's "Nuclear Safety Law", nuclear facilities refer to:

1) Nuclear power plants, nuclear thermal power plants, nuclear steam and heating plants, and other nuclear power plants and devices;
2) Research reactors, experimental reactors, critical assemblies, and other reactors outside of nuclear power plants.
3) Nuclear fuel cycle facilities, such as nuclear fuel production, processing, storage, and post-processing facilities;
4) Facilities for the treatment, storage, and disposal of radioactive waste.

Additionally, the "Nuclear Safety Law" defines radioactive waste as waste generated from the operation and decommissioning of nuclear facilities, which contains radioactive nuclides or is contaminated by radioactive nuclides, and its concentration or specific activity exceeds the clean release level determined by the state, and is expected to be no longer used.

(3) Nuclear Materials and Spent Fuel

In China's "Nuclear Safety Law", nuclear materials refer to:

1) Uranium-235 materials and their products;
2) Uranium-233 materials and their products;
3) Plutonium-239 materials and their products;
4) Other nuclear materials that need to be regulated as stipulated by laws and administrative regulations.

The IAEA's interpretation of nuclear materials is: Pu (except when the concentration of ^{238}Pu exceeds 80%); ^{233}U-; enriched ^{233}U or enriche^{235}U; uranium in forms other than ore or slag that contains naturally occurring isotopes; any material containing one or more of the above components.

Spent fuel refers to nuclear fuel that has been irradiated in the reactor core and permanently removed from the core.

(4) Inherent Safety

Inherent safety is a concept related to safety.

In IAEA terminology, inherent safety refers to the elimination of specific hazards through the selection of materials and design concepts, such as the use of fire-resistant materials to achieve fire safety. For reactors, inherent safety means that when the reactor experiences abnormal conditions, it relies solely on the reactor's natural safety, backup safety, and passive safety, without any mandatory intervention from external equipment or human operation, to remove core heat and control reactivity, ensuring safe operation and shutdown of the reactor.

Active safety refers to the need for active equipment (powered equipment) to achieve the functionality of the safety system, i.e., it requires external conditions to achieve the expected functionality. For example, after an accident shutdown, the active residual heat removal system needs the cooling pump to continue running during the core cooling phase, and the operation of the pump depends on the power supply. Similarly, the electric auxiliary feedwater pump is powered by an emergency diesel engine, and its function also relies on external power to be realized.

Redundant safety refers to the setting up of multiple systems through backup redundancy to prevent the failure of a safety system. Any one of these systems can achieve the predetermined safety function.

Passive safety refers to the use of natural scientific laws or principles such as the principle of inertia, the law of gravity, the law of heat transfer, and the negative reactivity feedback characteristics of the core to achieve the safety function of the reactor. For example, the passive residual heat removal system on the secondary side, in the event of a total power outage, uses the natural circulation of the steam generator to remove the residual heat from the core. This is passive safety. With passive safety features, the reactor can control reactivity and remove core heat using the natural properties and laws of the system design, without relying on human operation or mandatory intervention from external equipment, when abnormal conditions occur, leading the reactor towards normal operation or safe shutdown. The concept of passive safety was proposed in the 1980s and is one of the important markers of advanced reactors today.

(5) Operation, Shutdown, and Decommissioning of Nuclear Facilities

1) Operation

Operation can be divided into normal operation and abnormal operation (anticipated operational occurrences).

Normal operation: Operation within the prescribed operational limits and conditions. For nuclear power plants, this includes startup, power operation, shutdown, closure, maintenance, testing, and refueling.

Anticipated operational occurrences: Operational processes that are expected to occur at least once during the facility's operational lifetime, but will not cause significant damage to safety-important items or lead to accident conditions due to appropriate measures taken in the design. Examples of anticipated operational occurrences include normal power outages, turbine trips, and main cooling pump power outages.

2) Shutdown

Shutdown refers to when a nuclear facility has stopped operating and will not be restarted.

3) Decommissioning

Decommissioning refers to the measures taken to decontaminate, dismantle, and clear a nuclear facility so that the radiation dose at the site or equipment no longer in use meets the requirements of national standards.

Decommissioning of nuclear facilities is the action that should be taken after a nuclear facility has reached or exceeded its design life or has ceased operation for other reasons. Through decommissioning, the harmful effects of residual radioactive substances and other harmful substances can be reduced to an acceptable level, achieving limited use or unrestricted opening and use of the site.

Using a nuclear power plant as an example, there are three forms of decommissioning: first, immediate dismantling and removal of radioactive materials from the reactor after ceasing operations; second, sealing the reactor for several decades until its radioactivity naturally decays before dismantling; third, constructing a concrete shell around the reactor for long-term containment.

(6) Nuclear Emergency Response and Nuclear Safety Culture

1) Nuclear Emergency Response

In IAEA terminology, an emergency refers to unconventional situations that require immediate action, primarily to mitigate harm or adverse effects on human health and safety, quality of life, property, or the environment. A nuclear or radiological emergency refers to urgent situations caused by nuclear chain reactions or the decay energy or radiation exposure of chain reaction products that have caused or are expected to cause harm. The Chinese "National Nuclear Emergency Plan" describes it as "a state requiring immediate actions beyond normal work procedures to prevent accidents or mitigate their consequences, sometimes also referred to as an emergency state; it also broadly refers to immediate actions beyond normal work procedures."

2) Nuclear Safety Culture

The concept of nuclear safety culture was first proposed by the IAEA. After the Chernobyl nuclear accident, the International Nuclear Safety Advisory Group's "Summary Report on the Post-Chernobyl Review Meeting" (INSAG-1) first introduced the term "Safety Culture". In 1988, the IAEA further elaborated on the safety culture in its "Basic Safety Principles for Nuclear Power Plants" report, emphasizing it as a fundamental management principle for nuclear safety. In 1991, the IAEA's "Safety Culture" discussed the definition, characteristics, and essence of safety culture, elaborated on its requirements for decision-making, management, and individual levels, and proposed influencing issues and qualitative indicators to measure different levels of safety culture (IAEA, 1991). "Safety Culture" provided the first definition of nuclear safety culture as:

Nuclear safety culture is that assembly of characteristics and attitudes in organizations and individuals which establishes that, as an overriding priority, nuclear plant safety issues receive the attention warranted by their significance.

1.4 Development of Nuclear Safety

Due to the production of unstable isotopes and ionizing radiation resulting from changes in the internal structure of atomic nuclei in nuclear technology applications, there is potential for significant harm to human health, life, and the environment. Especially since nuclear technology was initially used for military purposes, the shadow of atomic bomb explosions has long lingered in people's memories. Therefore, nuclear and radiation safety has received special attention worldwide. To ensure nuclear and radiation safety and gain public trust, countries around the world have adopted a series of technical and management measures in the research, development, and application of nuclear technology, and in the design, construction, operation, and decommissioning of nuclear facilities. They have developed a complete set of theories and principles, promulgated and strictly implemented in the form of laws and regulations. Nuclear and radiation safety, as a new interdisciplinary field, can be defined as: all theories, principles, technical measures, and management measures taken to reduce the adverse effects of radiation produced during the application of nuclear technology or the operation and decommissioning of nuclear facilities on workers, the public, and the environment to an acceptable level, thereby gaining public trust. The above definition includes three aspects:

(1) Scope: Nuclear and radiation safety should be integrated into all stages of research, development, and application of nuclear technology, and all stages of design, construction, operation, and decommissioning of nuclear facilities.
(2) Purpose: The purpose of nuclear and radiation safety is to reduce the adverse effects of radiation produced during the application of nuclear technology or the operation and decommissioning of nuclear facilities on workers, the public, and the environment to an acceptable level, thereby gaining public trust.
(3) Content: The content of nuclear and radiation safety should cover all theories (such as deterministic and probabilistic), principles (such as defense in depth, nuclear safety culture), all technical measures (such as equipment grading and identification, operating limits and conditions) and management measures (such as reporting system, license application and supervision).

Nuclear safety and radiation safety have different emphases. Nuclear safety focuses on maintaining the normal operation of nuclear facilities, preventing accidents, and mitigating their consequences, thereby protecting workers, the public, and the environment from radiation damage. Radiation safety focuses on monitoring radiation levels, evaluating radiation effects, implementing radiation protection measures, and emergency response and intervention, to achieve optimized radiation protection and ensure radiation doses do not exceed specified limits.

The International Atomic Energy Agency collectively refers to nuclear safety, radiation safety, radioactive waste safety, and radioactive material transport safety as nuclear safety, which is the broad concept of nuclear safety. In this book, the concept of nuclear safety mostly refers to the broad concept of nuclear safety and radiation safety.

Nuclear safety and radiation safety, as an interdisciplinary field, involves more than ten specialties or disciplines, such as nuclear physics, radiochemistry, nuclear materials, nuclear equipment, nuclear instrumentation, radiation protection, nuclear medicine, geology, seismology, meteorology, hydrology, law, sociology, psychology, and others.

Currently, nuclear power has become the main industry of nuclear technology application. Therefore, this book often adopts many concepts of nuclear power safety when explaining nuclear safety concepts, but its key points are equally applicable to other nuclear industries.

Due to the potential harm of nuclear radiation to humans and the environment, many basic principles emphasizing safety were established at the beginning of nuclear power development, especially emphasizing the impossibility of nuclear accidents from the design and equipment reliability. For example, the principle of indepth defense against radioactive releases in design, inherent safety and fault safety principles, the single fault criterion of safety combinations, and the principles of redundancy and diversity of safety systems. At the same time, strict requirements have been put forward for design, construction, and operation, such as conservative design, strict site requirements, and strict quality assurance. In addition, the maximum credible accident is determined as the design basis accident, and dedicated safety facilities are set up. To ensure that the dedicated safety facilities can cope with these design basis accidents, conservative assumptions and calculation models are used in safety analysis, and acceptance criteria for various design basis accidents are stipulated.

The history of nuclear power development shows that the above design principles, especially the conservative design and emphasis on equipment reliability, are completely correct for ensuring nuclear power safety.

On March 28, 1979, a partial core meltdown occurred at Unit 2 of the Three Mile Island nuclear power plant in Pennsylvania, USA. The cause of the Three Mile Island nuclear accident was a common initial event—the loss of main feedwater to the steam generator. According to the design, after the loss of main feedwater to the steam generator, the auxiliary feedwater system automatically starts. Due to the closure of two valves on the auxiliary feedwater pipeline during maintenance, the water could not enter the steam generator after maintenance, resulting in the loss of cooling on the primary side of the steam generator, and the temperature and pressure in the reactor increased, causing the pressure stabilizer relief valve to automatically open for pressure relief. The pressure stabilizer relief valve should have automatically reseated after pressure relief, but due to equipment failure, it did not reseat, resulting in the loss of coolant in the reactor. At this time, the safety injection system automatically started. The operators made a wrong judgment based on the high water level signal of the pressure stabilizer and cut off the safety injection system, resulting in the fuel elements in the reactor not being cooled, leading to a partial core meltdown. The Three Mile Island nuclear accident shows that although there were maintenance errors and equipment failures, if there was no erroneous intervention by the operators, the safety system would have automatically started according to the design, and the partial core meltdown could have been avoided. After the Three Mile Island nuclear accident, in order to reduce human errors, countries around the world have made a lot of improvements in strengthening operator training, improving human–machine interfaces and control room design, and a lot of research has been carried out on the prevention and mitigation of severe accidents. Many of the results of these efforts have been incorporated into nuclear safety regulations and standards and improvements in nuclear power plants, greatly improving the level of nuclear safety.

On April 26, 1986, a strong explosion occurred at Unit 4 of the Chernobyl nuclear power plant in the Ukrainian Soviet Socialist Republic, and a large amount of radioactive material leaked from the nuclear power plant, spreading through smoke to parts of the Soviet Union and Europe, causing serious environmental pollution, a large number of people were evacuated and caused social panic, affecting the development of nuclear power in the world for a period of time. The main reason for the Chernobyl accident was serious design flaws in the type of nuclear reactor used at the power plant. The experimental program executed by the operators was not well thought out and violated operating procedures, which also contributed to the accident. However, the root cause should be attributed to the lack of safety culture in the Soviet nuclear power authorities, because the design flaws of this type of reactor were well known, had been discovered during the commissioning of other similar type nuclear power plants, and a special report had been written to the relevant authorities, but it did not attract the attention of the authorities and relevant parties. The lessons of the Chernobyl nuclear accident are profound. Since the 1990s, emphasizing the importance of safety culture has become an important issue for countries around the world to strengthen nuclear safety.

On March 11, 2011, a magnitude 9.0 earthquake occurred off the east coast of Honshu, Japan. After the earthquake, Units 1, 2, and 3 of the operating Fukushima Daiichi Nuclear Power Plant and all four units of the second nuclear power plant successfully achieved automatic "shutdown" by inserting control rods located at the bottom of the reactor core through the hydraulic system, indicating that the nuclear power plant had reached the design standard for beyond design basis earthquake accidents. The earthquake caused the loss of external power and plant power at the nuclear power plant, and the emergency diesel generator was immediately started to supply power to cool the fuel assemblies in the reactor core. The plant has a total of 13 emergency diesel generators, including 9 water-cooled generators and 4 air-cooled generators. Before the earthquake, 12 were on standby and 1 was under repair. After the earthquake, 2 air-cooled generators could not be used, 1 air-cooled generator and all water-cooled generator sets were operating normally. More than an hour after the diesel generator supplied power, a strong tsunami with a height of over 14 m submerged the generators located at an altitude of 13 m. The diesel generators located at a high altitude were completely submerged, rendering them unable to continue supplying power. The only source of power was a backup battery pack, but its power supply time was extremely limited. Subsequently, three mobile diesel generators were brought in, but they could not be connected due to incompatible interfaces. With the thermal cooling system shut down due to power failure, the entire reactor was in a state of high-temperature evaporation. The reaction of metal components with water produced hydrogen gas, which caused a chemical explosion and fire under appropriate concentration and high-temperature conditions. The safety protection facilities were severely damaged, the factory building was unrecognizable, and a large amount of radioactive material leaked to the outside world. The Fukushima nuclear accident was attributed to both the sudden major earthquake and tsunami, and human factors. It was a disaster caused by both "natural disasters" and "man-made disasters". The Fukushima nuclear power plant was nearing its lifespan, some components were aging, and there were safety risks. The operating unit did not properly inspect and maintain the nuclear power unit equipment and components, and some key equipment could not function after the accident. These three nuclear accidents have prompted people to analyze and think about nuclear power development planning, nuclear power plant site selection, nuclear power plant design and operation, nuclear technology research and development, nuclear safety culture, and nuclear power talent training.

From adopting conservative designs and emphasizing equipment reliability, to reducing human errors and considering the prevention and mitigation of severe accidents, to emphasizing the importance of safety culture and the application of probabilistic safety analysis methods, as well as continuous reflection on the understanding of nuclear safety and the safety design of nuclear power plants, new safety concepts, safety requirements, safety methods, and technologies have emerged. This reflects the continuous deepening and improvement of nuclear safety and radiation safety concepts.

The occurrence of the aforementioned three severe nuclear accidents has also continuously drawn people's attention to the importance of nuclear safety, deepened the understanding of reactor safety risks, and promoted the continuous improvement of nuclear safety technology. The safety performance of reactors has also been continuously improved. The progress of safety technology is mainly reflected in the design of safety systems.

(1) Development of Passive Safety Technology

Current passive safety technologies mainly include: natural circulation technology, passive residual heat removal technology, passive safety injection technology, passive containment cooling technology, etc.

1) Natural circulation technology refers to the circulation of fluid in a closed system under the action of gravity, without relying on external power sources, and only using the driving force generated by the density difference between the cold and hot source fluids. Natural circulation technology is an important basis for passive safety design. For naval reactors, by improving the natural circulation capability of the reactor, a certain power output can be maintained by using circuit natural circulation without starting the main coolant pump, maintaining natural circulation operating conditions.
2) Passive residual heat removal technology refers to the use of multiple circuit natural circulations to smoothly carry the decay heat of the reactor to the final heat sink without any external power, thereby achieving the safe cooling function of the reactor core in the event of total power loss and other extreme accidents. For example, the passive residual heat removal system of a marine nuclear reactor uses a circuit natural circulation to carry the decay heat of the reactor core to the steam generator, and uses the natural circulation circuit on the secondary side of the steam generator to carry energy to the passive residual heat removal cooler, and then uses the natural flow of seawater to carry energy into the sea.
3) Passive safety injection technology refers to the injection of cooling water into the reactor core by relying on gravity potential difference, energy storage pressure

difference, and other passive devices after the reactor has lost coolant, achieving re-submergence of the reactor core under water loss accidents.

4) Passive containment cooling technology uses the density difference formed by the heating or evaporation, cooling or condensation of the fluid (including gas) to form a driving pressure head or a gravity pressure head formed by the potential difference, without any external power, to achieve containment cooling.

(2) Application of Passive Safety Systems

The application of passive safety systems can greatly simplify system design, reduce system operation failure probability; improve system safety, reduce the probability of core meltdown by 1~2 orders of magnitude; and by reducing active equipment, reduce the requirements for emergency power sources, also reduce equipment in-service inspection and maintenance, improve system economy. Compared with nuclear power units of the same level, the AP1000 nuclear power unit, which extensively applies passive safety technology, has reduced its system valves by 50%, pumps by 35%, pipelines by 80%, seismic structures by 45%, cables by 70%, and the frequency of reactor core meltdown and large-scale radioactive release is 2 ~ 3 orders of magnitude lower than the second-generation nuclear power units. The time after a severe accident without the need for human intervention has been extended to 72 h, which can effectively reduce the probability of human error.

The superiority of passive safety technology is rapidly promoting its application in nuclear energy systems. Currently, almost all advanced pressurized water reactors being built and developed internationally have applied passive safety technology design.

1.5 Objectives and Basic Principles of Nuclear Safety

To protect humans and the environment from radioactive hazards and mitigate the adverse effects of nuclear safety issues on the public and society, it is necessary to establish a reasonable nuclear safety objective. The nuclear safety objective must be implemented throughout the entire process of site selection, design, construction, and operation of nuclear facilities, ensuring that site selection and plant status adhere to universally accepted health and safety principles. Nuclear safety objectives can be qualitative and quantitative. Considering the complex factors affecting nuclear facilities and their uncertainties, the higher the level of safety objectives, the more likely they are to be qualitative. The safety objectives of nuclear facilities are generally considered to have three: one is the overall objective of nuclear safety (basic safety objective), and the other two are auxiliary objectives that explain the overall objective, which are radiation protection objectives and technical safety (probabilistic safety) objectives. The three are not isolated from each other but are interconnected to ensure the integrity of nuclear safety objectives.

After the Three Mile Island nuclear accident and the Chernobyl nuclear accident, the IAEA published the "Basic Safety Principles for Nuclear Power Plants" in response to the two nuclear accidents, comprehensively describing nuclear safety objectives from the levels of basic safety objectives, radiation protection objectives, and technical safety objectives.

1.5.1 Safety Objectives

(1) Overall Nuclear Safety Objective

The overall nuclear safety objective refers to establishing and maintaining effective defense against radioactive hazards to protect workers, the public, and the environment from radioactive harm. The overall nuclear safety objective is supported by radiation protection objectives and technical safety objectives. These two objectives complement each other and, together with technical measures and managerial and procedural measures, ensure defense against radioactive hazards.

It should be noted that while the overall nuclear safety statement highlights radioactive hazards, this does not mean that nuclear facilities do not pose other hazards common to general industrial production activities, such as the impact of thermal emissions on the environment and economic losses caused by equipment damage. These hazards are also worth paying attention to, but to highlight the uniqueness of nuclear facilities, they are not within the scope of nuclear safety research, so the overall objective of nuclear safety does not focus on these hazards.

(2) Radiation Protection Objective

The radiation protection objective is to ensure that under all operating conditions of nuclear facilities, radiation exposure or radiation exposure caused by any planned discharge of radioactive substances remains below the specified limit and is as low as reasonably achievable. If a nuclear accident occurs, it should ensure that all accident radiation consequences are mitigated.

The radiation protection objective requires a complete set of radiation protection measures under normal operation of nuclear facilities, and a set of measures to mitigate accident consequences under anticipated operating event conditions and any accident conditions, including on-site and off-site countermeasures, to protect workers, the public, and the environment from radioactive hazards as much as possible.

(3) Technical Safety Objective The technical safety objective refers to taking all reasonably practicable technical means or measures to prevent accidents and mitigate their consequences once they occur; for all possible accidents considered during design, including those with a very low probability, it is necessary to ensure with high reliability that the consequences are as small as possible and below the specified limit, and that the probability of accidents with severe radioactive consequences is extremely low.

1.5.2 Quantitative Safety Objectives

The above are qualitative objectives. To better implement them, relevant countries have also established quantitative objectives.

(1) United States

On August 21, 1986, the U.S. Nuclear Regulatory Commission (NRC) issued a policy statement on nuclear power plant operation safety objectives, proposing two qualitative safety objectives:

1) The operation of nuclear power plants will not bring obvious additional risks to the life and health of the public;
2) The societal risk to the public's life and health from the operation of nuclear power plants is lower than or comparable to that seen from currently competitive other power generation technologies. Compared with other societal risks, it will not bring obvious additions.

At the same time, in support of the above two qualitative safety objectives, two quantitative safety objectives were proposed:

1) The average acute death risk to the public from the operation of nuclear power plants is less than 0.1% of the total acute death risk from other accidents;
2) The risk of cancer death to the public near the nuclear power plant from the operation of the nuclear power plant is less than 0.1% of the total public cancer death risk from all other causes.

To establish the design basis for advanced light water reactors, in 1985, with the participation of the U.S. Department of Energy (DOE), the Electric Power Research Institute (EPRI) compiled the U.S. Nuclear Power Plant User Requirements Document (URD), which was officially published in 1991. The user requirements document proposes: the probability of core damage is less than 10^{-5}/reactor-year; the probability of large-scale radioactive release is less than 10^{-6}/reactor-year.

(2) International Atomic Energy Agency

The operational safety objectives for nuclear power plants proposed in the report (INSAG-12) published by the International Atomic Energy Agency's Nuclear Safety Advisory Group in 1999 are:

1) The probability of core meltdown is 10^{-4} /reactor-year (for existing nuclear power plants) and 10^{-5} /reactor-year (for future nuclear power plants);
2) The probability of a large-scale radioactive release is 10^{-5} /reactor-year (for existing nuclear power plants) and 10^{-6} (for future nuclear power plants).

(3) China

In May 2002, the China Nuclear Safety Administration issued a policy statement "Technical Policies on Several Important Safety Issues in the Design of New Nuclear Power Plants", proposing the safety objectives for the operation of new nuclear power plants:

1) the probability of core meltdown is 10^{-5} /reactor-year;
2) the probability of a large-scale radioactive release is 10^{-6} /reactor-year.

1.5.3 Nuclear Safety Policies and Principles

Around the above nuclear safety objectives, a set of basic principles must be implemented throughout the entire process of site selection, design, construction, and operation of nuclear facilities. The basic principles of nuclear safety are universally applicable rules that must be followed to achieve nuclear safety objectives and are the foundation of specific safety principles.

(1) Safety Policy

Article 4 of Chapter 1 of China's "Nuclear Safety Law" stipulates that those engaged in nuclear activities must follow the policy of ensuring safety.

Only the development of nuclear energy that ensures safety is sustainable. Article 10 of the "Convention on Nuclear Safety" compiled by the International Atomic Energy Agency stipulates that "each contracting party should take appropriate steps to ensure that all organizations directly involved in activities related to nuclear facilities establish the necessary priority policies for nuclear safety". Both international laws and China's existing laws and regulations on nuclear safety emphasize that safety is the most important condition for the development and utilization of nuclear energy and technology. In practice, both the nuclear safety supervision and management departments and units involved in nuclear safety-related activities regard safety as a prerequisite for carrying out nuclear energy development and utilization activities. "Ensuring safety" has become a policy consistently adhered to in China's nuclear safety supervision and nuclear energy development and utilization activities.

The policy of "ensuring safety" emphasizes that the development and utilization of nuclear energy and technology must ensure safety as a prerequisite and place safety in the highest priority position. No individual or organization should sacrifice safety to develop nuclear energy and technology for profit. The policy of "ensuring safety" emphasizes that nuclear safety must be guaranteed in the development and application of nuclear energy and technology. This policy in China is also determined according to the spirit of President Xi Jinping's speech at the Fourth International Summit on Nuclear Security (Jinping 2014).

(2) Safety Principles

Article 4 of Chapter 1 of China's "Nuclear Safety Law" stipulates that the specific principles that nuclear safety work must adhere to are: safety first, prevention-oriented, clear responsibility, strict management, in-depth defense, independent supervision, and comprehensive guarantee.

1) Safety First

Safety first means always making conservative decisions to ensure that safety is not affected by any economic factors, schedule requirements, or other factors. The subjects involved in nuclear safety-related activities, including units involved in nuclear energy and technology utilization activities, nuclear safety equipment manufacturing units, units engaged in non-destructive testing of nuclear safety equipment, and their management and technical personnel, operators, as well as nuclear industry authorities, energy authorities, and nuclear safety supervision departments, should prioritize safety in decision-making, planning, management, production, manufacturing, research and development, inspection, testing, transportation, operation, etc., to ensure that nuclear safety is not affected by various interest trade-offs and that activities comply with nuclear safety requirements.

2) Prevention-Oriented

Prevention-oriented means prioritizing the prevention of nuclear accidents in nuclear safety work, preventing problems before they occur. The lessons from the Chernobyl nuclear accident in the Soviet Union and the Fukushima nuclear accident in Japan fully demonstrate that once a nuclear accident occurs, especially an accident involving a large-scale release of radioactive materials, the resulting damage, radioactive hazards, and radioactive pollution are very serious and irreversible, with high remediation costs and unpredictable remediation effects. The destruction caused by a nuclear accident is difficult to remedy with post-accident measures. Compared to "during the event" rescue and "post-event" remediation, "pre-event" prevention has the lowest cost and the best effect. Being prevention-oriented also means that once an accident occurs, emergency response measures should be taken to mitigate its consequences.

China's Radioactive Pollution Prevention Law, Work Safety Law, Environmental Protection Law, and other laws have established the principle of prevention-oriented. The nuclear safety licensing system stipulated by the Nuclear Safety Law is an important institutional arrangement for the principle of prevention-oriented.

(3) Clear Responsibility

"Clear Responsibility" refers to the specific and well-defined responsibilities undertaken by various responsible entities. Different responsible entities bear different safety responsibilities as stipulated by China's Nuclear Safety Law: nuclear facility operators, nuclear material holders, and carriers of radioactive materials are the main entities responsible for comprehensive nuclear safety; relevant government departments bear the safety management responsibilities stipulated in the "Three Determination Scheme" within their respective scopes of duty, with particular emphasis on the responsibility of the nuclear industry and energy departments to "manage safety in their respective industries"; contractors and suppliers bear "corresponding" responsibilities that match their rights and obligations within their respective scopes of duty.

(4) Strict Management

Strict management refers to the insistence of managers at all levels on strict standards and requirements to ensure the quality of all activities. On the one hand, it requires operators, nuclear material holders, and consignors of transport activities to strictly comply with the obligations stipulated by laws and regulations when engaging in nuclear safety-related activities, and to implement relevant nuclear safety standards and internal management systems; on the other hand, it requires relevant government departments to strictly carry out review, licensing, supervision, and law enforcement in accordance with relevant laws and regulations, and to severely crack down on illegal and irregular behaviors. The strict supervision of the nuclear safety regulatory department is a specific manifestation of strict management, reflected in the strict compliance with the nuclear safety requirements stipulated by current laws and regulations throughout the entire process of related activities (such as the manufacture, installation, non-destructive testing of nuclear safety equipment, and the site selection, design, construction, operation, and decommissioning of nuclear facilities).

(5) In-Depth Defense

In-depth defense refers to the prevention of nuclear accidents and mitigation of their consequences through the establishment of a series of progressive and independent protective measures or physical barriers. The concept of in-depth defense is implemented in all safety-related activities, including those related to organization, personnel behavior, technology, and management, to ensure that these activities are under multiple defensive measures. Even if a failure occurs, it will be detected, compensated, or corrected by appropriate measures.

(6) Independent Supervision

Independent supervision refers to the independent supervision of nuclear safety by the nuclear safety supervision and management department, free from interference and influence from any other entities, with independent decision-making and duty performance, ensuring the independence, authority, and effectiveness of China's nuclear safety supervision. Independent supervision is an internationally accepted practice, stipulated in relevant international conventions and national laws. However, it should be noted that nuclear safety work involves multiple departments, and each department should perform its own duties and cooperate with each other. The principle of independent supervision does not exclude the supervision of nuclear power enterprises by industry regulatory departments.

(7) Comprehensive Guarantee

Comprehensive guarantee refers to the formation of a comprehensive nuclear safety support system to provide effective support for ensuring nuclear safety. Firstly, the state establishes a nuclear safety strategic planning system to plan tasks to ensure nuclear safety over a longer time frame and promote their implementation; secondly, it clarifies the legal requirements for building a nuclear safety culture to enhance the level of nuclear safety from the subconscious level of personnel; thirdly, it strengthens the research and development of nuclear safety science and technology to promote innovation in nuclear safety means and methods; fourthly, it strengthens international cooperation in nuclear safety, fully drawing on the good practices formed by various countries in the use and production of nuclear energy and nuclear technology, improving its own safety level, and promoting the Chinese nuclear safety regulatory system.

Discussion Questions

(1) How many stages can the over 120-year history of nuclear energy and nuclear technology development be roughly divided into? Please briefly describe.
(2) Please briefly describe the fields and scope involved in nuclear technology and its applications.
(3) What is radiation? What are the two types? What are their similarities and differences?
(4) What are the common types of radiation? What are their characteristics?
(5) What are radioactive activity and specific activity?
(6) What are absorbed dose, dose equivalent, and effective dose? How are they different?
(7) What are the common types of nuclear decay? What are their characteristics?
(8) What are decay constant and half-life? What are they related to?
(9) What are the common types of nuclear reactions? What are their characteristics?
(10) What is a scattering reaction? What are the two types? What are their similarities and differences?
(11) What is fission yield? In the fission products of uranium-235, around what mass number is there a high fission yield?
(12) How many types of interactions between radiation and matter are there? Please briefly describe.
(13) What are activation analysis and labeling?
(14) What are radiation degradation and radiation sterilization (also known as radiation disinfection)?
(15) What is nuclear safety? Why is humanity more concerned about nuclear safety than ever before?
(16) What is the definition of a nuclear accident? What are its characteristics?
(17) What does the Chinese "Nuclear Safety Law" include in the term "nuclear facilities"? How many categories are there?
(18) What is spent fuel? What does the Chinese "Nuclear Safety Law" include in the term "nuclear materials"?
(19) What are active safety, redundant safety, and passive safety?
(20) What forms can the decommissioning of a nuclear power plant take?
(21) What are the definitions of nuclear emergency response and nuclear safety culture?
(22) What are the key differences between nuclear safety and radiation safety?
(23) Please briefly describe the time, location, and main causes of the Three Mile Island nuclear accident, the Chernobyl nuclear accident, and the Fukushima nuclear accident.
(24) What are the main components of passive safety technology?
(25) What is the overall goal of nuclear safety?

(26) What are the specific principles that must be adhered to in nuclear safety work? Please briefly explain.

References

International Atomic Energy Agency. (1991). Safety Culture, INSAG-4.

Xi Jinping. (2014, March 25). Speech at the Nuclear Security Summit in The Hague. People's Daily.

Li S, Zha LF, Zhao WY. (1994). Nuclear Energy and Nuclear Technology. Shanghai, Shanghai Scientific and Technical Publishers.

National People's Congress of China. (2017). Nuclear Safety Law of the People's Republic of China.. Beijing, Law Press China.

Wang CX. (2002). Applications of Nuclear Energy and Nuclear Technology. Beijing, Atomic Energy Press.

Chapter 2
Basic Knowledge of Nuclear Physics

2.1 Basic Properties of Atomic Nuclei

The basic properties of atomic nuclei usually refer to the static properties of atomic nuclei as a whole, which are closely related to the structure and changes of atomic nuclei (Yan et al. 2014).

2.1.1 *Composition and Mass of Atomic Nuclei*

(1) Composition of Atomic Nuclei

All atomic nuclei are composed of protons and neutrons, collectively referred to as nucleons. Protons carry a positive charge, represented by *p*. Neutrons carry no charge, represented by n. Therefore, nucleons exist in two states: charged state, which is the proton; and uncharged state, which is the neutron.

The proton is also a hydrogen nucleus, and its charge is equal to the basic charge (unit in Coulombs *C*), which is the absolute value of the electron charge: $e = 1.602\,189\,2 \times 10^{-19}\,C$.

The masses of protons and neutrons are respectively:

$$m_p = 1.672\,648\,5 \times 10^{-27}\text{kg}=938.279\,6\,\frac{\text{MeV}}{c^2} = 1.007\,277u$$

$$m_n = 1.674\,9543 \times 10^{-27}\text{kg}=939.573\,1\,\frac{\text{MeV}}{c^2} = 1.008\,665u$$

where, $c \approx 3 \times 10^8$(m/s) is the speed of light. u is the atomic mass unit, according to international standards, 1 atomic mass unit is defined as 1/12 of the mass of a ^{12}C atom.

L. Yu et al., *Nuclear Technology and Safety*, Nuclear Science and Technology,
https://doi.org/10.1007/978-981-95-2856-1_2

Free protons are stable and exist in large quantities in nature. Neutrons, however, are unstable and free neutrons basically do not exist in nature.

Inside a reactor, the average lifespan of neutrons released by fuel fission, from generation to absorption or leakage out of the reactor, is about 10^{-3} s for thermal neutron reactors, and only 10^{-6}~10^{-7} s for fast neutron reactors, much shorter than the existence time of free neutrons.

The number of protons is represented by Z, also known as the nuclear charge number, which is also the atomic number of the element. Neutrons are represented by N. The mass number of the atomic nucleus is represented by A, so $A = Z + N$, representing the total number of nucleons in the atomic nucleus (Chen et al. 2019).

(2) Mass of Atomic Nuclei

When the binding energy of the extranuclear electrons is very small and can be ignored, then the mass of the atomic nucleus m is equal to the mass of the atom of the element m_{A} minus the mass of the outer electrons of the atom m_{e}, that is,

$$m = m_A - Zm_e \tag{2.1}$$

where, Z is the charge number. m_A is equal to the atomic weight of the atom times the atomic mass unit. The mass of the electron is very small, $m_e \approx 5.4858 \times 10^{-41}$ u.

For atomic mass, when using u as the unit, it is close to the mass number A. Therefore, in approximate calculations, the following approximation formula can be used:

$$m = m_A - Zm_e \approx m_A \approx A \tag{2.2}$$

2.1.2 Nuclear Density

Definition: The number of atoms per unit volume is called nuclear density, represented by N, and the unit is $1/\text{m}^3$.

Since each atom contains only one nucleus, the calculation of nuclear density can be reduced to the calculation of atomic density, i.e., the calculation of the number of atoms per unit volume. Below, we introduce the calculation methods for nuclear density (atomic density) based on whether the substance is a single element, a mixture, or a compound.

(1) Calculation of Nuclear Density for Single Elements

When the molecules of a substance are composed of the same type of atom, this substance is called a single element. If the atomic weight of the atom is denoted as A, then A grams of this atom is called a "mole (gram) atom". Any substance contains the same number of atoms in each mole (gram) atom, denoted as N_A, N_A is known as Avogadro constant, its value is

$$N_A = 6.022\,045 \times 10^{23} (1/\text{mol})$$

Therefore, the number of atoms contained in each gram of this atom is N_A/A, and there are 1000 N_A/A atoms per kilogram. If the density of this substance is denoted as ρ (kg/m^3), then 1000 N_A/A is the number of atoms contained per cubic meter, hence the formula for calculating nuclear density is

$$N = 1000\rho N_A/A \ (1/\text{m}^3) \tag{2.3}$$

(2) Calculation of Nuclear Density for Compounds

In the molecules of a compound, each element's atoms have a definite number. If the molecular weight of the compound is denoted as M, then M grams of this compound molecule are called a "mole molecule", which contains N_A molecules. If the density of the compound is denoted as ρ (kg/m^3), and the number of the i-th type of atom in each compound molecule is n_i, then the density of the i-th type of atomic nucleus is calculated as

$$N_i = \frac{1000\rho N_A n_i}{M} (1/m^3) \tag{2.4}$$

(3) Calculation of Nuclear Density for Mixtures

A substance formed by mixing two or more different substances together is called a mixture. If the average density of the mixture is denoted as ρ (kg/m^3), the atomic weight of the i-th type of atom is A_i, and its atomic percentage is W_i, then the average atomic weight of the mixture is $A = \sum_i W_i A_i$, hence the formula for calculating the nuclear density of the i-th type of atom is

$$N_i = \frac{1000\rho N_A W_i}{\sum_i W_i A_i} (1/m^3) \tag{2.5}$$

2.2 Energy of Atomic Nuclei

The energy produced when atomic nuclei undergo fission or fusion is the result of changes in binding energy caused by changes in nuclear structure. Therefore, the binding energy of atomic nuclei is the source of nuclear energy utilization, and the binding energy is a sign of the stability of atomic nuclei.

2.2.1 Binding Energy

(1) Mass Defect

The atomic nucleus formed by the combination of *Z* protons and (*A-Z*) neutrons, denoted as ${}_Z{}^AX$, has a mass m that is always less than the total mass of the *Z* protons and (*A-Z*) neutrons that make it up. This difference in mass is called the "mass defect". The mass defect is denoted as Δm, so we have

$$\Delta m = Zm_p + (A - Z)m_n - m > 0 \tag{2.6}$$

Equation (2.2) can also be written as

$$\Delta m = Z(m_p + m_e) + (A - Z)m_n - (m + Zm_e) \tag{2.7}$$

If the mass defect of electrons binding to atoms can be ignored, then $m_p + m_e$ is approximately equal to the mass of a hydrogen atom m_H, and $m + Zm_e$ is approximately equal to the atomic mass of the nucleus under discussion m_A, so the above equation can be rewritten as

$$\Delta m \approx Zm_H + (A - Z)m_n - m_A \tag{2.8}$$

Using Eq. (2.8), the mass defect of an atomic nucleus can be calculated with considerable precision from its atomic massm_A. For example, when a proton and a neutron combine to form a deuterium nucleus, the atomic mass of deuterium is $m_D = 2.014\,102\,u$, the atomic mass of hydrogen is $m_H = 1.007\,825\,u$, and the mass of a neutron is $m_n = 1.008\,665\,u$. With $Z = 1$ and$A = 2$, the mass defect of a proton and a neutron combining to form a deuterium nucleus, according to Eq. (2.8) is:

$$\Delta m \approx 1.007\,825 + 1.008\,665 - 2.014\,102 = 0.002388 \text{ u} \tag{2.9}$$

(2) Binding Energy

The famous relationship between mass and energy proposed by Einstein is

$$E = mc^2 \tag{2.10}$$

In the equation, *c* is the speed of light, $c = 2.997\,924\,58 \times 10^8$ m/s. Therefore, one atomic mass unit is equivalent to 931.5016 mega electron volts (MeV) of energy, and 1 MeV = $1.602\,189\,2 \times 10^{-13}$ J. Equation (2.10) indicates that when nucleons combine to form an atomic nucleus, there is always a loss of mass, i.e., some energy is released from the atomic nucleus system during the combination process.

$$\Delta E = \Delta mc^2 \tag{2.11}$$

Equation (2.11) equals the energy released when nucleons combine to form the atomic nucleus. Conversely, to separate all nucleons in the atomic nucleus, this much energy must be provided. This energy is called the binding energy of the atomic nucleus. Thus, substituting Eq. (2.6) gives,

$$\Delta E = [Zm_p + (A - Z)m_n - m]c^2 \tag{2.12}$$

or

$$\Delta E \approx [Zm_H + (A - Z)m_n - m_A]c^2 \tag{2.13}$$

This is the formula for calculating the binding energy of an atomic nucleus. Using Eq. (2.13), we can calculate the energy released when a proton and a neutron combine to form a deuterium nucleus as mentioned earlier:

$$\Delta E = \Delta mc^2 = 0.002\,388 \times 931.501 \approx 2.224\,\text{MeV} \tag{2.14}$$

2.2.2 Specific Binding Energy

The binding energy of different nuclides varies greatly. Generally, atomic nuclei with a large number of nucleons A also have a large binding energy. We call the ratio of the binding energy of an atomic nucleus to the number of nucleons the "specific binding energy", which is the average binding energy per nucleon, denoted by ε.

$$\varepsilon = \Delta E/A \tag{2.15}$$

The specific binding energy represents the work that must be done on average for each nucleon to break the atomic nucleus into free nucleons. Therefore, the size of the specific binding energy can be used to indicate the strength of the atomic nucleus binding. Atomic nuclei with a larger ε are more tightly bound and therefore more stable. Atomic nuclei with a smaller ε are more loosely bound and therefore more likely to split, unstable.

Based on experimental and calculated results, the specific binding energy of different atomic nuclei and the corresponding number of nucleons are represented by a curve, as shown in Fig. 2.1. This curve is called the specific binding energy curve.

As can be seen from Fig. 2.1, only light nuclei and heavy nuclei have smaller specific binding energies. The specific binding energy of light nuclei shows periodic fluctuations and several extreme values. Before $A \approx 60$, the specific binding energy increases with A. After $A \approx 60$, the specific binding energy decreases smoothly with A. Because medium nuclei have a larger specific binding energy, these nuclei are the most stable. This characteristic of the specific binding energy curve gives

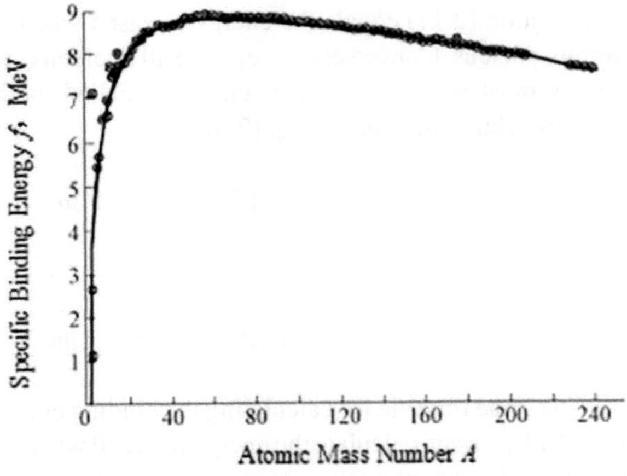

Fig. 2.1 Specific binding energy varies with atomic mass number

us an important insight: if we can transform atomic nuclei with smaller specific binding energies into atomic nuclei with larger specific binding energies, energy can be released. For example, heavy atomic nuclei with relatively small specific binding energies can be split into medium atomic nuclei with larger specific binding energies, or extremely light atomic nuclei with small specific binding energies can be fused into light atomic nuclei with relatively large specific binding energies. The energy released in the former is called fission energy, and the latter is called fusion energy. Representative examples include the fission of uranium-235 into two medium-mass nuclei, and the fusion of two deuterium nucleons into a helium nucleus, etc.

2.3 Nuclear Decay and Decay Laws

Nuclear decay is one of the ways an atomic nucleus can transform into another atomic nucleus. This section will introduce the basic laws governing this nuclear decay process.

2.3.1 Nuclear Decay Laws

At a given time t, the number of radioactive nuclei is $N(t)$. These $N(t)$ nuclei do not decay simultaneously. Experiments have shown that radioactive decay of atomic nuclei is a statistical process (Yan 2021). For a single atomic nucleus, the moment of decay is uncertain, but the percentage of a large number of similar atomic nuclei decaying within a small-time interval is certain. Let the probability of a single atomic nucleus of a certain nuclide decaying within a unit time interval be λ, then whether it is α decay or β decay, the corresponding λ is a certain constant, and it is only related to the characteristics of the nucleus itself, and has nothing to do with the chemical and physical conditions that affect the properties of the extranuclear electrons, such as temperature, pressure, electromagnetic field, etc. This λ is called the decay constant. It should be pointed out that each radioactive nucleus has its own inherent decay constant. Among the known radioactive nuclides, there are no two nuclides with exactly the same decay constant.

Suppose at $t = 0$ there are $N(0)$ identical radioactive atomic nuclei, and at the end of t there are $N(t)$ undecayed nuclei, then according to the physical meaning of the decay constant λ, the average number of nuclei decaying in the interval $t \to t + \mathrm{d}t$ is $\lambda\, N(t)\mathrm{d}t$. Therefore, the decrease in the number of undecayed nuclei at time t ($-\mathrm{d}N$) is given by:

$$-\mathrm{d}N = \lambda N(t)\, \mathrm{d}t \tag{2.16}$$

Note that λ is independent of time. Because $N(t)$ decreases with time, a negative sign is prefixed to the derivative. After integrating Eq. (2.16), we get

$$\int_{N(0)}^{N(t)} \frac{\mathrm{d}N(t)}{N(t)} = -\int_{0}^{t} \lambda \mathrm{d}t \tag{2.17}$$

So

$$N(t) = N(0)e^{-\lambda t} \tag{2.18}$$

This is the universal law followed by the nuclear decay process, that is, radioactive decay decays according to the exponential law.

2.3.2 Half-Life and Mean Lifetime

We call the time it takes for a radioactive nucleus to decay by half the half-life, denoted by $T_{1/2}$. It is not difficult to obtain from Eq. (2.18):

$$\frac{N(0)}{2} = N(0)e^{-\lambda T_{1/2}} \tag{2.19}$$

That is

$$T_{1/2} = \frac{\ln 2}{\lambda} = \frac{0.693}{\lambda} \tag{2.20}$$

It can be seen that the half-life is inversely proportional to the decay constant. The time required for a radioactive nucleus with a large decay constant to decrease by half is short, and vice versa.

The actual lifetime of each radioactive nucleus varies and can take any value between 0 and ∞. However, the average lifetime of a large number of radioactive nuclei (denoted by T_{av}) is certain and is an important physical quantity. From the above analysis, we know that $\lambda N(t)$ represents the decay rate of the radioactive nucleus at time t, therefore, $\lambda N(t)\,dt$ is the number of radioactive nuclei that decayed in the time interval from t to $t + dt$, and the lifetime of these nuclei is t. Substituting into Eq. (2.18), it is not difficult to obtain the average lifetime of radioactive nuclei as

$$T_{av} = \frac{1}{N(0)}\int_0^{\infty} t\lambda N(t)\mathrm{d}t = \frac{1}{N(0)}\int_0^{\infty} t\lambda N(0)e^{-\lambda t}\mathrm{d}t = \frac{1}{\lambda} \tag{2.21}$$

It can be seen that the average lifetime of a radioactive nucleus is longer than the half-life, and it also represents the time required for the radioactive nucleus to decrease to $N(0)/e$. The relationship between the average lifetime of a radioactive nucleus and the half-life can be obtained from Eqs. (2.20) and (2.21):

$$T_{av} = \frac{T_{1/2}}{0.693} \approx 1.443T_{1/2} \tag{2.22}$$

$T_{1/2}$ and T_{av} are both characteristic quantities describing radioactive decay. Depending on the needs, either can be used to describe the decay characteristics of radioactive nuclei.

If the daughter nucleus generated after the decay of the parent nucleus is still radioactive, this is a case of continuous decay. For example, A → B → C, where A, B, and C represent successive generations of radioactive atomic nuclei. The longer the decay chain, the more differential equations need to be solved. Naturally, the differential equation at this time is different from Eq. (2.16).

2.4 Nuclear Reactions

2.4.1 Laws of Nuclear Reactions and Reaction Energy

(1) Laws of Nuclear Reactions

Nuclear reactions involve the transformation of a target atomic nucleus into a new nucleus and another particle by being struck by particles or photons. These processes can be represented by the following equation:

$$a + x \rightarrow y + b \tag{2.23}$$

Here, a is the incident particle, b is the particle emitted after the reaction, known as the outgoing particle. x is the atomic nucleus being bombarded, known as the target nucleus. y is the newly formed nucleus. This type of nuclear reaction can also be represented as x(a, b) y. For example, the first nuclear reaction discovered in history was:

$${}_2^4\text{He} + {}_7^{14}\text{N} \rightarrow {}_8^{17}\text{O} + {}_1^1\text{H} \tag{2.24}$$

It can also be expressed as ${}_7^{14}\text{N}(\alpha,\text{p}){}_8^{17}\text{O}$.

In nuclear reactions, outgoing particles can be of the same type as the incoming particles or different. Nuclear reactions in which the outgoing particles are of the same type as the incoming particles are called scattering. Although the type of atomic nucleus remains unchanged during scattering, it is broadly referred to as a nuclear reaction.

All nuclear reactions obey the following conservation laws:

(1) Conservation of nucleon numbers: The number of nucleons remains the same before and after the reaction.
(2) Conservation of charge: The total charge of all particles remains the same before and after the reaction.
(3) Conservation of momentum: The momentum of the system is conserved before and after the reaction.
(4) Conservation of energy: The energy of the system (including rest energy) is conserved before and after the reaction.

These four conservation laws are the most fundamental. In addition to these, nuclear reactions must obey other conservation laws, the details of which can be found in relevant reference books.

(2) Reaction Energy

In the nuclear reaction x(a,b)y, let m_a, m_b, m_x, m_y represent the rest masses of the respective particles, and their kinetic energies be E_a, E_b, E_x, E_y. According to the law of energy conservation:

$$m_a c^2 + E_a + m_x c^2 + E_x = m_b c^2 + E_b + m_y c^2 + E_y \tag{2.25}$$

Since the target nucleus is generally at rest, setting $E_x = 0$, Eq. (2.25) can be rewritten as:

$$m_a c^2 + m_x c^2 + E_a = m_b c^2 + m_y c^2 + E_b + E_y \tag{2.26}$$

This equation can be rearranged as:

$$Q = [(m_a + m_x) - (m_b + m_y)]c^2 = E_b + E_y - E_a \tag{2.27}$$

Here, Q is called the reaction energy, which equals the difference in rest energies before and after the reaction and also equals the change in kinetic energy of the particles.

From Eq. (2.27), it can be seen that when $Q > 0$, the rest mass of the reactants decreases while the kinetic energy increases. Such reactions are called exoergic nuclear reactions. When $Q < 0$, the rest mass of the reactants increases while the kinetic energy decreases. Such reactions are called endoergic nuclear reactions.

2.4.2 Neutron-Induced Nuclear Reactions

Since neutrons carry no electric charge, they do not experience Coulomb repulsion when interacting with atomic nuclei and can directly penetrate the electron cloud to interact with the nucleus. Nuclear reactions in reactors are primarily the result of neutrons interacting with various materials.

(1) Neutron-Producing Nuclear Reactions

Here, four major types of neutron-producing nuclear reactions are introduced.

1) (α, n) reaction: Bombarding certain light nuclei such as ^{7}Li, ^{9}Be, ^{11}B and ^{19}F with α-particles can yield neutrons with energies ranging from about 1 to 13 MeV. The corresponding nuclear reactions are: ^{7}Li (α,n) ^{10}B, ^{7}Be (α,n) ^{12}C, ^{11}B (α,n) ^{14}N, ^{19}F (α,n) ^{22}Na.
2) (γ, n) reaction: Some atomic nuclei emit neutrons when irradiated with γ-rays, and the neutrons produced in this way are often referred to as photoneutrons. Since the vast majority of γ-ray sources cannot emit γ-rays with energy greater than 4 MeV, the target material is limited to beryllium and deuterium, which have low neutron binding energy. Their nuclear reactions are ^{9}Be (γ,n) ^{8}Be, ^{2}H (γ,n) ^{1}H.
3) (p, n) and (d, n) reactions: Neutrons can also be produced by bombarding certain light atomic nuclei with protons or deuterium nuclei. Their nuclear reactions include^{3}H (p,n) ^{3}He, ^{3}H (d,n) ^{4}He, etc.
4) (n, f) reaction: Bombarding fissionable atomic nuclei (such as ^{235}U) with neutrons causes the target nucleus to split into two medium-mass atomic nuclei (known as fission fragments), releasing an average of 2 to 3 secondary neutrons per nuclear fission. If F_1 and F_2 denote the fission fragments, and v represents the average number of neutrons released per nuclear fission, then the nuclear fission reaction can be represented as

$$\mathrm{n}+{}^{235}\mathrm{U} \rightarrow \mathrm{F}_1 + \mathrm{F}_2 + \nu \mathrm{n} \tag{2.28}$$

(2) Nuclear reactions induced by neutrons

Depending on the energy of the incident neutron and the nature of the target nucleus, the neutron and target nucleus can undergo one or several of the following reactions.

1) Elastic scattering: Represented by the (n, n) reaction. After the neutron interacts with the target nucleus, the neutron is re-emitted and the internal energy of the target nucleus remains unchanged. This reaction is known as elastic scattering. In this reaction, in addition to momentum conservation, kinetic energy is also conserved, which can be described by the collision between elastic balls.
2) Inelastic scattering: Often represented by the (n, n′) reaction. This reaction process is similar to elastic scattering, with the emitted particle still being a neutron, but the internal energy of the target nucleus changes after the interaction, and the target nucleus is in an excited state. When it returns from the excited state to the ground state, γ-rays are often emitted.
3) Radiative capture: Often represented by the (n, γ) reaction. The reaction in which the target nucleus captures a neutron and emits γ-rays is called radiative capture. In this reaction, the target nucleus captures a neutron to form an unstable compound nucleus, which emits γ-rays after a very short time (about 10^{-14} s) and returns to the ground state of the daughter nucleus. The atomic number of the daughter nucleus is the same as that of the target nucleus, but the mass number of the daughter nucleus increases by 1.

The (n, γ) reaction can occur in many nuclei from light to heavy, especially under the action of thermal neutrons, almost all elements can undergo this reaction. Therefore, in a thermal neutron reactor, the ^{238}U absorption of neutrons with energy near a few electron volts, the ^{238}U (n,γ)^{239}U reaction, and the control rod (such as cadmium rod) reaction to low-energy neutron absorption ^{113}Cd (n,γ) ^{114}Cd all belong to this type of reaction.

When the incident neutron forms a compound nucleus with the target nucleus, the neutron binding energy ΔE exists within the compound nucleus. If the total kinetic energy of the neutron and the target nucleus in the center-of-mass system is E_c, then the excitation energy of the compound nucleus is $E_c + \Delta E$. If there is an excited state of the compound nucleus near $E_c + \Delta E$, the probability of forming a compound nucleus is particularly high. This phenomenon is called "resonance absorption".

1) Reactions that emit charged particles: Reactions in which a neutron interacts with a target nucleus to form a new nucleus and emit charged particles such as protons, α particles, etc. For example, in a pressurized water reactor, nuclei such as ^{16}O and ^{17}O emit a proton after absorbing a neutron, undergoing ^{16}O (n,p) ^{16}N, ^{17}O (n,p) ^{17}N reactions, etc. The half-life of ^{16}N is 7.3s, and it emits β and γ rays. This reaction is the main source of radioactivity in the water of a pressurized water reactor.
2) Reactions that emit several neutrons: When high-energy neutrons bombard a target nucleus, sometimes two or more neutrons can be produced, which can be represented as (n, 2n), (n, 3n), etc.
3) Fission reaction: When a heavy nucleus is acted upon by a neutron, it splits into two different fragments, releasing 2 to 3 neutrons and a large amount of energy. This process is called a fission reaction, which is the (n, f) reaction introduced earlier. Therefore, it is both a neutron-producing nuclear reaction and a neutron-induced nuclear reaction.

Nuclear fission is the most important nuclear reaction in a reactor. Isotopes ^{233}U, ^{235}U, ^{239}Pu, and ^{241}Pu can undergo fission under the action of neutrons at various energies, and they are usually referred to as easily fissionable nuclei. For ^{238}U and ^{232}Th, fission can only be induced by fast neutrons, and they are usually referred to as fissionable nuclei.

(3) Classification of Neutrons by Energy

There are various ways in which neutrons interact with nuclei as introduced earlier, and the energy of the resulting neutrons can vary greatly (Zhang 2023). The type of nuclear reaction that occurs and the energy of the resulting neutrons mainly depend on the energy of the incident neutron, the nature of the target nucleus, and the type of reaction. Therefore, when studying the interaction of neutrons with matter, neutrons are often classified according to their energy range.

1) Slow Neutrons ($0< E \leq 1$ keV)

Slow neutrons can be further divided into the following three categories:

Cold neutrons ($0< E \leq 0.002$ eV), thermal neutrons (0.002 eV$< E \leq 1$ eV) and epithermal neutrons (1eV$< E \leq 1$ keV).

The term "cold neutrons" refers to the fact that its average energy is lower than that of thermal neutrons.

Neutrons in thermal equilibrium with the surrounding medium are called thermal neutrons. The distribution of thermal neutrons approximately follows Maxwell's distribution law. At 20 °C (or 293 K), the most probable speed of thermal neutrons is 2200 m/s, and the corresponding energy is 0.0253 eV. These two values are usually taken as representative values for thermal neutrons.

The term "epithermal neutrons" refers to the fact that its average energy is higher than that of thermal neutrons. Epithermal neutrons can undergo strong resonance absorption when interacting with heavy nuclei, so neutrons in this energy region are also called resonance neutrons.

The nuclear reactions caused by slow neutrons, apart from elastic scattering, mainly involve absorption by the nucleus, such as the (n, γ) reaction, (n, α), (n, p) reaction, and (n, f) reaction. The most common is the (n, γ) reaction, which can occur in many nuclei from light to heavy. As for the (n, α) and (n, p) reactions, they are limited to a few light nuclei (such as ^{6}Li, ^{10}Be, ^{14}N, and ^{27}Al). The (n, f) reaction only occurs in a few heavy nuclei (^{233}U, ^{235}U, ^{239}Pu, and ^{241}Pu).

2) Intermediate Energy Neutrons (1 keV $< E \leq 0.5$ MeV)

The main interaction of intermediate energy neutrons with atomic nuclei is elastic scattering, followed by radiative capture.

3) Fast Neutrons (0.5 MeV $< E \leq 10$ MeV)

In addition to elastic scattering with atomic nuclei, fast neutrons can also produce inelastic scattering.

3) High Energy Neutrons ($E > 10$ MeV)

In addition to elastic and inelastic scattering with atomic nuclei, high energy neutrons can also produce nuclear reactions that emit two or more particles, such as the (n, 2n) and (n, 3n) reactions.

Inside a nuclear reactor, the energy of neutrons ranges from 0 to 17 MeV, but high energy neutrons with energies above 10 MeV are very rare. Therefore, the nuclear reactions caused by neutrons inside a nuclear reactor mainly fall into two categories: scattering (elastic and inelastic) reactions and absorption (radiative capture and fission) reactions.

2.4.3 Neutron Cross Section and Nuclear Reaction Rate

In the physical calculations of a reactor, concepts such as neutron cross section and reaction rate are often used to quantitatively describe the interaction of neutrons with matter. Before introducing the neutron cross section and nuclear reaction rate, let's first introduce a few commonly used basic physical quantities.

(1) Neutron Density and Neutron Flux Density

Neutron Density: The number of free neutrons per unit volume is called neutron density, denoted by n, and the unit is $1/m^3$.

Intensity of Neutron Beam: Consider a uniform, unidirectional neutron beam with a neutron density of n, with a speed of v, refers to the number of neutrons passing through a unit area perpendicular to the direction of v per unit time. This is known as the intensity of the neutron beam, represented by I, and its unit is $1/(s{\cdot}m^2)$.

Neutron flux density: The number of neutrons entering a small sphere centered at a certain point in space per unit time, divided by the maximum cross-sectional area of the sphere, is known as the neutron flux density at that point. It is represented by ϕ, and its unit is the same as that of I, $1/(s{\cdot}m^2)$.

According to the definition of ϕ, the neutron flux density is the total distance traveled by all neutrons in a unit volume per unit time. Therefore, $\phi = nv$, which is the physical meaning of ϕ. In reactor theory, neutron flux density ϕ is often used to replace neutron density n, or neutron density n is used to replace neutron flux density ϕ.

(2) Neutron Nuclear Reaction Cross Section

The physical meaning of the nuclear reaction cross section is the probability of a certain type of nuclear reaction occurring. It is mainly divided into microscopic cross sections and macroscopic cross sections.

1) Microscopic Cross Section

Assume there is a unidirectional, uniform, and monoenergetic neutron beam with an intensity of I. This neutron beam strikes perpendicularly on a thin target with a unit area. The thickness of the target is Δx, and the number of atomic nuclei per unit

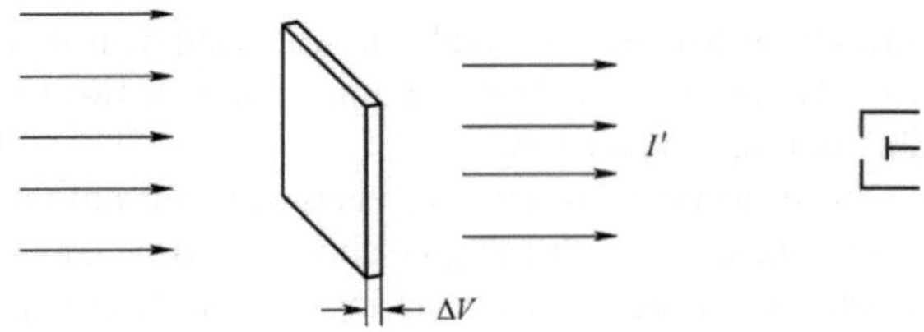

Fig. 2.2 Attenuation of a parallel neutron beam after passing through a thin target

volume in the target is N. A neutron detector is placed at a certain distance behind the target (see Fig. 2.2). If the neutron beam intensity measured without the target is I, and the neutron beam intensity measured with the target is $I\prime$, then $I\prime - I = \Delta I$ equals the number of neutrons that interact with the target nuclei. Because once a neutron interacts with a target nucleus (whether it is scattering or absorption), it will disappear from its original flight direction, and these neutrons cannot be detected by the neutron detector. Experiments show that when the target area remains unchanged, ΔI is proportional to the neutron beam intensity I, target thicknessΔx, and nuclear density of the target N, i.e.,

$$\Delta I = -\sigma I N \Delta x \tag{2.29}$$

In this equation, σ is a proportionality constant, which we call the "microscopic cross section". It is related to the properties of the target nucleus and the energy of the neutron. The above equation can be rewritten as

$$\sigma = \frac{-\Delta I}{I N \Delta x} = \frac{-\Delta I/I}{N \Delta x} \tag{2.30}$$

In this equation, $-\Delta I/I$ is the proportion of neutrons in the parallel neutron beam that interact with the target nucleus, and $N\Delta x$ is the number of target nuclei per unit incident area. Therefore, σ is the interaction rate of the incident neutron beam of unit intensity relative to each target nucleus.

It is easy to understand that the number of neutrons interacting with the target nucleus, in addition to being related to the intensity of the incident neutron beam I, the target thickness Δx, and the nuclear density of the target N, is also related to the energy of the incident neutron. When the thickness is Δx tending to infinitesimally small, Eq. (2.30) can also be written in differential form:

$$\sigma = \frac{-\mathrm{d}I}{I N \mathrm{d}x} \tag{2.31}$$

According to Eq. (2.31), σ is a measure of the probability of the average incident neutron interacting with the target nucleus. Its unit is m^2. Since the values of microscopic cross sections are generally very small, in practical applications, "barn" (abbreviated as b) is often used as the unit, where 1b equals 10^{-28} m^2 or 10^{-24} cm^2.

For a monoenergetic neutron beam with an initial intensity of I_0, the intensity after interacting with the target nucleus can be easily obtained by integrating Eq. (2.31) to find the remaining neutron beam intensity.

$$I(x) = I_0 e^{-N\sigma x} \tag{2.32}$$

Since neutrons interact with matter in various ways, different microscopic cross-sections are used to describe different types of interactions. We use subscripts to denote the cross-sections of various nuclear reactions. For example, the subscripts *s, e, i, γ, f, a,* and *t* represent the scattering, elastic scattering, inelastic scattering, radiative capture, fission, absorption, and total reaction cross-sections of neutron-nucleus interactions, respectively.

According to the definition of cross-section, we have:

$$\sigma_s = \sigma_e + \sigma_i \tag{2.33}$$

$$\sigma_a = \sigma_\gamma + \sigma_f + \cdots \tag{2.34}$$

$$\sigma_t = \sigma_s + \sigma_a \tag{2.35}$$

2) Macroscopic Cross-Section

Since each cubic meter of the target material contains N nuclei, $N\sigma$ represents the total cross-sectional area of target nuclei per cubic meter. This quantity frequently appears in formulas of reactor physics and is referred to as the macroscopic cross-section, denoted as Σ, and its unit is 1/m. Its physical meaning is a measure of the average probability of a neutron interacting with the atomic nuclei in a unit volume. For example, the macroscopic cross-sections of scattering, elastic scattering, inelastic scattering, radiative capture, fission, absorption, and total nuclear reactions can be represented by Σ_s, Σ_e, Σ_{in}, Σ_γ, Σ_f, $\Sigma_a$andΣ_t, respectively.

Since the macroscopic cross-section Σ is the product of the microscopic cross-section and the nuclear density, to calculate it, in addition to obtaining σ, we must also know the number of atomic nuclei N per unit volume. For single-element materials, the nuclear density is calculated according to Eq. (2.3) while for compounds and mixtures, it is calculated according to Eqs. (2.4) and (2.5), respectively. Thus, the macroscopic cross-section $\Sigma_x(x = s, a, f, \ldots)$ of a compound or mixture can be written as:

$$\Sigma_x = \sum_i N_i \sigma_{xi} \tag{2.36}$$

where: σ_{xi} is the microscopic cross-section of a certain nuclear reaction of the ith element; N_i is the number of atomic nuclei of the ith element per unit volume of the medium.

With the concept of macroscopic cross-section, Eq. (2.32) can also be written as:

$$I(x) = I_0 e^{-\Sigma x} \tag{2.37}$$

3) Mean Free Path

The distance a neutron travels between two consecutive interactions is called the free path, and its average value is called the mean free path, denoted as λ (note, do not confuse with the radioactive decay constant), and its unit is m. For different types of neutron-target nucleus interactions, different types of mean free paths can be defined. For example, the mean free paths for neutrons undergoing successive interactions—such as scattering, elastic scattering, inelastic scattering, radiative capture, fission, absorption, and total reaction—can be denoted as λ_s, λ_e, λ_{in}, λ_γ, λ_f, $\lambda_a$andλ_t, respectively.

From Eq. (2.37), we have

$$I(x)/I_0 = e^{-\Sigma x} \tag{2.38}$$

Because $I(x)/I_0$ represents the proportion of the initial neutron beam intensity remaining after the incident neutron has traveled a distance x within the target, $e^{-\Sigma x}$ is the probability that the incident neutron I_0 travels a distance x within the target without interacting with the target nucleus. Differentiating Eq. (2.37) gives

$$-\frac{\mathrm{d}I(x)}{I_0} = e^{-\Sigma_x} \Sigma \mathrm{d}x \tag{2.39}$$

Clearly, Eq. (2.39) represents the fraction of the initial neutron beam intensity that undergoes a change while traveling a distance x within the target and interacting within an infinitesimal interval dx near x. In other words, it describes the probability of the first interaction occurring with the target nuclei. Therefore, the mean free path can be determined by the following equation:

$$\lambda = \int_0^\infty x e^{-\Sigma x} \Sigma \mathrm{d}x = \frac{1}{\Sigma} \tag{2.40}$$

It can be seen that the mean free path of a neutron for a specific reaction is the reciprocal of the macroscopic cross-section of this reaction. Similarly, depending on the different types of interactions between the neutron and the target nucleus, we can define the scattering mean free path $\lambda_s = 1/\Sigma_s$, the absorption mean free path $\lambda_a = 1/\Sigma_a$, the total nuclear reaction mean free path $\lambda_t = 1/\Sigma_t$, etc., and we have

$$\frac{1}{\lambda_t} = \frac{1}{\lambda_s} + \frac{1}{\lambda_a} \tag{2.41}$$

1) Nuclear Reaction Rate

Assuming that the neutron moves randomly within the target nucleus at the same rate v, the macroscopic cross-section of the target nucleus is Σ, and the mean free path is λ, then the statistical average number of nuclear reactions that a neutron has with the target nucleus per unit time is $v/\lambda = v\Sigma$. We have previously introduced the concept of neutron density n, which represents the number of free neutrons per unit volume, so the total number of nuclear reactions between neutrons and target nuclei per unit time per unit volume is $R = nv\Sigma$, which is called the nuclear reaction rate. It is a commonly used and important physical quantity in reactor physics analysis, and its unit is $1/(\mathrm{s} \cdot \mathrm{m}^3)$. Corresponding to the macroscopic cross-section, depending on the different types of interactions between the neutron and the target nucleus, we can define the scattering reaction rate $R_s = nv\Sigma_s$, the absorption reaction rate $R_a = nv\Sigma_a$, the total nuclear reaction rate $R_t = nv\Sigma_t$, etc.

According to the definition of the nuclear reaction rate, for a uniform, unidirectional neutron beam, the decrease in the neutron beam intensity after the neutron has traveled a distance x within the target nucleus and has undergone a nuclear reaction within a distance $\mathrm{d}\,x$ is the nuclear reaction rate of the neutron interacting with the target nucleus. Therefore, from Eq. (2.31), we have

$$R = \frac{-\mathrm{d}I}{\mathrm{d}x} = \sigma NI = \Sigma I \tag{2.42}$$

In general, if the neutron is not moving in one direction, but is moving randomly, the neutron flux density ϕ at the location of the target material can replace the neutron beam intensity I, and Eq. (2.42) still holds, i.e.,

$$R = \sigma N\phi = \Sigma\phi \tag{2.43}$$

This is because the longer the total distance traveled by the neutron per unit time and per unit volume ($\phi = nv$), and the more target nuclei (N) per unit volume, the more opportunities there are for the neutron to collide with the target nucleus, and thus the higher the nuclear reaction rate (R). Equation (2.43) is widely used in reactor physics. For different types of nuclear reactions, by replacing σ or Σ in the equation with the corresponding microscopic or macroscopic cross-section, we can obtain the reaction rates for different types of nuclear reactions, such as fission (reaction) rate, scattering (reaction) rate, etc.

2.4.4 Doppler Effect

The cross-section is usually related to neutron energy, mainly indicating that the cross-section depends on the relative energy of the neutron and the nucleus involved in the interaction (Wu et al. 2014). Thus, even for monoenergetic neutrons, the energy relative to the target nucleus will vary within the range of the single neutron energy measurement. This phenomenon is called the Doppler effect, which is similar to

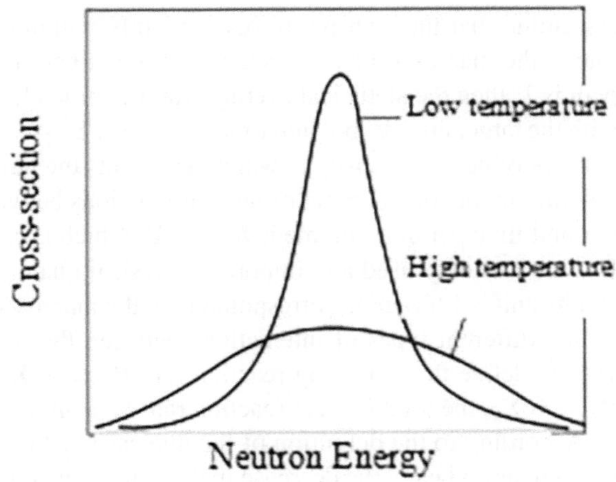

Fig. 2.3 Doppler broadening of resonance as the temperature increases

the observed wavelength change of a moving light or sound source with a fixed frequency.

As the vibrational energy of the target nucleus increases with rising temperature, the range of relative energy between the neutron and the nucleus also increases with temperature. Thus, due to the Doppler effect, the width of the resonance peak increases with rising temperature, a phenomenon often encountered in reactor physics analysis known as Doppler broadening. The broadening of the peak is accompanied by a decrease in its height, while the area under the resonance curve remains constant. For isolated clear resonances, Doppler broadening occurs as the temperature increases, as shown in Fig. 2.3.

For thermal neutron reactors, Doppler broadening affects the neutron absorption rate. If the average flux density of resonant neutrons remains constant, the neutron absorption rate in the resonance will not be affected by Doppler broadening, as the area under the cross-section curve remains constant. However, in actual thermal neutron reactors, the average flux density of resonant neutrons passing through the resonance region during the neutron slowing down process increases due to the broadening of the resonance peak, i.e., the total absorption rate of neutrons in the resonance region (which depends on the product of neutron flux density and cross-section) increases with rising temperature.

2.5 Nuclear Fission and Fusion

2.5.1 Fission Critical Energy and Critical Mass

Nuclear fission is the most important nuclear reaction occurring between neutrons and nuclear fuel in a reactor. A large amount of fission energy is released in the process of nuclear fission, and neutrons are also emitted. This makes it possible for the neutrons produced by fission to cause further nuclear fission reactions, releasing more neutrons and energy. Thus, nuclear fission can continue as required under human control, achieving the purpose of utilizing atomic energy. Based on the liquid drop model of atomic nuclei, we can understand some characteristics of the nuclear fission process. Consider the process of a liquid drop changing due to force (see Fig. 2.4): the liquid drop starts as a sphere A, then is pulled into an ellipsoid B. Although the volume does not change, the surface area increases. As long as the volume energy exceeds the surface energy, the liquid drop will return to its original shape. If the deformation force is large enough, the liquid drop will reach the dumbbell shape C. In this state, the surface energy generally exceeds the volume energy that provides cohesion for the liquid drop, and the liquid drop will no longer be able to return to its original shape, but will split into two small liquid drops, which initially deform somewhat, as in D, and finally become spherical E.

The nuclear fission process can be considered similar to the above process. The target nucleus absorbs a neutron to form an excited compound nucleus, whose excitation energy equals the binding energy of the neutron plus the kinetic energy of the neutron before it was captured. Due to this excess energy, it can be assumed that the compound nucleus undergoes a series of deformation processes, i.e., it undergoes a process similar to Fig. 2.4. If the excess energy is not sufficient to cause further deformation beyond B, it will force the nucleus to return to its original shape. The excess energy is then released by the excited compound nucleus by emitting a gamma-ray photon. If the compound nucleus has enough energy to enter the dumbbell state C, then because the surface energy (plus electrostatic repulsion energy) exceeds the volume energy, it cannot return to the initial state A, but changes from C to D, then to E, indicating fission into two separate nuclei—fission fragments. Due to electrostatic repulsion, they fly away in opposite directions. The excess energy that the compound nucleus must have to deform to state C is called the critical energy of fission. If it can obtain so much critical energy, fission usually occurs. If it cannot obtain so much critical energy, fission is impossible. To illustrate the possibility of fission of some heavy nuclei under neutron action, the critical energy of a compound nucleus formed

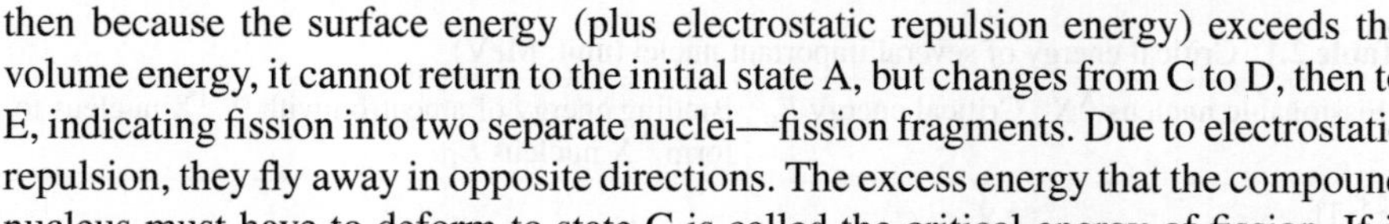

A B C D E

Fig. 2.4 Liquid drop model of heavy nucleus fission

by a heavy nucleus absorbing a neutron, and the binding energy of a neutron are listed in Table 2.1.

From Table 2.1, it can be seen that after ^{233}U, ^{235}U, and ^{239}Pu capture a neutron, they form compound nuclei ^{234}U, ^{236}U, and ^{240}Pu. Because the binding energy of the last neutron in the compound nucleus $E_B > E_c$, ^{233}U, ^{235}U, and ^{239}Pu can undergo fission after capturing a thermal neutron. However, the binding energy E_B of the compound nuclei formed after ^{232}Th and ^{238}U capture a thermal neutron is $E_B < E_c$, therefore, it cannot cause fission. Only when the kinetic energy of the incident neutron is raised above 0.6MeV, fission can possibly occur. Thus, in the calculation of neutron balance in nuclear reactors, it is not necessary to consider the fission of ^{238}U caused by thermal neutrons, but only the fission caused by fast neutrons.

Heavy nuclei such as ^{233}U, ^{235}U, ^{239}Pu can undergo fission after absorbing thermal neutrons, and we usually refer to them as fissile nuclei. ^{241}Pu is also a fissile nucleus, but due to its short half-life, large fission cross-section, and low accumulation in reactors, it is costly to produce, so it is rarely extracted and used as fuel separately. The only naturally occurring fissile nucleus is ^{235}U. In addition, we refer to nuclei that require higher energy neutrons, such as fast neutrons, to cause fission as fissionable nuclei, including ^{232}Th and ^{238}U. ^{232}Th and ^{238}U can transform into fissile nuclei ^{239}Pu and ^{233}U after absorbing neutrons under certain conditions, as follows:

$$^{232}\text{Th}(\text{n},\gamma)^{233}\text{Th} \underset{23.5\text{Minutes}}{\overset{\beta^-}{\rightarrow}} {}^{233}\text{Pa} \underset{27.4\mathit{Days}}{\overset{\beta^-}{\rightarrow}} {}^{233}\text{U} \tag{2.44}$$

$$^{238}\text{U}(\text{n},\gamma)^{239}\text{U} \underset{23\text{Minutes}}{\overset{\beta^-}{\rightarrow}} {}^{239}\text{Np} \underset{2.3\mathit{Days}}{\overset{\beta^-}{\rightarrow}} {}^{239}\text{Pu} \tag{2.45}$$

Therefore, ^{232}Th and ^{238}U are also referred to as regenerative materials or convertible materials. ^{233}U, ^{235}U, and ^{239}Pu have all been used as nuclear fuels, but only

Table 2.1 Critical energy of several important nuclei (unit: MeV)

Fissionable nucleus ^{A}X	Critical energy E_c	Binding energy of a neutron with $^{A-1}$X nucleus to form ^{A}X nucleus E_B
^{232}Th	5.9	–
^{233}Th	6.5	5.1
^{233}U	5.5	–
^{234}U	4.6	6.6
^{235}U	5.75	–
^{236}U	5.3	6.4
^{238}U	5.85	–
^{239}U	5.5	4.9
^{239}Pu	5.5	–
^{240}Pu	4.0	6.4

Table 2.2 Microscopic cross-sections of thermal neutrons in fissile and fissile nuclides (unit: b)

Nuclide	σ_f	σ_c	σ_a	σ_f/σ_a
^{233}U	531	48	579	0.917
^{235}U	581	99	681	0.855
^{238}U	—	2.7	2.7	–
Natural uranium	4.2	3.4	7.6	0.553
^{239}Pu	743	269	1 012	0.734

^{235}U occurs naturally and is currently the most widely used nuclear fuel. However, in natural uranium ore, ^{238}U is the most abundant, accounting for approximately 99.28%, while ^{235}U only accounts for about 0.714%.

Table 2.2 provides the cross-sections (i.e., thermal neutron cross-sections) for several important fissionable nuclei at an energy of 0.0253 eV: σ_f is the fission cross-section, σ_c is the capture cross-section, σ_a is the total absorption cross-section, which equals the sum of σ_f and σ_c, and $a=\sigma_f/\sigma_a$ is the fraction of fissions caused by the absorption of a neutron by a given fissionable nucleus. Therefore, in ^{235}U, 85% of the absorbed thermal neutrons cause fission, while in ^{239}Pu, only 73% do.

Under certain conditions, the minimum mass of nuclear material required to achieve a chain fission reaction is called the critical mass.

2.5.2 *Fission Energy and Reactor Thermal Power*

(1) Fission Energy

We know that when nuclear fuel undergoes fission, it releases an enormous amount of energy, which can be calculated using the binding energy formula. When the nuclei of ^{235}U, ^{233}U, or ^{239}Pu absorb thermal neutrons and fission, they produce fragments while emitting two to three neutrons, β-rays, γ-rays, and neutrinos. The fission energy is initially manifested as the kinetic energy of these products. These particles gradually lose their kinetic energy by colliding with the nuclei of various materials in the reactor core, transferring energy to the impacted nuclei and causing the temperature to rise continuously until thermal equilibrium is achieved between the fission fragments, neutrons, and the impacted nuclei.

During nuclear fission, approximately 90% of the energy is released instantaneously, while the remaining portion—primarily the β and γ decay energy from fission products and neutron capture products—is released over time. The release period varies: some energy is emitted within seconds, while other portions may take years to be released. This delayed fission energy poses significant challenges to reactor safety. For example, after a reactor is shut down, continuous cooling is still necessary; otherwise, the reactor could overheat and eventually lead to core meltdown.

Table 2.3 Fission energy of several fission isotopes (unit: MeV/fission)

Isotope	E_f	Isotope	E_f
^{232}Th	196.2 ± 1.1	^{238}U	208.5 ± 1
^{233}U	199.0 ± 1.1	^{239}Pu	210.0 ± 1.2
^{235}U	201.7 ± 0.6	^{241}Pu	213.8 ± 1.0

The energy released per fission event depends on the fissile nucleus and varies slightly with core design, as neutron capture products are influenced by the materials used in the reactor. Table 2.3 lists the fission energy values for several common fissile isotopes. However, it is generally accepted that each fission of a ^{235}U, ^{233}U, or ^{239}Pu nucleus releases approximately 200 MeV of energy.

The kinetic energy of the fission fragments accounts for about 84% of the total fission energy. The flight path of the fragments in the fuel core is very short, about 0.0127 mm, so it can be assumed that the kinetic energy of the fission fragments is converted into heat energy at the site of nuclear fission. In a clean and evenly loaded reactor, the spatial distribution of this part of the heat energy is basically the same as the distribution of the thermal neutron flux in the fuel core. Fission neutrons collide with the nuclei of the moderator, converting their kinetic energy into heat energy, with a range varying from a few centimeters to several tens of centimeters. The gamma rays produced during fission have strong penetration power, and their kinetic energy is converted into heat energy in the reactor core, reflector, thermal shield, and biological shield, with a very small portion of gamma rays penetrating beyond the biological shield. The heat distribution produced by gamma rays depends on the specific design of the reactor. High-energy beta particles have a range of about 4 mm in the cladding and less than 0.254 mm in the uranium fuel core, so most of the energy of high-energy beta particles is converted into heat energy within the fuel element. A small portion of high-energy beta particles penetrate the fuel element and enter the moderator, but they do not reach outside the reactor core. The approximate distribution of ^{235}U fission energy is shown in Table 2.4, so in actual calculations, it is often assumed that each fission event releases 200 MeV of energy.

From the above analysis, it can be seen that the majority of the fission energy is converted into heat within the fuel elements. The distribution of fission energy is related to the specific design of the reactor and the control rod operation program. In the absence of precise data, for thermal neutron reactors, it can be assumed that over 90% of the fission energy is converted into heat within the elements, about 5% of the fission energy is converted into heat within the moderator, and less than 5% of the fission energy is converted into heat within the reflector and thermal shield. In the design of large pressurized water reactors, the heat release of the fuel elements is often taken as 97.4% of the reactor's heat release, while in boiling water reactors it is taken as 96%.

(2) Reactor Thermal Power

The heat energy released by the reactor per unit time is called the reactor's thermal power. Taking the energy released by each fission of ^{235}U as $E_f = 200$ MeV, and 1

Table 2.4 Approximate distribution of fission energy

Type		Source	Energy / MeV	Range	Location of conversion to heat energy
Fission	Prompt	Kinetic energy of fission fragments	168	Extremely short	Within fuel elements
		Kinetic energy of fission neutrons	5	Medium	Mostly within the moderator
		Kinetic energy of prompt γ	7	Long	Throughout the reactor
		Neutrinos	12	Long	Outside the reactor
	Delayed	Energy of β rays from fission product decay	8	Short	Mostly within the elements; a small part within the moderator and throughout the reactor
		Energy of γ rays from fission product decay	7	Long	
(n, γ) reactions caused by excess neutrons	—	γ decay energy produced by non-fission reactions caused by excess neutrons plus (n, γ) reactions	≈3~12	Varies from short to long	Throughout the reactor
Total		—	≈210~219	—	198~207 MeV can be utilized

MeV = 1.602×10^{-13} J. Therefore, the energy released by each fission of ^{235}U is approximately 3.204×10^{-11} J. Thus, to release 1J of energy, 3.12×10^{10} fissions of ^{235}U are required. If the fission rate of ^{235}U in the reactor is $R_f = \Sigma_f \phi$, then the power per unit volume at any point in the reactor, $q_v(r)$, is.

$$q_v(r) = E_f \Sigma_f \phi(r) = \frac{\Sigma_f \phi(r)}{3.12 \times 10^{10}} W/m^3 \tag{2.46}$$

where: Σ_f is the macroscopic cross-section of thermal neutron fission of ^{235}U in the reactor core; $\phi(r)$ is the neutron flux density at location r within the reactor core.

If the volume of the reactor core is $V(\text{m}^3)$, then the thermal power P of the reactor is.

$$P = \frac{R_f V}{3.12 \times 10^{10}} = \frac{\Sigma_f \overline{\phi} V}{3.12 \times 10^{10}} W \tag{2.47}$$

where: $\overline{\phi}$ The average thermal neutron flux density of the reactor core is represented as

$$\overline{\phi} = \frac{1}{V}\int \phi(r)\mathrm{d}V \tag{2.48}$$

Equation (2.47) indicates that the thermal power of the reactor is proportional to the fission rate. In a well-designed reactor, Σ_f is determined, which means that the thermal power of the reactor is proportional to the average thermal neutron flux density of the core. The higher the neutron flux density in the reactor, the greater the power, and vice versa. Therefore, we often use the level of neutron flux density in the reactor to represent the power level of the reactor.

Since 1 MW = 86 400 × 10^{10} J/d, the number of fissions per day for a reactor with a thermal power of P(MW) is

$$R_f = \frac{P \times 10^6 \times 86\,400\mathrm{J/d}}{200 \times 1.6 \times 10^{-13}\mathrm{J/Fission\ number}} = 2.68 \times 10^{21} P\mathrm{Fission\ number/d} \tag{2.49}$$

That is, it consumes 2.68 × $10^{21} P$ atoms of ^{235}U per day. After multiplying by the gram atomic weight of ^{235}U and dividing by Avogadro's constant, the above equation can be converted into the number of grams of ^{235}U fissioned per day as.

$$\text{Fuel consumption rate} \approx 2.68 \times 10^{21} P \times \frac{235}{6.02 \times 10^{23}} = 1.05\mathrm{Pg/d} \tag{2.50}$$

This indicates that the thermal power produced by the fission of 1.05 g of ^{235}U per day is 1MW. In reality, not all ^{235}U nuclei that absorb neutrons undergo fission; some are captured by neutrons and consume ^{235}U nuclei. Therefore, to obtain 1 megawatt-day of energy, the consumption of ^{235}U is 1.17 times the above value, i.e., $\sigma_a/\sigma_f = 1.17$. Hence, the actual consumption rate is 1.05 × 1.17 P = 1.23 P g/d.

The energy released per unit mass of nuclear fuel is called the burnup depth, usually expressed in "MW·d/t". 1t = 10^3 kg = 10^6 g. As an example, assuming the nuclear fuel is pure ^{235}U, if all ^{235}U undergoes fission, the energy released should be about 1 MW·d/g, equivalent to 10^6 MW·d/t. However, due to radiation capture, the maximum energy released by ^{235}U is about 8 × 10^5 MW·d/t. The burnup depth reached by the fuel unloaded from the reactor core is called the discharge burnup depth, which is one of the important indicators of power reactor design. Currently, the average discharge burnup depth of pressurized water reactors can reach more than 3 × 10^4 MW·d/t. For example, for a pressurized water reactor with a thermal power of 2800MW, the daily consumption of ^{235}U is 1.23 P = 3444 g/d, then the mass of ^{235}U consumed in one year is 1257.1 kg.

2.5.3 *Fission Fragments and Fission Neutrons*

(1) Fission Fragments

During the nuclear fission process, fission fragments and neutrons are produced. There are many ways of nuclear fission, most of which produce two fragments during fission. The product yield curve of fission fragments is shown in Fig. 2.5. From the figure, it can be seen that the shape of the curve varies with the energy of the neutron causing the fission. The figure shows the product yield curves of fission fragments when ^{235}U nuclei are fissioned by 14 MeV neutrons and thermal neutrons. For thermal neutron fission, more than 80 types of fission fragments have been found, indicating that there are more than 40 different fission pathways for ^{235}U nuclei. The mass numbers of the fission fragments are distributed approximately between 72 and 161. The curve also shows that the fission process is generally not symmetrical. The yield of symmetrical fission (where the two fragments have approximately equal masses, such as $A = 117, 118$) is only 0.01%, while the maximum yield of asymmetrical fission (where the two fragments have mass numbers of 95 and 139, respectively) is 6%.

In almost all cases, these fission fragments have a high neutron-proton ratio. They usually undergo a series of beta decays, converting excess neutrons into protons to become stable nuclei. We refer to the fission fragments and their decay products as fission products. Of course, some fragments emit neutrons during the beta decay process, and we usually call these fission products that emit neutrons during the decay process precursors of delayed neutrons. Due to the radioactive decay and accumulation of fission fragments, over 300 types of fission products will eventually be produced in the reactor. Some of these fission products, such as xenon and samarium, have a large thermal neutron absorption cross-section. They

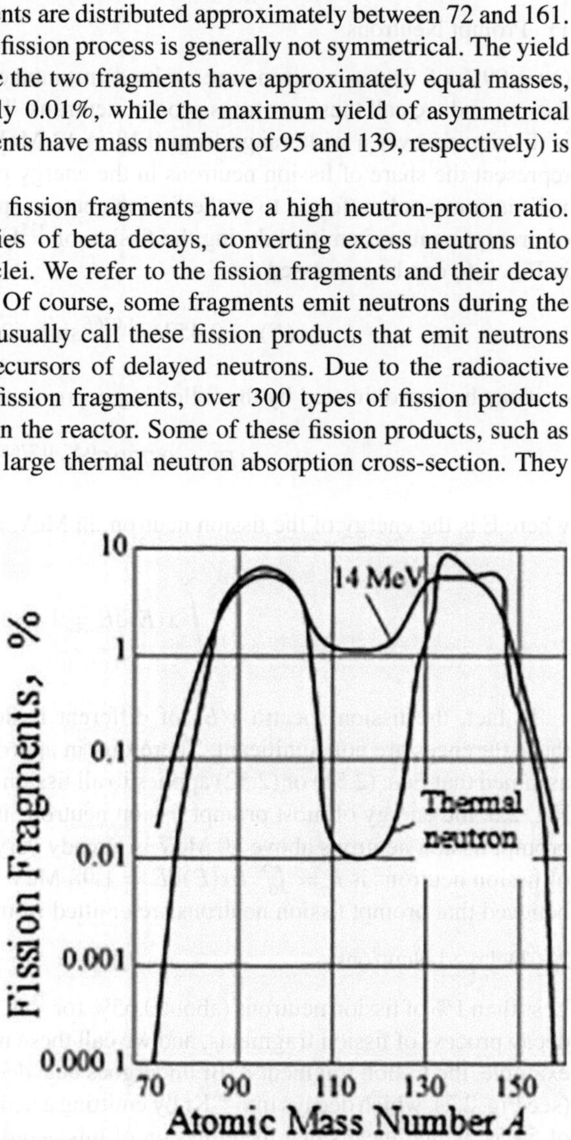

Fig. 2.5 The product yield curve of fission fragments

consume neutrons in the reactor, and we call these fission products with large neutron absorption cross-sections poisons.

(2) Fission Neutrons

Neutrons emitted from heavy nucleus fission are called fission neutrons, which are the most important fission products. We usually classify them into two types based on their characteristics, namely prompt neutrons and delayed neutrons, which are introduced as follows.

1) Prompt Neutrons

Over 99% of fission neutrons are emitted instantaneously (about 10^{-14} s) during fission, and we call these neutrons prompt neutrons. The energy of prompt neutrons is distributed over a wide range from 0.18 to 12 MeV. We usually use $x(E)$ dE to represent the share of fission neutrons in the energy range between E and E + dE. $x(E)$ is commonly referred to as the fission neutron spectrum. The energy spectrum of prompt neutrons emitted during the fission of ^{235}U (as shown by the solid curve in Fig. 2.6) can be expressed as:

$$x(E) = 0.453e^{-1.036E}\sinh\sqrt{2.29E} \tag{2.51}$$

or it can be approximated by the following expression (see dashed line in Fig. 2.6)

$$x(E) = 0.770E^{0.5}e^{-0.776E} \tag{2.52}$$

where E is the energy of the fission neutron, in MeV, and it satisfies

$$\int_0^\infty x(E)\mathrm{d}E = 1 \tag{2.53}$$

In fact, the fission spectra $x(E)$ of different fission isotopes are different, but the differences are not significant. Therefore, in approximate calculations, it can be assumed that Eqs. (2.51) or (2.52) applies to all fission isotopes. As can be seen from Fig. 2.6, the energy of most prompt fission neutrons is around 1 MeV. The share of prompt fission neutrons above 10 MeV is already very small. The average energy E of fission neutrons is $E = \int_0^\infty Ex(E)dE = 1.98$ MeV ≈ 2 MeV, and it is generally believed that prompt fission neutrons are emitted isotropically.

2) Delayed Neutrons

Less than 1% of fission neutrons (about 0.65% for ^{235}U fission) are emitted during the decay process of fission fragments, and we call these neutrons delayed neutrons. For example, the fission fragment 87Br undergoes beta decay to produce an excited ^{87}Kr (see Fig. 2.7), which decays into ^{86}Kr by emitting a neutron. This decay has a half-life of 54.5s, which means that the emission of this neutron is delayed compared to the instant of fission. Fission fragments like ^{87}Br, which can produce delayed neutrons

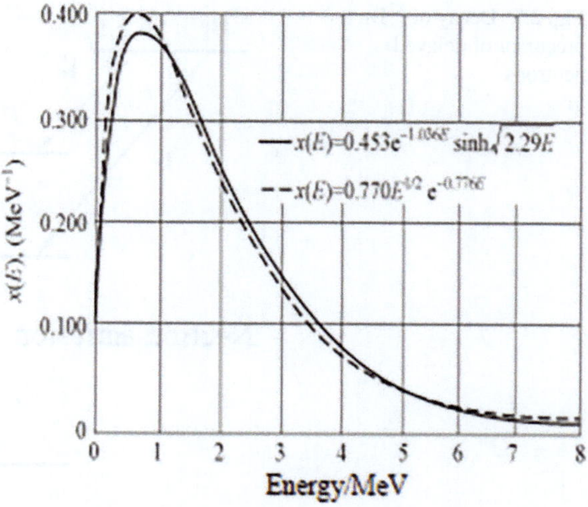

Fig. 2.6 Energy spectrum of fission neutrons during thermal neutron fission of ^{235}U nucleus

during the beta decay process, are the precursors of delayed neutrons mentioned earlier. Based on the length of the half-life of the precursor nuclei, they are usually divided into six groups, with the relevant parameters listed in Table 2.5, where $T_{1/2}$, λ_{i}, and E_{i} represent the half-life, decay constant, and energy of the *i*th group of delayed neutron precursors, respectively; v_i represents the average number of delayed neutrons produced in the *i*th group from a single fission, known as the yield of the *i*th group of delayed neutrons; β_{i} represents the proportion of the *i*th group of delayed neutrons in all fission neutrons (prompt and delayed), known as the fraction of the *i*th group of delayed neutrons; $\beta = \sum_i \beta_i$ represents the total fraction of delayed neutrons produced in each fission, which is also the proportion of delayed neutrons in all neutrons produced in each fission (prompt and delayed), often used in subsequent neutron dynamics analysis; β_i/β represents the proportion of the *i*th group of delayed neutrons in all delayed neutrons, known as the relative fraction of the *i*th group of delayed neutrons; t_{i} represents the average lifetime of the precursor nuclei of the *i*th group of delayed neutrons, where $t_{\mathrm{i}} = 1/\lambda_{\mathrm{i}}$, then the average lifetime of all delayed neutrons is represented by $\bar{t}$, similarly for the thermal fission of ^{235}U, there is $\bar{t} = \sum_i \beta_i t_i / \sum_i \beta_i$.

The energy spectrum of delayed neutrons is different from that of prompt neutrons; it is discontinuous. As can be seen from Table 2.5, the average energy of delayed neutrons is much lower than that of prompt neutrons. Although delayed neutrons account for a very small proportion of all fission neutrons (less than 1%), it is precisely because of these less than 1% delayed neutrons that control of the fission rate or power within the reactor is possible.

Fig. 2.7 Decay of ^{87}Br, a precursor of delayed neutrons

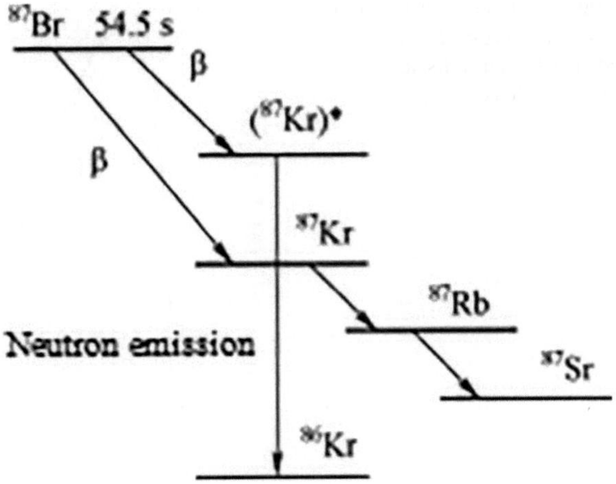

Table 2.5 ^{235}U thermal neutron fission delayed neutron data

Group	1	2	3	4	5	6
Half-life $T_{1/2,\,i}$ /s	55.72	22.72	6.22	2.30	0.61	0.23
Energy E_i/ keV	250	560	430	620	420	430
Fraction β_i	0.00021	0.00142	0.00128	0.00257	0.00075	0.00027
Average lifetime t_i /s	80.39	32.78	8.97	3.32	0.88	0.33
Decay constant λ_i/(1/s)	0.0124	0.0305	0.111	0.301	1.14	3.01
Product yield v_i	0.00052	0.00346	0.00310	0.00624	0.00182	0.00066
Relative fraction β_i/β	0.032	0.219	0.196	0.395	0.116	0.042
Total fraction β	0.0065					
Total product Yield v	0.0158					

2.5.4 Nuclear Fusion

Nuclear fusion, also known as thermonuclear reaction, involves combining light nuclei with small binding energies (mass numbers generally less than 16), such as deuterium and tritium, under certain conditions, into a medium-mass nucleus with a larger average binding energy. This process releases a tremendous amount of energy (also resulting in mass loss, which is converted into energy output, hence the term "nuclear fusion energy"). This reaction is one of the important ways to obtain nuclear energy. Due to the strong electrostatic repulsion between atomic nuclei, the probability of nuclear fusion under normal conditions is very small. Only at temperatures of tens of millions of degrees Celsius do light nuclei have enough kinetic energy to overcome electrostatic repulsion and undergo sustained nuclear fusion, hence the term "thermonuclear reaction" or "hot nuclear reaction". Since the electrostatic repulsion of atomic nuclei is directly proportional to the product of their charges, the smaller the atomic number, the lower the kinetic energy (i.e., temperature) required for fusion. Some lighter atomic nuclei, such as hydrogen, deuterium, tritium, helium, lithium, etc., are most likely to release fusion energy. The average binding energy of ^{4}He nuclei is much larger than that of some nearby light nuclei (see Fig. 2.1), so its fusion reaction is more probable and releases more energy. Specifically, the fusion of two deuterium nuclei ${}_1{}^2$H into a helium nucleus ${}_2{}^4$He is an example. Because the specific binding energy of ${}_1{}^2$H nucleus is 1.11 MeV, and that of ${}_2{}^4$He nucleus is 7.07 MeV. Therefore, separating two deuterium nuclei into two protons and two neutrons requires $1.11 \times 4 = 4.44$ MeV of energy; while fusing into a helium nucleus can release $7.07 \times 4 = 28.28$ MeV of energy. Therefore, the fusion of two deuterium nuclei into a helium nucleus can release 28.28–4.44 = 23.84 MeV of binding energy. Comparing the fission of ^{235}U with the fusion of deuterium nuclei, it can be seen that for fission, the energy released per nucleon is approximately $200/235 \approx 0.85$ MeV; while for fusion, the energy released per nucleon is approximately $23.84/4 \approx 6$ MeV. Therefore, when the same weight of material participates in the reaction, the energy released by the fusion reaction is much greater than that of the fission reaction, which is a major reason why controlled thermonuclear reactions are of particular interest.

However, it has been found that deuterium and tritium are more likely to undergo fusion. The process of a deuterium nucleus and a tritium nucleus fusing to form a helium nucleus and a neutron is as follows:

$$ {}_1{}^2\mathrm{H} + {}_1{}^3\mathrm{H} \rightarrow {}_2{}^4\mathrm{He} + {}_0{}^1\mathrm{n} \tag{2.54} $$

This fusion reaction can release 17.2 MeV of energy. Deuterium and tritium are generally referred to as fusion nuclear fuels. The water compound of deuterium is called heavy water. Deuterium is abundant in nature, with 0.03 g of deuterium in every liter of water, and about 4 billion tons of deuterium in the water on Earth, so the fusion fuel on Earth is much more abundant than fission fuel. Tritium is rare in nature, with only a few kilograms, mainly from cosmic rays. However, tritium can be produced from lithium. The amount of lithium on Earth is relatively abundant and can meet human needs for fusion energy.

Nuclear fusion is one of the significant ways to harness nuclear energy. Fusion energy is not only abundant but also safe and clean. We now know that in the universe, the primary source of energy is nuclear fusion. The sun and many other stars in the universe can emit light and heat for a long time, continuously radiating energy to the outside world due to light nuclear fusion. Of course, to achieve such fusion reactions, a certain amount of energy must first be provided to the system. If we can control nuclear fusion reactions and gradually release energy, then we can use nuclear fusion energy for power generation (we call the device that releases energy through nuclear fusion reactions a fusion reactor). This is also a significant topic in the field of high technology.

We now know that to achieve controlled nuclear fusion reactions, we first need to heat nuclear fuels such as deuterium and tritium to high temperatures. At such high temperatures of tens of millions of degrees, the deuterium–tritium mixed fuel has exceeded the three familiar states of matter: solid, liquid, and gas, and has become the fourth state of matter, known as plasma. So, what is plasma? Simply put, plasma is a conductive gas. If the gas is heated to tens of thousands of degrees, the gas atoms will all ionize, turning into positively charged ions and negatively charged free electrons. This gas composed of ions and electrons is plasma. The higher the temperature of the plasma, the greater the density, the longer the confinement time (the time to maintain high temperature), the more energy is released. When the temperature reaches a critical point, the released energy is sufficient to heat the next addition of deuterium–tritium fuel, and the fusion reaction can continue. This is called the "ignition" of controlled nuclear fusion. There are currently three major difficulties to be solved to achieve "ignition". The first is heating, that is, how to heat the plasma to such a high temperature. The second is confinement, even if the heating temperature is reached, the high-temperature plasma cannot contact the wall of the container holding it, otherwise, the plasma will cool down, and the container will be burned. The third is cleanliness, that is, how to prevent impurities from mixing into the plasma, because impurities will greatly increase radiation loss and cool the plasma. If these difficulties can be overcome and ignition is achieved, the next step is to establish a fusion reactor and achieve the peaceful use of fusion energy. The main components of the fusion reactor should include a high-temperature fusion plasma core, a position layer, a screen layer, a magnet, auxiliary systems, and a plasma system, as well as a heat exchange system and power generation device similar to a nuclear power plant. For more than 60 years, people have been striving to explore ways to achieve "controlled fusion reactions". Due to different heating and confinement methods, there are several types of experimental devices. Tokamak (an abbreviation of the Russian words for "toroidal", "chamber", and "magnetic"), magnetic mirror, magnetic compression device, and laser fusion device are all experimental devices designed and manufactured for the study of controlled fusion reactions.

2.6 Reactivity and Its Control

2.6.1 Definition of Reactivity

The size of neutron generation and disappearance in a nuclear reactor determines the state of the nuclear reactor, and the size of neutron generation and disappearance in the nuclear reactor is reflected in the effective multiplication factor k (the ratio of neutron generation to neutron disappearance). When the effective multiplication factor k is greater than 1, neutron generation is greater than neutron disappearance, and the number of neutrons in the reactor will increase over time, and the nuclear reactor is in a supercritical state; when the effective multiplication factor k is less than 1, neutron disappearance is greater than neutron generation, and the number of neutrons in the reactor will decrease over time, and the nuclear reactor is in a subcritical state; when the effective multiplication factor k is equal to 1, neutron

Table 2.6 Description of reactor states

Reactor State	Subcritical	Critical	Supercritical
k	<1	=1	>1
k_{ex}	<0	=0	>0
ρ	<0	=0	>0

generation equals neutron disappearance, and the number of neutrons in the reactor will not change over time, they can balance at any value, and the nuclear reactor is in a critical state.

However, balance is relative, and imbalance is absolute. For various reasons, the effective multiplication factor k cannot always be equal to the critical value of 1. The difference between the effective multiplication factor and 1 is usually called the excess multiplication factor, represented by k_{ex}:

$$k_{ex} = k - 1 \tag{2.55}$$

Then the ratio of k_{ex} to k is called reactivity, represented by ρ.

$$\rho = k_{ex}/k = (k - 1)/k \tag{2.56}$$

Equation (2.56) is the definition and calculation formula of reactivity. Since k is always greater than zero, so ρ and k_{ex} has the same sign, either positive or negative, both of which are measures or parameters describing the degree to which a reactor deviates from the critical state. In the study of reactor dynamics, reactivity is commonly used to characterize the degree to which a reactor deviates from the critical state, and it is a very important dynamic physical quantity.

Since reactivity represents the relative change in k, it is often expressed as a percentage, or measured by the fraction of delayed neutrons β. When the reactivity unit represented by the β value is called a "dollar", 1 β is a "dollar", and one percent of it is called a "cent". The relationship between k, k_{ex}, and ρ is shown in Table 2.6.

In summary, the sign of reactivity indicates the state of the reactor; and the magnitude of reactivity will determine the rate of change in neutron flux density (or reactor power) within the reactor.

2.6.2 *Factors Influencing Reactivity*

For a newly built reactor, in order to ensure that it can operate at the required power for a long time, it must have sufficient reactivity reserve. The reactivity reserved at cold state (20 °C) before a newly built reactor is put into operation, that is, the remaining reactivity of the clean cold core, is called the backup reactivity of the reactor, denoted as ρ_0. After the nuclear reactor starts operation, the core temperature changes from cold to hot, the temperature change range is large, and the temperature change will cause reactivity changes; in addition, during the operation of the reactor, nuclear fission continuously produces fission fragments, some of which absorb neutrons, thereby causing reactivity changes; furthermore, the fission in the reactor continuously changes, and the fuel composition changes accordingly. These changes cause reactivity changes, therefore, for pressurized water reactors, the main effects affecting reactivity changes are the following three.

(1) Temperature Effect

After the reactor is put into operation, the core temperature will rise, causing the core material density to decrease, and the neutron temperature to rise and uranium nucleus thermal motion to intensify, these changes lead to reactivity changes. We call the phenomenon of reactivity changes caused by core temperature changes as the temperature effect, the increase in reactivity caused by the rise in core temperature is called the positive temperature effect (positive feedback), and the decrease in reactivity caused by the rise in core temperature is called the negative temperature effect (negative feedback). Reactors are usually designed to have a negative temperature effect. The temperature effect of different types of reactors is different, the temperature effect of pressurized water reactors is the most significant, and it is a negative effect. Thus, we call the reactivity change caused by the temperature effect as temperature reactivity. For a pressurized water reactor operating at rated parameters, this change is about 2–12%.

(2) Poisoning Effect

During the operation of a nuclear reactor, fission fragments and their decay products are continuously produced and accumulated. Among these products, some have a large thermal neutron absorption cross-section, the most prominent of which are ^{135}Xe and ^{149}Sm. We usually call these harmful fission products and their decay products as nuclear poisons. We call the phenomenon of reactivity reduction caused by nuclear poisons capturing neutrons as the poisoning effect, and the reactivity reduction caused by nuclear poisons is called poisoning reactivity. For different nuclear reactors, different operating powers and operating times, the size of poisoning reactivity is different. Generally, for a pressurized water reactor at rated conditions, its equilibrium poisoning reactivity is about 2–5%.

(3) Burnup Effect

After a nuclear reactor is put into operation, nuclear fission fuel will continuously be consumed. This consumption, which reduces the number of fissionable nuclei and causes a decrease in reactivity, is known as the burnup effect. The reactivity reduction caused by burnup is referred to as burnup reactivity. The magnitude of burnup reactivity varies greatly depending on the type of nuclear reactor, its operating power, and operating time. Generally, the higher the reactor power and the longer the operating time, the deeper the burnup and the greater the reduction in reactivity. For a typical pressurized water reactor, this reduction is approximately 3–12% at the end of its operation.

In different operating conditions and at different operating times, the aforementioned three effects (temperature, poisoning, and burnup) have varying degrees of prominence. For example, when a nuclear reactor transitions from a cold state to a hot state or experiences a significant change in operating temperature, the temperature effect is dominant. When the reactor operates at high power or experiences a significant change in power, the poisoning effect becomes significant. When the nuclear reactor operates for a long period, the burnup effect becomes increasingly apparent.

In summary, to ensure that a nuclear reactor, especially a nuclear power reactor, can operate for a sufficient number of days at its design power, it must have enough reactivity reserve before it starts operating to compensate for the reactivity reduction caused by temperature, poisoning, and burnup effects. For a typical pressurized water reactor, the value of the backup reactivity is usually in the range of 15–25%.

2.6.3 Reactivity Control Tasks

After the nuclear reactor starts, due to effects such as temperature and poisoning, the reactivity within the reactor changes. To achieve the goals of starting the reactor, increasing or decreasing power for stable operation, and shutting down, external control methods must be adopted to control reactivity.

(I) Common Physical Quantities

1) Excess Reactivity

At any moment, the maximum reactivity that can be obtained by adjusting the reactivity of control poisons is called the "excess reactivity" of the reactor at that moment, denoted as ρ_{ex}. Control poisons refer to all substances in the reactor used for control, such as control rods, burnable poisons, and chemical compensation poisons. The magnitude of excess reactivity is related to the operating time and conditions of the reactor. The excess reactivity of a new reactor is called the "initial excess reactivity". For a new reactor core, the initial excess reactivity is at its maximum in a cold, poison-free state.

2) Shutdown Margin

When all control poisons are inserted into the reactor core, the negative reactivity achieved by the nuclear reactor is called the shutdown margin, denoted as ρ_s. Obviously, the shutdown margin is also related to the operating time and conditions of the reactor. To ensure the safety of the nuclear reactor, it is required to have a sufficiently large shutdown margin under hot, equilibrium xenon poisoning conditions. Otherwise, as the reactor core gradually cools and ^{135}Xe decays, the reactivity of the reactor will gradually increase, and the shutdown margin will gradually decrease, potentially causing the reactor core to return to a critical or supercritical state. In the design of pressurized water reactors, the shutdown margin is generally specified to be 3–5% in a cold, poison-free state, and it must be ensured that the shutdown margin is greater than 2–3% when a set of control rods is stuck and cannot be inserted into the reactor core in a cold, poison-free state.

3) Total Control Equivalent

The total control equivalent equals the sum of excess reactivity and shutdown margin, denoted as $\Delta\rho$. That is,

$$\Delta\rho = \rho_{ex} + \rho_s \tag{2.57}$$

The reactivity change caused by a set of control rod assemblies fully inserted into the reactor core is defined as the reactivity equivalent of that set of control rod assemblies.

(2) Main Tasks

The main tasks of reactivity control are: to adopt various effective control methods to control the excess reactivity of the reactor, ensuring safety while meeting the needs of long-term reactor operation; to maintain a relatively flat power distribution throughout the reactor core life by controlling the spatial arrangement of poisons and the optimal rod withdrawal program, making the power peak factor as small as possible; to adjust the reactor in response to external load changes, enabling it to adapt to these changes; and to quickly and safely shut down the reactor in the event of an accident, maintaining an appropriate shutdown margin.

2.6.4 Basic Principles and Methods of Reactivity Control

(1) Basic Principles

In principle, the reactivity of a thermal neutron reactor can be controlled by manipulating six factors in the effective multiplication factor, namely by changing neutron absorption within the reactor, altering neutron moderation properties, adjusting fuel content, and modifying neutron leakage. In reality, once the components of a thermal neutron reactor are determined, the fast neutron multiplication factor and the yield of neutrons absorbed per event can be considered essentially constant. Moreover, controlling the probability of escaping resonance capture is not very effective, while controlling the thermal neutron utilization factor and the non-leakage probability is more effective and easier to implement.

Moving control rods can control the number of neutrons absorbed by the fuel, thereby changing the value of the thermal neutron utilization factor. Additionally, moving control rods can cause some form of distortion in the neutron flux density, changing the neutron leakage value of the reactor core, thereby altering the non-leakage probability. Typically, power reactors use control rods to control rapid changes in reactivity.

(2) Control Methods for Pressurized Water Reactors

Due to the large initial residual reactivity of pressurized water reactors, the total control equivalent is substantial. The grid of a pressurized water reactor is relatively dense, and the efficiency of control rods is relatively low. If all reactivity is controlled by control rods, a large number of control rods would need to be arranged, which is challenging. Currently, power station pressurized water reactors use three control methods: control rods, solid burnable poisons, and boron acid solution in the coolant. Naval pressurized water reactors use a combination of control rods and solid burnable poisons.

The advantages of using control rods to control a reactor are speed, reliability, flexibility, and maneuverability. The disadvantage is that it causes significant disturbance to the neutron flux density distribution within the reactor, leading to uneven distribution of neutron flux density. Additionally, a large number of control rods can pose difficulties for reactor structural design and manufacturing.

(3) Solid Burnable Poison Method

To reduce the number of control rods, burnable poisons such as boron steel tubes can be inserted into a new reactor core in a certain distribution. This method takes full advantage of the fast "burnup" of boron, which significantly reduces the neutron absorption capacity of the boron steel tubes as the reactor burnup deepens. This reactivity compensation is automatic and does not require external control. If solid burnable poisons can be reasonably arranged within the reactor, they can automatically release reactivity as they burn up, precisely compensating for burnup reactivity. This not only reduces the number of control rods but also helps to flatten the neutron flux density distribution within the reactor, improving fuel utilization and extending the core life.

Discussion Questions

1) What is an atomic nucleus composed of, and what are its characteristics?
2) What are mass defect and binding energy?
3) What is specific binding energy? What is the pattern of its change with atomic mass number?
4) What is nuclear decay? What laws does nuclear decay follow?
5) How is the average life of a radioactive nucleus defined? What is its relationship with half-life?
6) What conservation laws should nuclear reactions follow?
7) What is reaction energy? Why can it be positive or negative?
8) What is a neutron nuclear reaction? How many types of nuclear reactions produce neutrons?
9) What are fission and fusion reactions? What are their differences?
10) How many types of neutrons are there based on energy range?
11) What are the definitions of neutron density and neutron flux density? What are their differences?

12) What are microscopic cross-section and macroscopic cross-section? What are their units? What are their differences?
13) What are the definitions of nuclear reaction rate and Doppler effect?
14) What are fission critical energy and critical mass?
15) What is a fissile nucleus? What are the main types of nuclei?
16) What is fission energy, and what are its release characteristics? How much energy is approximately released when a ^{235}U nucleus undergoes fission?
17) The neutron density of energy $E_1 = 0.0253$ eV is $n_1 = 10^5$ neutrons/cm^3, while the neutron density of energy $E_2 = 1$ keV is $n_2 = 10^3$ cm^{-3}, which has a greater neutron flux density?
18) According to calculations, the thermal power produced by the fission of 1.23 g of ^{235}U per day is 1MW. How much ^{235}U needs to undergo fission to produce 1 kW·h of energy?
19) If the thermal power of a reactor is 330 MW, how many ^{235}U nuclei undergo fission per second?
20) What are fission fragments and fission neutrons?
21) What are prompt neutrons and delayed neutrons? Briefly explain their differences.
22) Why does fusion reaction release much more energy than fission reaction when the same weight of material is involved?
23) What is reactivity? Why is the sign of reactivity indicative of the state of the reactor?
24) Briefly explain what is meant by temperature effect, poisoning effect, and burnup effect.
25) What are residual reactivity and shutdown margin?
26) What are the main control methods for pressurized water reactors?

References

Chen WZ, Zhang LM, Xiao HG. (2019) Principles of Nuclear Reactor Engineering (Volume I). Beijing, China Atomic Energy Press.

Wu HC, Cao LZ, et al. (2014). Nuclear Reactor Physics. Beijing, China Atomic Energy Press.

Yan FJ. (2021). Atomic Physics. Beijing, Higher Education Press.

Yan CQ, Cao XR. (2014). Nuclear Reactor Engineering. Harbin, Harbin Engineering University Press.

Zhang DH. (2023). Advanced Fast Neutron Reactor Technology. Shanghai, Shanghai Jiao tong University Press.

Chapter 3 Applications of Nuclear Technology

Nuclear energy encompasses three types of technologies: technologies required to produce nuclear explosions; technologies that enable nuclear reactors to generate energy and apply it; applications of nuclear technology. The first two technologies use the fission of uranium and plutonium atoms to release energy. The public knows little about the applications of nuclear technology, but these technologies are diverse and have a profound impact on all aspects of daily life (Li et al. 1994). A study in the United States in 1992 pointed out that non-power peaceful applications related to nuclear technology involve $357 billion in revenue and 3.7 million jobs each year. Now, nuclear technology applications in the United States seem to be four times larger than the nuclear power industry. It seems necessary to pay attention to the application of nuclear technology (Wang 2002).

Decades ago, atomic energy aroused various lofty and yearning dreams in people. They dreamed that one day the electricity generated by atomic energy would be so cheap that people all over the world could enjoy cars, trains, and aircraft that could run without refueling; seawater could be desalinated; barren deserts could become oases. Mr. Winston Churchill at the time saw the peaceful use of atoms as a "long stream of prosperity for the world". Although these early dreams, under the influence of deep fear and harsh reality in the post-World War II years, outlined the necessity, public image, and expectations of peaceful development of atomic energy, not all dreams had withstood the test of time, and some had burst prematurely. However, many dreams had been realized in research laboratories, hospitals, factories, and farmers' fields. They had produced lasting results that benefit today's society. Especially the achievements from 2005 to 2020 have given people hope to solve many problems on the planet using nuclear energy technology. Facts have also shown that nuclear technology plays an important role in many fields.

In the context of fostering emerging industries and supporting the transformation and upgrading of the national economy in China, it is even more important to carry out research on the application of nuclear technology in different fields.

L. Yu et al., *Nuclear Technology and Safety*, Nuclear Science and Technology,
https://doi.org/10.1007/978-981-95-2856-1_3

3.1 Industrial Applications of Nuclear Technology

The industrial application of nuclear technology is extremely wide (Wei and Jia 2011). By 2016, nearly 150 countries and regions around the world had carried out research and development on the application of nuclear technology in the industrial field. The United States has listed nuclear technology as one of the 22 major technology development directions to be prioritized. The industrial application of nuclear technology has become a global industry. Many emerging countries are developing their radiation processing industries with great enthusiasm and practical steps; some countries are exploring the use of comprehensive technologies, including nuclear technology, to transform their resource advantages into economic advantages. The rapid development of industrial applications of nuclear technology has become one of the driving forces for innovative development of new technologies, new materials, new processes, and new methods. In the field of industrial application of nuclear technology, the technology of developed countries and regions such as the United States and Europe is in a leading position.

The application of nuclear technology in industry is mainly reflected in analysis and detection during the production process, radiation processing, isotope tracing, etc. Various isotope monitoring instruments integrated based on the principle that the radiation emitted by radioactive nuclides can be used as an information source to obtain control parameters and other information in industrial processes, such as nuclear logging in the development and production of oil fields, coal fields, mines, etc., have been widely used. It plays an irreplaceable role in mineral development and industrial production control, such as thickness gauges, nuclear scales, radiographic flaw detectors, container inspection systems, ion smoke fire alarms, etc., which can be used to monitor production processes, achieve non-destructive testing or component analysis, detect fire situations, etc., and are important components of industrial process control and automated monitoring.

Some countries have developed and deployed ^{60}Co (Cobalt-60) container CT inspection systems, neutron imaging detection systems, as well as explosive detection devices, high-energy electron beam sterilization systems, and radioactive material detection equipment based on nuclear technology. These are concrete achievements in the application of nuclear technology, providing powerful tools for combating smuggling and counterterrorism while playing an increasingly vital role in safeguarding lives, property, and national security. For example, radiation processing, an important manifestation of industrial applications of nuclear technology, has achieved significant results in fields such as cross-linked cables, heat shrink materials, rubber vulcanization, positive temperature coefficient (PTC) devices of polymer resistors, sterilization of medical supplies, food irradiation preservation, and wastewater and exhaust gas treatment, forming an industrial scale. China's radiation processed products include more than 20 categories and over 300 types, forming three pillar products: heat shrink materials, radiation cross-linked wires and cables, and radiation emulsion polymerization.

Sterilization of medical supplies has become the main method for disinfecting large quantities of medical supplies. Radiation sterilization can be applied to thousands of medical supplies, including metal products, plastic products, and disposable high polymer material medical supplies. Both traditional Chinese and Western medicines and cosmetics can also be sterilized by radiation, with broad development prospects.

Approximately 45 countries worldwide have approved over 200 types of irradiated food, with an annual market sales volume reaching over 300,000 tons. Moreover, food irradiation has been used abroad as an effective means to prevent foodborne diseases and carry out international agricultural product quarantine. In addition, isotope irradiation devices and complete sets of equipment developed to meet the needs of radiation processing, high-power irradiation accelerators and complete sets of equipment, as well as large-scale industrial online detection, safety detection equipment for hazardous materials, radiation therapy equipment, etc., using isotopes and accelerators as radiation sources, have formed an industrial scale. The annual total output value of the radiation processing application field is growing at a rate of about 20% per year.

Radiation processing (or modification) based on ion implantation provides a powerful technical platform for the upgrading of semiconductor materials and the development of the electronic information industry. There are more than 3000 ion implanters used for integrated circuits worldwide. The radiation modification of silicon rectifier devices (thyristors) with a 12 MeV electron accelerator has been widely used; metal ion sources implanted into material surfaces enhance the mechanical strength and wear resistance of the material surface and have been widely used in tool processing, the automotive industry, and the aerospace industry; N-type high-resistance silicon manufactured by neutron transmutation doping on nuclear reactors is suitable for preparing high-quality high-resistance materials. Ion implantation has become one of the means of microelectronic circuit processing. Ion beam etching has the advantages of high resolution and short exposure time compared to light beam and electron beam etching. The low-temperature plasma etching method will replace the high-temperature deposition method and become the only economically feasible method for fine pattern high-fidelity transfer in large-scale integrated circuit production. Implanting oxygen ions into semiconductor silicon wafers forms an insulating SiO_2 layer under a very thin single-crystal silicon layer. This type of semiconductor material has the advantages of low power consumption, fast time response, and small volume, and will have good development prospects. As of 2019, more than 80 countries worldwide have carried out applications of nuclear technology, with more than 1000 electron accelerators used for radiation processing, with a total power of about 45 MW; there are about 250 irradiation devices of ^{60}Co radiation sources, with an intensity of 9.25×10^{18} Bq.

3.1.1 Application of Tracer Technology

Nuclear tracer technology is an important part of nuclear technology applications, including stable isotope tracer technology and radioactive isotope tracer technology.

Nuclear tracer technology is also known as isotope tracer technology or tracer atom method, simply referred to as tracer method. It refers to the technical method of studying the movement and transformation rules of the traced substance using tracers (marked atoms or marked compounds). With the advancement of science and technology, nuclear tracer technology has gone through the process of birth, development, and perfection, experiencing the research process from simply tracing elements to tracing complex compounds. It provides unique and precise research methods for understanding objective things. It helps humans understand many phenomena and processes that are difficult to reveal by other methods. In addition, the rapid development of nuclear technology provides more precise detection instruments for nuclear tracer technology, especially more available stable and radioactive isotopes and their marked compounds, greatly promoting the development of nuclear tracer technology and expanding its application range.

Radioactive tracing technology has a wide range of applications in industry, demonstrating unique advantages in solving problems that are difficult to address by other methods. For example, it is used to study the wear and tear of mechanical parts such as engines, bearings, machine tools, and cutting tools, providing a scientific basis for improving the wear resistance of parts, studying wear mechanisms, and selecting wear-resistant materials. For instance, the internal combustion engines of large power plants generally operate continuously, with their pistons and crankshaft connecting rods working non-stop. How can we determine whether the piston or piston rings need to be replaced without stopping the engine? In addition to regular maintenance and replacement, neutron irradiation can be used to transform the iron in the piston rings into radioactive iron isotopes. During engine operation, the metal particles produced by wear and tear will detach from the piston rings and fall into the machine's lubricating oil. If the lubricating oil is connected to a detector, it can easily measure the amount of radioactive iron, thereby determining whether the piston and piston rings need to be replaced. In the metallurgical industry, radioactive isotope tracing technology is crucial for determining the corrosion of refractory materials in steelmaking blast furnaces, the operation time of steelmaking, the content and distribution of impurities in products, and for studying the solidification process, dust movement patterns, and distribution characteristics. For example, on-site measurement of the corrosion of refractory furnace lining materials is challenging because it is impossible to inspect at high temperatures. If the corrosion and wear of the furnace lining exceed the limit and continue to be used without detection, it can cause significant damage. Tracing technology effectively solves this problem. In operation, radioactive isotopes (usually cobalt-60) can be embedded at a certain depth (the limit depth of wear and corrosion) in the refractory materials of the furnace lining. During production, the radioactive intensity can be measured at regular intervals. Based on the changes in radioactive intensity, the degree of corrosion and wear of the furnace lining can be determined. Currently, tracing technology has become a mature technology for on-site monitoring of the wear of refractory materials in blast furnaces in the metallurgical industry. In casting processes, the application of radioactive isotope tracing technology can conveniently and accurately determine the crystallization rate of steel ingots and continuous casting billets, impurities, and the causes of defects. Underground oil pipelines in airports, refineries, and oil fields often leak due to corrosion or quality issues, causing not only oil loss but also environmental pollution and fires. Tracing technology can effectively solve this problem. In operation, aviation oil containing radioactive ^{131}I-benzene can be added to the pipeline to be inspected, followed by the addition of non-radioactive aviation oil, and then a detector (leak detector) is suspended in the aviation oil. When the radioactive oil leaks out through the hole and seeps into the soil, the leak detector moving with the oil can detect the location of the leak. This method is simple, accurate, and economical. Similarly, this method is also suitable for finding blockages in oil pipelines and inspecting underground pipelines in the petroleum and chemical industries. In the catalytic industry, radioactive isotope tracing technology can be used to study catalysts, catalytic mechanisms, and reaction mechanisms, creating conditions for controlling the production and use of catalysts and developing excellent catalysts.

3.1.2 Application of Radiation Processing Technology

The United States is a major country in the development of the world's nuclear technology application industry, with an industry scale reaching 600 billion dollars in 2013. The industry's characteristics are particularly evident in radiation processing: for example, high-performance heat shrink materials and products, irradiated astronaut food being marketed for civilian use; irradiation devices are large-scale and highly specialized, with a single source capacity exceeding 12 million curies; research is in-depth and application fields are broad. For instance, in food irradiation processing, the United States has conducted systematic research on safety and quarantine treatment, and has introduced a series of regulations, making it one of the most advanced countries in the world for commercial irradiation treatment of agricultural products and food. India, Thailand, Vietnam, and other countries use irradiation technology to treat tropical fruits, achieving large-scale exports; many countries use irradiation technology for quarantine treatment of imported and exported grains to ensure the safety of their species.

Radiation processing is an important application field of civilian non-power nuclear technology, mainly referring to the process of improving the quality or performance of the processed object by using gamma rays, electron beams, and X-rays. Radiation processing can be applied to medical and health products, drugs, food, biological products, agricultural and sideline products, chemical products, polymer materials, and "three wastes", for disinfection, sterilization, insect killing, preservation, cross-linking polymerization, modification, degradation, and harmless treatment. Due to its strong penetration, no pollution residue, and high efficiency, radiation processing has gradually penetrated into various fields of the national economy and is known as "green" processing technology. The irradiation devices used for radiation processing mainly include gamma irradiation devices, electron accelerator irradiation devices, and X-ray irradiation devices.

According to statistics from the International Atomic Energy Agency, there are approximately 300 gamma irradiation devices in over 50 countries worldwide, with a total source capacity of about 400 million curies. In addition, there are over 2000 industrial irradiation electron accelerators with a total power exceeding 80 MW. The main countries producing cobalt sources for industrial irradiation include Canada, Russia, China, Argentina, and India. The primary manufacturers of industrial gamma irradiation devices are Canada, China, India, and Hungary. The main producers of industrial irradiation accelerators are the United States, Canada, Belgium, Russia, China, Japan, France, and Switzerland.

Material improvement is one of the important applications of radiation processing. Material improvement mainly includes applications such as radiation polymerization, radiation cross-linking, radiation degradation, radiation grafting, and radiation curing. Electron or gamma irradiation can alter various properties of polymers such as polyethylene. The original material is composed of long parallel molecular chains, and radiation causes these chains to link together. This process is called cross-linking. Irradiated polyethylene has good heat resistance and is an excellent material for wire insulation. Radiation can bind suitable polymers to a fiber substrate. This method can be used to produce various dust-resistant fabrics.

To improve the reliability and safety of wire and cable products, modern industry has higher requirements, such as temperature resistance, environmental aging resistance, and crack resistance, which conventional wires and cables cannot meet. By cross-linking the insulation layer of wires and cables, the working temperature, solvent resistance, environmental aging resistance, and crack resistance of the wires and cables can be greatly improved. Radiation cross-linking can also cause changes in the electrical properties of insulation materials and improve mechanical performance. Radiation cross-linked cables are currently an important area of radiation processing, such as photovoltaic halogen-free flame-retardant cables, F40 radiation cross-linked cables for aircraft, cables for nuclear power plants, and low-smoke halogen-free flame-retardant cables resistant to 150 °C. These cannot be produced by conventional chemical methods.

Currently, heat shrinkable material products are widely used in sectors such as electricity, electronics, communications, telecommunications, transportation, construction, petroleum, chemical, automotive, shipbuilding, aerospace, etc., mainly for insulation, corrosion prevention, moisture proofing, sealing, labeling, wire and cable splicing, branching, wire harnessing, terminals, machinery, etc. The radiation processing industry of heat shrinkable materials has become an indispensable part of the national economy. The use of irradiation technology in heat shrinkable materials was developed by the American company Raychem as early as 1960. It has undergone half a century of technological and process development, is relatively mature, and often uses a method of molding before irradiation in production.

In terms of tire irradiation applications, the United States and Japan began research in 1957 and 1977 respectively. Currently, the popularity of Japan's meridian tire irradiation pre-vulcanization technology exceeds 90%, and many Japanese tire companies have set up low-energy electron accelerators on their tire production lines in China for tire irradiation vulcanization. China began in 2000 to carry out technical transformations on existing tire production lines, conducted experimental research on tire irradiation pre-vulcanization technology, and developed embedded electron accelerators for tire irradiation pre-vulcanization, promoting the development of China's irradiation pre-vulcanization technology.

3.1.3 Radioactive Isotope Logging and Gamma Ray Flaw Detection

(1) Radioactive Isotope Logging

Radioactive isotope logging has been widely used in coalfield geological exploration and oil exploration, and has become an indispensable means in exploration. We know that the main strata in coal seams are siltstone, sandstone, limestone, etc., all of which have a higher density than coal, and the density variation of coal produced anywhere is similar. When gamma rays irradiate light materials, they are minimally absorbed and mostly reflected back; when gamma rays irradiate heavy materials, they are absorbed more and reflected back less. Based on these characteristics of gamma rays, it is possible to distinguish between coal and other rocks of different densities and masses. During exploration, a hole is first drilled with a drill, then a cobalt-60 gamma ray source and a detector are inserted. The detector is separated from the radiation source to avoid direct gamma ray entry, and the detector only detects scattered gamma rays. The position of the coal seam can be determined based on the amount of scattered gamma rays measured. Using this method in combination with other logging methods, the thickness of the coal seam can also be accurately measured.

The oil industry widely uses neutron logging technology to explore for oil and natural gas. The principle is similar to radioactive isotope gamma ray exploration for coal, but the gamma ray source is replaced with a neutron source, using the neutron lifetime or neutron-gamma of pulsed neutrons. Well logging, this method can be used to estimate the oil reserves in geological formations and delineate the oil–water interface. Using pulsed neutron lifetime logging, the distribution of mineralized water and oil layers in rock formations can be determined. Another method, known as neutron-induced gamma-ray spectroscopy logging, can be used to measure the carbon/oxygen ratio in geological formations, thereby distinguishing between oil and non-oil layers. Neutron logging technology can also be used to explore for mineral deposits such as boron, copper, silver, manganese, tungsten, mercury, and rare earth elements.

(2) Gamma Ray Inspection

Gamma rays can be used to detect defects such as damage and cracks in industrial equipment, especially large-scale equipment (including airplanes, ships, satellites, rockets, spacecraft, etc.), making it one of the main methods of non-destructive testing. Gamma ray inspection devices are simple to use, convenient, and highly intuitive, making them particularly suitable for inspecting large castings, forgings, and welds. A gamma ray inspection device mainly consists of a controller, a detector system, and a power container. The controller includes a power supply, an electronic control system, an electric lock, and a timer. The power container includes a display, relays, a power motor, a wire storage bottle, and protective equipment. The detector system includes a radiation source and a detector. Normally, the radiation source is kept inside a protective container. During operation, the radiation source can be automatically controlled to move into the working position to minimize the operator's exposure to radiation.

Cobalt-60 and iridium-192 are commonly used in gamma ray inspection devices. When the thickness of the steel being inspected is greater than 50 mm, cobalt-60 is often used as the radiation source. When the steel thickness is between 10 and 50 mm, iridium-192 is commonly used. When the steel thickness is less than 10 mm or for other light metal materials (such as aluminum), thulium-170 is generally used. If the steel thickness reaches 400 to 500 mm, an accelerator must be used as the radiation source. In gamma ray inspection technology, the correct choice of radiation source is important for improving sensitivity, clarity, and reducing exposure time.

Currently, the number of countries using gamma ray inspection technology is increasing, and the range of applications is also expanding. It is estimated that about 40% of X-ray inspections have been replaced by gamma ray inspections.

3.2 Application of Nuclear Technology in Agriculture

The main applications of nuclear technology in agriculture are radionuclide tracing technology and nuclear radiation applications (Li and Chen 2005). That is, using radionuclides and nuclear radiation with biology to explore the characteristics of radionuclides and nuclear radiation, as well as the physical, chemical, and biological effects produced by their interactions with matter. This reveals scientific phenomena in the field of agriculture and solves technical problems in agricultural production. Therefore, the development of nuclear science and technology provides the material and technical basis for the application of nuclear technology in agriculture. The application of nuclear technology in agriculture is an important part of nuclear science and technology, and it is also the continuation and development of nuclear science and technology, gradually forming a new discipline of nuclear agriculture. Nuclear agriculture is characterized by using nuclear physics and radiochemistry as research methods, with living matter, life activities, and the living environment as research objects. It mainly includes: plant radiation-induced mutation breeding, agricultural product irradiation processing, agricultural radionuclide tracing, radiation biology, and insect radiation sterility research.

Radiation-induced mutation breeding: This mainly involves using radiation to treat plant seeds or other organs (tissues), inducing mutations in their genetic material, causing changes in their traits, and then selecting beneficial mutations to create new germplasm and breed new varieties. Radiation-induced mutation breeding has the technical advantages of breaking bad gene linkage, innovating gene traits, increasing mutation frequency, shortening breeding cycles, and improving breeding effects.

Agricultural product irradiation processing: This mainly involves using gamma rays, X-rays, or electron beams to treat agricultural products in order to achieve effects such as inhibiting sprouting, delaying ripening, killing insects, sterilizing and disinfecting, and degrading harmful substances for preservation and storage. This maintains the nutritional quality and flavor of the products, extends their shelf life, and improves the safety of agricultural products for consumption. This technology is a food safety assurance technology strongly promoted by the United Nations Food and Agriculture Organization, the International Atomic Energy Agency, and the World Health Organization. It has the advantages of being safe, green, efficient, low-carbon, and environmentally friendly. It does not require heating, does not damage packaging, and can solve technical problems that conventional technologies cannot solve in the field of preservation and storage.

Agricultural radionuclide tracing: This mainly involves using isotopic tracers to study the migration, transformation, transport, absorption, and metabolic laws of the traced substances in plants or animals and their surrounding environment. Because tracing technology can reveal the absorption, metabolism, and transformation laws of substances from a microscopic perspective, it is widely used in basic and applied research in disciplines and fields such as nutritional physiology, soil chemistry, plant protection, animal ecology, and environmental protection. It provides scientific basis and technical support for high yield, high efficiency, high quality, and water-saving cultivation of major crops such as wheat and corn, scientific fertilization, irrigation, pest and disease control, and healthy animal production.

Radiobiology: This field primarily studies the effects, action patterns, and mechanisms of nuclear radiation on biological characteristics, including the basic concepts of radiobiology, the initial and secondary reactions of radiation on organisms, and the physical, chemical, and biological changes. Research in radiobiology provides a theoretical basis for radiation protection, radiation-induced mutation, insect radiation sterilization, and the irradiation preservation and freshness of agricultural products.

Insect Radiation Sterilization: This involves using a certain dose of radiation to irradiate insects, damaging their reproductive system while maintaining their mating and survival abilities, thus rendering them sterile. These sterilized insects are then released into the natural environment where they mate with wild insects but cannot produce offspring, thereby reducing insect population density and achieving pest control.

3.2.1 Application of Tracer Techniques

By using radioactive isotope tracer techniques, we can study the metabolism of crops, and the roles of roots, stems, leaves, fruits, and other parts in this process. Questions such as how to fertilize soil, what is the optimal timing for fertilizer application, the mechanism of photosynthesis, pesticide residues, and drug usage standards can be addressed. The principle of tracer techniques in agriculture involves replacing non-radioactive atoms in compounds with radioactive atoms of the same element, such as replacing the ^{12}C atoms in organic compounds with ^{14}C atoms. This allows us to track the movement of these radioactive isotopes during chemical and biological reactions or migration processes.

In agricultural production, the application of isotope tracing techniques aims to optimize the relationship between fertilizer, soil, and crops to achieve high crop yields. Isotope tracing is very useful for measuring the effective nutrients in soil and is the simplest and most sensitive method. After nitrogen enters the soil, it undergoes a series of physical, chemical, and biological changes. The transformation of soil nitrogen is part of the natural cycle. Therefore, using isotope tracing to measure the utilization rate of nitrogen fertilizer (not limited to nitrogen fertilizer) is an important means of improving fertilizer utilization. For example, using ^{15}N tracing to study the timing and pathways of nitrogen fertilizer loss, and the transformation of soil nitrogen, can provide information on the loss and improved utilization of nitrogen fertilizer in soil. Tracer techniques are also widely used in biological nitrogen fixation research. Over the past 50 years, ^{15}N or ^{13}N tracing techniques have been mainly used to confirm nitrogen fixation and study nitrogen metabolism during the fixation process. When confirming nitrogen fixation, the basic method is to grow nitrogen-fixing organisms in systems containing ^{15}N and ^{13}N for a certain period, then sample and analyze. If ^{15}N is detected in their products or derivatives using mass spectrometry, or if ^{13}N is detected using a radiation counter, it indicates that the organism can fix nitrogen.

Phosphorus in crops is absorbed from soil containing phosphate through the roots. However, the effectiveness of phosphorus as a fertilizer varies widely according to different experimental results. By applying a solution of phosphate containing radioactive phosphorus to the soil and observing the growth process of crops, samples can be taken at different times and from different parts. Based on the changes in radioactive phosphorus in different parts, we can observe how phosphorus is absorbed by the roots of crops, how it rises along the stem, and how it is distributed in leaves and fruits through the leaf veins. Finally, a picture of phosphorus absorption can be given, indicating which parts of the crop contain more radioactive phosphorus, and the effect of time on the amount of phosphorus absorbed by different parts of the crop can be calculated. Using this method, research has been conducted on crops such as tomatoes and tobacco. It was found that most of the phosphorus in tomatoes is concentrated in the seeds, and the phosphorus content in diseased tobacco leaves is higher than in healthy ones. Another successful example is the study of crop roots and leaves, which found that the leaves of crops are as important as the roots. They not only serve as a source of carbon dioxide—carbon, but also absorb mineral salts, helping the roots supply crops with nitrogen, phosphorus, and other necessary elements. This discovery has changed the traditional view that fertilization should only be applied to the roots, and provides a basis for fertilizing crops not only at the roots. Using this new fertilization method, cotton yields have increased by 10–20%, potato yields by 7%, and bean yields by 17%.

In photosynthesis research, oxygen-18 is used as a tracer element in water components. It was discovered that oxygen is derived from the decomposition of water during photosynthesis, rather than from carbon dioxide as previously thought. The released hydrogen combines with the carbon in carbon dioxide molecules to further produce carbohydrates. Using carbon-14 as a tracer element, it was found that roots can also absorb carbon dioxide from the soil and transport it to the green parts of crops. Previously, it was believed that carbon could only enter crops from the air. These results can lead to modifications of traditional fertilization methods, and a new approach to applying inorganic and organic fertilizers to enrich the soil with carbon dioxide.

Understanding the persistence of pesticides on plant surfaces, their accumulation in fruits and soil, the elimination of excessive pesticide residues, and their effects on pests are all important issues. Currently, scientists have successfully identified the most effective and least toxic methods of pesticide application by using labeled compound formulations. This provides a basis for pest control, improvement of pesticide application techniques, and prevention of pesticide pollution.

In livestock farming, the use of radioactive isotope tracing technology can study various issues in livestock and poultry farming, such as nutrition and metabolic processes in livestock and poultry, genetic changes in livestock populations, etc. This is of great significance for the rational feeding and development of livestock. For example, in livestock feed, protein plays an important role. For ruminant mammals (such as cows and sheep), a lack of 20–25% protein can lead to a reduction in yield of more than 20%, and the cost will increase by 50%. Radioactive isotope tracing technology can be used to study the absorption and digestion process of ruminant mammal feed, such as determining the secretion rate of various acids produced by microbial life activities in the stomach, the nitrogen deficiency in microbial protein synthesis, the conditions required for protein absorption, and identifying the appropriate feed additives under various conditions.

Tracing technology can also be used to study the formation process of milk in cows. It was found that milk fat is mainly synthesized from low molecular weight fatty acids, which are formed by the activity of microbes in the cow's stomach. Other components in milk fat (such as phospholipids) are synthesized in the liver, while the main protein in milk comes from amino acids formed by glucose itself. This result would be difficult to obtain without the use of tracing technology. Of course, the examples listed above do not cover all applications of radioactive isotope tracing technology, and the scope of this technology is constantly expanding. These examples are helpful for understanding the characteristics of radioactive isotope tracing technology itself.

3.2.2 *Application of Radiation Technology*

(1) Pest Sterilization by Radiation

The insect sterilization technology using nuclear radiation (SIT technology), which was developed in the early 1950s, uses nuclear radiation to alter the genetic material of pests, reducing their potential reproductive capacity. It is a genetic control method that uses the instinct of pests to perpetuate their species to eliminate the species itself.

The nuclear radiation sterilization method involves irradiating a certain stage of the pest to be controlled with a certain dose of ionizing radiation. From a biological principle, the body cells of the irradiated pests are basically not damaged, but the chromosomes of their reproductive cells are induced to break and transpose, causing asymmetric combinations, forming gametes with dominant lethal mutations, and causing the zygotes formed after mating with normal insects to die. These dominant lethal mutations are not gene point mutations, but chromosome breaks, leading to abnormal nuclear division later. Ultimately, all the formed zygotes die. Therefore, pests treated with a certain radiation dose can maintain normal life activities and the ability to find mates, but they can make the eggs they produce after mating unable to hatch, lose reproductive ability, and cannot reproduce offspring.

Of course, there are many technical biological issues in the irradiation treatment, which are not detailed here. However, the usual practice is to artificially breed a large number of a certain pest, then use radiation to sterilize it, and release a large number in the area where the pest is harmful. All natural population pests that mate with irradiated sterile individuals lose their ability to reproduce offspring. After several generations of release, the number of natural populations decreases, and the pests in this area are eventually completely eliminated.

The large-scale release of effective radiation-sterilized insects is the foundation of sterilization technology. However, insect breeding factories have corresponding production capabilities, and the large-scale production of radiation-sterilized insects is limited and sometimes economically unfeasible. Therefore, scientists propose four scenarios for using insect radiation sterilization technology to control pests: (1) when established pest populations are at low levels or their spread and distribution are limited during certain periods of the normal recurrence cycle; (2) to eliminate newly invaded pest populations in new areas; (3) to prevent pests from establishing populations in new areas; (4) to first reduce the pest population using other methods, then release sterile insects to control this pest. For example, if chemical pesticides are first used to reduce the number of pests, and then sterile individuals are released multiple times, and the release quantity only needs to be 1/10 of the amount required when using the sterilization method alone, this can lead to the extinction of the pest species and can be completed one generation earlier. Therefore, the prominent feature of using insect radiation sterilization technology to control pests is to completely eradicate pests on a large scale and solve the problem once and for all. This avoids a series of drawbacks brought about by general chemical control, which played an important role in the development of pest control in the past, such as pests developing resistance to chemical pesticides, residual toxins causing harmful effects on ecological balance, killing of natural enemies leading to rampant pests, and secondary pests rising to become major pests. Environmental pollution is also a global key issue, and chemical pesticides are an important source of pollution. Using as few chemical pesticides as possible is an important measure for humans to protect their living conditions. Therefore, the use of insect radiation sterilization technology to eliminate important pests is an important reason that has attracted worldwide attention.

From an economic benefit perspective, most people believe that using insect radiation sterilization technology is very costly and troublesome, and its investment is much higher than the cost of general pesticide spraying. This statement is not wrong, because large-scale insect breeding requires a lot of manpower, equipment, and materials, such as feed, radiation sources, etc. The release of sterile insects also requires airplanes, ground staff, etc., which are indispensable procedures. However, if you carefully calculate, the economic benefits brought by completely eradicating pests are incalculable. For example, the United States used this technology to eradicate the Mediterranean fruit fly on Curaçao Island and in Florida. Although it cost tens of millions of dollars, this cost is only equivalent to the cost of the losses suffered in 3–5 years when the Mediterranean fruit fly pest was not eradicated, but afterwards, there is no need to control the Mediterranean fruit fly ever again. This also shows that the economic benefits of eradicating pests with insect radiation sterilization are considerable.

(2) Biological Growth Stimulation

Nuclear radiation is ubiquitous on Earth, coming from the Earth itself, the sun, and other outer space. Organisms have always lived in an environment with a certain dose of nuclear radiation and other radiation. From the generation of embryos, the formation of juveniles, the development of adults, the reproduction of offspring to the death of life, they have never left an environment with nuclear radiation. Therefore, nuclear radiation has become one of the indispensable conditions for biological growth and development. Some people put protozoa in a lead shielding room. Due to the reduction of nuclear radiation in the environment, their growth and development were significantly inhibited. When a γ-ray source equivalent to the background dose was placed in the lead room, their growth returned to normal.

The annual radiation dose equivalent received by organisms is 2×10^{-4} to 6×10^{-4} Sv. Different regions and environments have differences and will have different impacts on organisms. The growth and development of organisms are restricted by many factors. Although the differences caused by environmental radiation factors are not easy to detect, they objectively exist. Exploring this impact, changing the irradiation environment, providing the best irradiation conditions, can stimulate biological growth and development, and improve the yield, quality, and benefits of agricultural production.

Experiments have shown that direct or indirect ionizing radiation, with a very low dose, once or multiple times irradiating biological cells, tissues, organs, or bodies, not only will not change the genetic basis and affect offspring, but this physical factor, like hormones, will stimulate their growth and development, and can have beneficial effects on the vitality, morphology, and physiological and biochemical properties of the current generation. This effect is called the low-dose stimulation effect. For example, irradiating crop seeds before sowing with an appropriate low dose of nuclear radiation can break dormancy earlier, improve seed germination rate, promote growth and development, increase sprouting, improve quality, enhance disease resistance and vitality, and increase the yield of chloroplasts and grains. In addition, sometimes beneficial morphological changes can occur. Also, irradiating the fertilized eggs, embryos, juveniles, or young animals of farmed animals with an appropriate low dose can accelerate metabolism, promote growth and development, enhance immune function, reduce mortality, shorten the time to maturity, and improve product yield and quality.

Research on the stimulatory effects of low-dose ionizing radiation has a history of nearly a century abroad, and has been widely applied in countries such as the Commonwealth of Independent States, Canada, and Hungary. Pre-sowing irradiation of seeds can increase the yield of cereals and legumes by 10–20%, vegetables by 20–30%, and potatoes by 15–20%. The Commonwealth of Independent States has built various types of mobile irradiation devices. Reports indicate that large irradiation vehicles are equipped with a 3500 Ci ^{137}Cs gamma-ray source and an electric conveyor system, capable of processing 750–1000 kg of seeds per hour. China has also begun to promote its application, for example, irradiating oak silkworm eggs with a radium-beryllium neutron source significantly stimulates increased production. Ionizing radiation of freshwater fish eggs and fry yields similar results. Neutron irradiation technology for shrimp increases production by an average of 15%. Low-dose gamma-ray irradiation of eggs improves the resistance of eggs during incubation, increases the hatching rate by 6.86%, the healthy chick rate by 5.52%, and increases the egg production of the current generation of hens by 8.03%.

In agriculture, low-dose irradiation technology is generally used to increase production by irradiating seeds before sowing. Seeds generally do not need special treatment, but it is necessary to ensure that the timing of irradiation is appropriate, and the dose is suitable, accurate, and uniform. Therefore, it is required: to calibrate monitoring instruments regularly; to irradiate seeds before sowing, it is best to use a conveyor belt to pass them uniformly through the irradiation field. If irradiation is carried out in container packaging, the volume of the container should not be too large, and the container should be turned over when the irradiation dose reaches half to ensure uniform irradiation of the seeds; seeds irradiated should be sown within a few days, as prolonged storage may affect the effect; as the germination rate of irradiated seeds increases, the number of seeds sown can be appropriately reduced; because seedlings grow well and develop quickly, the requirements for water and fertilizer increase, and corresponding management should keep up to ensure increased production.

Low-dose irradiation growth stimulation technology applied to crops can promote seed germination and budding, especially for some seeds that are difficult to germinate and have a low germination rate. After irradiation treatment, the dormant state can be relieved, the germination rate can be increased, seedling emergence can be accelerated, and the uniformity of seedling emergence can be improved.

In addition, it can stimulate crop growth and increase yield. Irradiation can not only accelerate crop growth, but also increase the yield of chloroplasts and seeds, fruits, and tubers. For example, the Soviet Union sowed buckwheat in fields irradiated with gamma rays, subjected it to low-dose irradiation for two months, and accumulated irradiation of 8.6 to 1100 mGy. As a result, within the dose range of 41 and 204 mGy, the yield of crop chloroplasts increased significantly compared to before irradiation. Some people used 2.9 Gy of gamma rays to irradiate cucumber seeds, increasing the yield by 10–15%. Irradiating rice with 4.8–5.9 Gy gamma rays increased the yield by 5.8–10.7%. And irradiating wheat increased the yield by 12.8–14%.

Nuclear radiation stimulation has a greater effect on increasing the production of oak silkworms and mulberry silkworms. This technology has the effects of increasing production, disease resistance, and improving silk quality. According to statistics, the general increase in production is more than 10%. In Chinese provinces such as Liaoning, Henan, Shaanxi, Gansu, Heilongjiang, Jilin, and Shandong, there is a certain scale of application, especially welcomed by sericulturists.

Low-dose irradiation growth stimulation technology has also been applied in the aquaculture industry and has achieved great success. This technology can not only increase the yield of biological products but also improve the quality of biological products. For example, in fish and shrimp farming, tests have shown that irradiated shrimp have a 6.7% increase in edible parts and a 1.3% increase in protein. Various inorganic salts, such as calcium, phosphorus, copper, and selenium, have increased to varying degrees.

In the breeding industry, neutron source irradiation is generally used to irradiate the young animals. They also have a detailed and strict operating procedure to ensure the increased production of irradiated animals.

3.3 The Application of Nuclear Technology in Medicine

Nuclear medicine is the product of the combination of nuclear technology with medicine and related disciplines (Luo 2015). It is an interdisciplinary and marginal discipline. Nuclear medicine uses tracers labeled with nuclides (including radioactive nuclides and stable nuclides) for medical and biological (in vivo and in vitro) medical (mainly including diagnosis and treatment) and research purposes. Its development can be traced back to the early twentieth century. In 1901, shortly after the discovery of the radiation phenomenon, French physicians D'Arlo and Bloch placed radium salts on the skin surface to treat tuberculous skin damage, pioneering the use of nuclides to treat human diseases; in 1920, Hungarian radiochemist Hevesy used ^{201}Pb to study the movement of lead in plants, discovered the metabolic turnover rate, and used ^{214}Bi to study the metabolism of bismuth in rabbits, which can be regarded as the early biomedical application of isotope tracing. In 1925, American clinical nuclear physician Blumgart injected ^{214}Bi into the cubital vein to measure the time taken for radioactivity to appear in the contralateral limb, thereby determining the difference in blood flow velocity between healthy individuals and patients. In 1934, after the discovery of artificial radioactivity by Joliot-Curie couple, Hevesy used ^{32}P to study the metabolic behavior of phosphorus in mice. In 1937, the Lawrence brothers, American physicists, built a cyclotron that could produce milli-curie level ^{32}P, which was then used to treat leukemia. In 1938, American chemists Livingood and Seaborg discovered ^{131}I, which was quickly used to treat thyroid cancer. In 1939, Hamilton/Solly and Evans from the University of California, Berkeley, conducted in vitro simulation studies on the metabolic behavior of radioactive iodine. In 1941, Pecher first used ^{89}Sr to treat metastatic bone cancer. The establishment of nuclear reactors in the 1940s expanded the range of available radionuclides for nuclear medicine. During this period, scintillation detectors, scanners, and autoradiography were successively developed, while radiolabeled compounds such as ^{3}H, ^{14}C, ^{32}P, ^{125}I, and ^{131}I were widely used in life sciences research. Among these radionuclides, ^{131}I became the most extensively applied in medicine—^{131}I-rose Bengal for hepatobiliary imaging, ^{131}I-ortho-iodohippurate for renal function testing, and Na-^{131}I for treating thyroid disorders. In the 1960s, American scientists Berson and Yalow pioneered radioimmunoassay (RIA), and RIA kits were rapidly commercialized and adopted worldwide. Concurrently, the development of the ^{99}Mo-^{99m}Tc generator enabled hospitals far from reactors and accelerators to conveniently use ^{99m}Tc-labeled radiopharmaceuticals for clinical diagnostics. In the second half of the twentieth century, the rapid development of electronics technology, computer technology, and image reconstruction technology provided strong technical support for the development of nuclear medicine, and the invention and continuous improvement of gamma cameras and emission computed tomography brought nuclear medicine into a period of rapid development.

With the development of computer technology, humans have developed technologies to fuse CT, MRI, and ECT images, which can integrate information obtained from various imaging technologies, accurately determine the size and range of lesions and their relationship with surrounding tissues, thereby obtaining more physiologically meaningful functional parameter maps, giving nuclear medicine a unique advantage in clinical diagnosis of diseases. Entering the new century, the development of nuclear medicine still depends on the development of imaging technology, radioactive nuclides, radioactive drugs, and molecular biology and other related technologies.

Nuclear medicine includes two parts: diagnosis and treatment. Diagnosis is divided into in vivo and in vitro examinations, and in vivo examinations are further divided into radioactive nuclide imaging and non-imaging examinations.

3.3.1 Radioactive Nuclide Diagnosis

Organ diagnostic examinations include functional examinations and localization examinations (such as tumors). Here, the different uptake and exclusion effects of radioactive drugs labeled with radioactivity by healthy organs and diseased organs are used. Since the small amount of substance applied generally does not affect metabolism, it is possible to grasp the disease at the molecular level before morphological changes are manifested. The radioactive drug is usually injected into the patient's blood vessels, and after a certain period of time, the gamma rays emitted by the examined organ are measured. Based on the spatial distribution of the emitted rays, a color scintillation map of the organ is made by the computer.

This process has two methods: SPECT (Single Photon Emission Computed Tomography) and PET (Positron Emission Tomography).

SPECT is a nuclear medical diagnostic device developed by combining gamma camera and electronic computer technology, used to obtain three-dimensional distribution images of radioactive nuclides in the human body. Compared with the plane images of the gamma camera, SPECT has obvious advantages, overcoming the disadvantage of plane imaging that masks small lesions due to organ and tissue overlap, and improving the resolution and positioning accuracy of deep lesions.

Almost all human organs and bone marrow can be diagnosed with suitable radioactive nuclides. For example, renal function examination with ^{131}I labeled hippuric acid, by recording the change in radiation emitted from the kidney over time, the function of the kidney can be judged. Similar to this is the ventilation examination: patients inhale air mixed with a small amount of short-lived ^{81m}Kr (half-life: 13 s). The recorded radiation allows assessment of lung lobe function and can also aid in tumor localization.

The radioactive isotopes used in SPECT imaging should ideally only emit single-energy gamma rays and not charged particles, as the latter not only do not contribute to imaging but also increase unnecessary internal radiation exposure to the patient. The energy of gamma rays should ideally be between 100 and 300 keV. If the energy is too low, the absorption loss from the emission point exiting the body increases. If the energy is too high, the required thickness of the collimator increases.

Like other nuclear medicine imaging techniques, PET also uses tracer principles to display the body's metabolic activity. However, PET has two important features that distinguish it from other nuclear medicine imaging techniques: First, the radioactive tracers it uses are labeled with positron-emitting isotopes. Commonly used positron isotopes in PET include ^{18}F, ^{11}C, ^{15}O, ^{13}N, etc., which are isotopes or similar elements that make up the human body. These isotopes replace the isotopes in the tracer molecules without changing their original biological characteristics and functions, thus more objectively and accurately displaying the body's metabolic information. Second, it uses coincidence detection technology. By replacing collimators with coincidence detection, both sensitivity and spatial resolution, which were originally mutually restrictive, are significantly improved.

PET is also an imaging device that reflects the genetic, molecular, metabolic, and functional status of lesions. It uses positron-emitting isotopes to label human metabolites as imaging agents, reflecting metabolic changes through the lesion's uptake of the imaging agent, thereby providing clinical information about the disease's metabolic activity. This is a new milestone in the development of life science and medical imaging technology. After introducing compounds or metabolic substrates such as glucose, fatty acids, amino acids, and water, labeled with positron-emitting isotopes into the body, chemical images of the body are obtained using a positron scanner. Because it can display the metabolic activity of organs or tissues and the function and distribution of receptors, PET has received widespread clinical attention and is also known as "live biochemical imaging". The advent of PET has brought medical imaging technology to a new level, making it possible to non-invasively, dynamically, and quantitatively evaluate the physiological and biochemical changes in cellular metabolic activity of living tissues or organs under physiological conditions and during disease processes, and to obtain molecular-level information, which is currently impossible with any other method. PET can also be used to reveal the relationship between the concentration of a drug in plasma and its local concentration in the brain. This shows that, in addition to medical diagnosis, PET can also be used to evaluate drugs and toxins. PET is widely used in clinical practice in developed countries and has become the most effective means of diagnosing and guiding the treatment of tumors, coronary heart disease, and brain diseases, the three major threats to human life. Currently, the most commonly used PET imaging agent is ^{18}F-FDG (an ^{18}F-labeled glucose analog—fluorodeoxyglucose).

The radioactive isotopes used in PET imaging should ideally only emit $\beta+$ particles and not gamma rays, as the latter would increase accidental coincidence counts and reduce the signal-to-noise ratio. The half-life of the isotope should ideally be between 10 s and 80 h. If the half-life is too short, it is difficult or even impossible to label it on the carrier molecule; if the half-life is too long, the residual radioactive activity in the body after imaging is too high, causing additional radiation to the patient, which limits the total activity of the radioactive drugs used for imaging. Isotopes with shorter half-lives can be injected in larger quantities, and after collecting sufficient data in a short time, they decay quickly, which is beneficial for obtaining high-quality images.

3.3.2 Radiation Therapy

Radiation therapy is a local treatment technique for malignant tumors that uses rays (such as X-rays, gamma rays, electron beams, etc.) produced by various radioactive isotopes. The goal of radiation therapy is to strive to improve the radiation therapy gain ratio, that is, to maximize the concentration of radiation dose in the lesion area, kill tumor cells, and minimize or avoid unnecessary radiation to surrounding normal tissues and organs. Radioactive isotope therapy is based on the foundations of radiation physics, radiation biology, radiation therapy technology, and clinical oncology. These substances or facilities capable of emitting radiation include radioactive isotopes (such as ^{60}Co, ^{192}Ir, ^{137}Cs which can produce γ-rays, and ^{252}Cf which can generate neutron beams), superficial and deep X-ray machines (capable of producing X-rays), and accelerators (which can generate X-rays, electron beams, proton beams, neutron beams, meson beams, and heavy particle beams). Radiation therapy, which began in the late nineteenth century and has a history of over a hundred years, has become one of the three main methods for treating malignant tumors, alongside chemotherapy and surgery. Prior to the 1940s, the effectiveness of radiation therapy was not fully demonstrated due to rudimentary equipment and a lack of understanding of the mechanisms of radiation. However, since the 1950s, the rapid development of contact X-ray therapy machines, superficial X-ray therapy machines, deep X-ray therapy machines, ^{60}Co teletherapy machines, various medical electron accelerators, and brachytherapy machines, along with the use of ultra-high voltage therapy machines, improvements in auxiliary tools, and accumulated experience, have significantly improved the effectiveness of radiation therapy. It can now be used for almost all types of cancer treatment. According to statistics, over 70% of cancer patients in China and over 50% in the United States require radiation therapy.

In radiation therapy for cancer, high energy is deliberately given to the tumor to destroy it, while the surrounding healthy tissue is allowed to receive as low a dose as possible. Currently, γ-radiation (either from ^{60}Co or accelerator-induced radiation) or high-energy electrons (emitted by an electron accelerator) are commonly used. These radiation sources cannot avoid causing a relatively high radiation dose to the surrounding tissue. If neutrons or π-mesons are used, the surrounding radiation dose can be reduced. However, π-meson therapy can only be performed near accelerators with sufficient power, making it inconvenient to promote, and it can only be used to treat a few types of cancer.

A type of irradiation method that falls between external and internal irradiation is called neutron capture tumor therapy. In this method, a non-toxic boron compound (containing enriched ^{10}B) is provided to the patient. This compound can be enriched by malignant tumors, so its concentration in the tumor is higher than in the blood and surrounding tissues. When irradiated with thermal neutrons, the high-energy α-particles and ^{7}Li nuclei produced by the $^{10}B(n,\alpha)^{7}Li$ reaction kill the cells they encounter. Slow neutrons themselves do not actually cause damage to the body, as the probability of them being captured by various atoms constituting the cells is very low. The biological effect of neutrons depends on their interaction with hydrogen atoms.

Compared to other radiation therapies, the advantage of neutron capture therapy is that neutron radiation only acts on the boron-containing points to be destroyed. The range of the products after boron fission is very short, and the destruction is limited to the cells where boron is enriched. It is important to find boron compounds that are enriched in malignant tumors. Currently, boron compound neutron capture therapy is used in the treatment of neurofibromas and malignant melanoma. Many research reactors and accelerators can be used as neutron sources for this modern method. This method can potentially treat deep tumors, but it requires high-energy neutrons to penetrate deep into the tissue to reach the tumor.

The radionuclides used for radiation therapy should meet the following conditions: they emit α, β, Auger electrons, or only a small amount of weak γ-rays; their half-life is several hours to several tens of days; their decay products are stable nuclides; they can obtain high specific activity radiopharmaceuticals.

The linear energy transfer of α-particles is high, about 1000 times that of β-particles. The range of α-particles with energies of 4–8 MeV in tissue is about 25–60 μm, which is comparable to the diameter of a cell. Therefore, α-particles are the most concentrated energy accumulation for internal radionuclide therapy for tumors. β-particles have a certain range in tissue. For example, the maximum range of 1 MeV β-rays in tissue is about 4 mm, which is about the sum of the diameters of 100 cells. If the drug molecules can selectively enter tumor cells, the β-particles they emit are sufficient to kill the tumor cells.

In recent years, precise radiation therapy techniques have rapidly developed, and various stereotactic radiosurgery techniques (such as X-knife, γ-knife, three-dimensional conformal radiation therapy (3D-CRT), intensity-modulated radiation therapy (IMRT), etc.) are gradually replacing traditional radiation therapy techniques. Four-dimensional conformal radiation therapy techniques that can solve the problems of organ movement, anatomy, and tumor volume changes during radiation therapy (such as respiratory gating) are also under development. At the same time, neutron and proton therapies have already been carried out in a few countries around the world, and China has already purchased a proton accelerator for treatment. Biological conformal radiation therapy techniques are also being explored and developed.

3.4 Application of Nuclear Technology in the Environment

Nuclear technology is not only widely used in industries such as industry, agriculture, and medicine, but it has also seen rapid development in the fields of environmental science and environmental protection (China Association for Science and Technology 2018). China has conducted research on the treatment of waste gas, wastewater, and solid waste using nuclear technology, such as the use of electron beams to remove SO_2/HO_x, etc. Additionally, the use of nuclear technology for the qualitative and quantitative analysis of environmental sample elements has many irreplaceable features of conventional non-nuclear technology, such as high sensitivity, accuracy, precision, high resolution (including spatial resolution and energy resolution), non-destructiveness, multi-element measurement capabilities, etc., and has been widely used in the environmental field. The nuclear analysis techniques used in the fields of environmental science and environmental protection mainly include radionuclide tracing technology, radionuclide activation analysis, radiation decomposition technology, and isotope measurement technology. For example, radioactive tracing methods are widely used in the study of environmental conditions, as well as the study of atmospheric diffusion behavior in the environment, the application of ^3H in stratigraphic structure analysis, groundwater dynamics research, and surface water level change analysis, etc. The application of nuclear technology in the environment can be roughly divided into two categories: radiation technology and nuclear analysis technology. In terms of the environment, irradiation technology is mainly used for the treatment of environmental waste, and nuclear analysis technology is used for the analysis and detection of environmental samples.

3.4.1 Application of Tracing Technology

As pointed out earlier, isotope tracing is a microanalysis method that uses radioactive nuclides (or stable nuclides) as tracers to label the object of study. The radioactive nuclides (or stable nuclides) and their compounds used in isotope tracing, and the chemical and biological properties of the corresponding common elements and their compounds existing in nature are the same, only with different nuclear physical properties. Therefore, isotopes can be used as a kind of marker to replace the corresponding non-labeled compounds with labeled compounds containing isotopes (such as labeled food, drugs, and metabolites). Using the nuclear physical property of radioactive isotopes continuously emitting characteristic rays, nuclear detectors can be used to track location, quantity, and changes at any time. Stable isotopes do not emit rays, but they can be measured by mass spectrometers, gas chromatographs, nuclear magnetic resonance, and other mass analysis instruments by using the mass difference between them and the corresponding common isotopes. Both radioactive isotopes and stable isotopes can be used as tracers, but stable isotopes as tracers have lower sensitivity, fewer types available, higher prices, and their application range is limited. Using radioactive isotopes as tracers not only has high sensitivity, but also has the characteristics of simple measurement methods, accurate quantification, and accurate positioning.

For environmental engineering, agricultural environmental protection, and environmental chemistry, isotope tracing technology has outstanding advantages. By adding labeled compounds or tracers to the system under study, and with the help of isotope measurement technology, the movement and change rules of this substance with similar substances can be discovered. Due to the special radiation performance of isotopes, the measurement of samples using isotope tracing technology often does not require separation to achieve extremely high measurement sensitivity and accuracy. This method has long been widely used in the study of pollutant migration behavior in soil, surface water, and groundwater. In addition, tracer technology is often used in the study of pollutant transfer rules in the biological chain and the mechanism of pollutant treatment. The research carried out using tracer methods for the migration and transformation of key nuclides has played a very important role in the treatment and disposal of nuclear waste.

By measuring the difference in the content of tracer isotopes in samples under different conditions, it is possible to infer the processes that have occurred in the environmental matrix in nature. For example, using isotopes as accurate time markers, through the study of uranium series imbalance in seawater, a series of marine environmental changes can be inferred and predicted. In the field of environmental hydrogeology, the use of ^{14}C, tritium, and several stable isotope tracing technologies for the study of surface water and groundwater has long accumulated systematic experience. The use of radioactive nuclide tracing technology in agricultural environmental protection also has a long history. For example, using labeling technology can comprehensively track the application, absorption, degradation, transfer, and accumulation of pesticides and chemical pollutants in the ecosystem. Because tracer technology can reveal the movement rules of atoms and molecules and phenomena that are difficult to discover by other methods, it plays an important role in the field of high technology. For example, in the field of environmental chemistry, it can be used to identify intermediate products of reactions. When studying the metabolism of biological organisms, tracer methods can not only quantitatively measure the transfer and change rules of metabolites, but also determine the quantitative distribution of metabolites in various organs, which is a powerful weapon for ecological research. Because tracers with extremely high radioactive specific activity can be obtained artificially, the sensitivity and accuracy of radioactive measurement are greatly improved. This feature continues to expand the application range of tracer technology in the environmental field.

(1) Radioactive Isotope Tracer Method

Radioactive tracing technology is unique in its application to the study of groundwater flow direction, reservoir monitoring, dam fissures, the velocity and migration of sediments in rivers, and more.

Isotope dating technology in macro-environmental research is irreplaceable by other methods. Since the intensity of radioactivity is a function of time, the half-life can be simply used as the sole basis for groundwater age. For example, by measuring the content in groundwater and its source snow water in the Altai Mountains of Xinjiang, it was determined that it takes nearly two thousand years for the snow water from the Altai Mountains to transform into groundwater in the Shihezi region. This provides an important basis for estimating the limit of available groundwater in the region and predicting the consequences of overuse. Similarly, the use of isotope analysis technology to study the origin of Qinghai Lake in China concluded that the water in Qinghai Lake will not dry up for a considerable length of time. This conclusion plays a very important role in the development of the Qinghai Lake region. For regions near nuclear facilities or where nuclear energy is used, the concentration of certain radioactive nuclides in the environment directly reflects the quality of the environment. For example, the migration of radon in the environment has always been a concern for environmental scholars worldwide.

Radioactive nuclide tracing technology has the advantages of high sensitivity, simplicity, and immunity to environmental and chemical factors, and is widely used in various disciplines. In tracer studies in earth science and environmental science, naturally occurring radioactive nuclides are commonly used. For example, using ^{14}C to study the circulation patterns of the world's oceans, using ^{10}B to trace the source of volcanic magma to verify plate subduction theory, using ^{36}Cl to trace groundwater permeability, etc. Using ^{129}I to trace nuclear leaks has become an important means of nuclear inspection. Tracer studies in chemistry, biology, and medicine often use radioactive nuclide-labeled compounds, the most common of which are compounds labeled with ^{3}H, ^{14}C, ^{32}P, ^{125}I, ^{131}I, etc. Tracer technology can also be used to study the distribution, migration, and transformation of trace elements in crops, the loss and residue of fertilizers and pesticides in soil, and agricultural ecological problems such as soil erosion and grassland degradation. Statistics show that the use of tracer technology in fertilization can increase nitrogen fertilizer utilization by 10% to 20%.

(2) Stable Isotope Tracer Method

Since the 1980s, due to the development of research in environmental management, environmental quality assessment, environmental impact assessment, and pollution trend prediction, mathematical models that can accurately reflect objective situations are often needed. After proposing a model, its reliability needs to be verified. The improvement in accuracy and precision of environmental isotope measurement techniques such as tritium, ^{14}C, and ^{18}O has greatly facilitated the verification of models. Stable isotope measurement technology has further expanded the application range of nuclear measurement technology in environmental science.

Hydrogen and oxygen isotopes can be used to study the water cycle. Water on Earth forms a cycle through evaporation, condensation, precipitation, infiltration, and runoff. Some thermodynamic properties of water molecules are related to the mass of the hydrogen and oxygen atoms that make up them, thus isotope fractionation occurs during the water cycle. Since there are three stable oxygen isotopes and two stable hydrogen isotopes, ordinary water molecules exist in nine different isotopic combinations. The vapor pressure of these different isotopic water molecules is inversely proportional to the mass of the molecule, so the vapor pressure of water molecules with smaller molecular weight is much higher than that of water molecules with larger molecular weight. Thus, the water vapor generated by the evaporation of liquid water is enriched with lighter H and ^{16}O, and the residual water is enriched with heavier D and ^{18}O. This leads to the fractionation of stable hydrogen and oxygen isotopes during the water cycle. Therefore, the high and low content of hydrogen and oxygen isotopes in water can be used to study the water cycle.

Stable isotopes are a class of isotopes that naturally exist in organisms and are non-radioactive. Carbon and nitrogen each have two stable isotopes with different numbers of neutrons, namely ^{13}C and ^{12}C, and ^{14}N and ^{15}N, respectively. The content of heavy isotopes (such as ^{13}C and ^{15}N) in organisms is very low. In this way, the difference in the ^{15}N isotope ratio of nitrates from different sources can effectively determine whether the pollutant comes from fertilizers, urban sewage, or minerals in the soil. Similarly, due to the different isotope abundances of ^{18}O in deep water and surface water, the age and replenishment relationships between the two can be determined by measuring the ^{18}O in the water. Typically, the absolute content of stable isotopes is not of great significance, but rather it is compared with international standards, that is, the abundance of stable isotopes is compared. The abundance of stable isotopes is represented as the ratio between the two most abundant isotopes in the sample and the corresponding ratio in the international standard, denoted by the symbol *Q*. Because the ratio difference between the sample and the standard reference is small, the abundance of stable isotopes is expressed as the deviation in parts per thousand between the sample and the standard. Q^3C and Q^5N are generally used to represent the abundance of carbon and nitrogen stable isotopes in a substance, respectively.

The slight differences in mass of isotopes cause minor differences in their physicochemical properties (such as conduction rate in the gas phase, bond energy strength, etc.), hence there are differences in isotope composition before and after a material reaction. This characteristic was successfully introduced into several research fields such as biology in the early 1970s, such as the study of photosynthesis pathways, light energy utilization rate, environmental pollution, plant water utilization rate, mineral metabolism, climate effects, and biomass changes, etc.

In addition, molecular isotope tracing technology, also known as monomer compound isotope analysis, is becoming an important and effective environmental tracer due to its characteristic and stability. Stable isotope tracing can track the source of petroleum pollution in the environment and its evolution in the environment. Different crude oils and their products have certain differences in chemical composition. By distinguishing the petroleum pollution in the environment through chemical composition and biomarker characteristics, the so-called "chemical fingerprint" technology is used. Normal alkanes, polycyclic aromatic hydrocarbons, and isoprenoids are all commonly used chemical fingerprints. Scholars have used total petroleum hydrocarbons, normal alkanes, benzene series, and polycyclic aromatic hydrocarbons as chemical fingerprint technologies to study petroleum pollution in Brazil and the Black Sea. However, the chemical composition is volatile, leachable, and biodegradable, so the application of "chemical fingerprints" has certain limitations. Isotope fingerprints, due to their characteristic and relative stability, are also widely used in environmental tracing research. For example, foreign scholars have used the total carbon isotopes of crude oil to distinguish two crude oil leak events that occurred in Alaska 25 years apart; they have used isotope fingerprint technology to track the famous crude oil leak residue that occurred in Alaska, USA; by analyzing the crude oil residues and the crude oil molecular isotope composition on the contaminated bird feathers, they have analyzed the environmental impact of the Erika oil spill on the Atlantic coast of France, and so on. The Guangzhou Institute of Geochemistry of the Chinese Academy of Sciences used chemical and stable isotope fingerprint tracing to track two small-scale heavy oil leak events that occurred in Guangdong Province and found the source of pollution. These are all successful applications of tracing technology.

3.4.2 Application of Radiation Technology

Radiation technology utilizes the interaction between rays and matter, the activated atoms and molecules produced by ionization and excitation, to cause a series of physical, chemical, and biological changes in the matter, leading to the degradation, polymerization, cross-linking, and modification of the matter. Radiation technology provides new purification pathways for some pollutants that are difficult to treat with conventional methods. In environmental protection, the main use of radiation technology is to use ionizing radiation to cause changes in environmental pollutants, thereby achieving the purpose of treatment and reuse. For example: transforming harmful substances into harmless or useful substances, killing bacteria and pathogens, accelerating the degradation rate of difficult-to-degrade substances, etc. Radiation treatment is widely applicable to the treatment of exhaust gas, wastewater, and solid waste, and is becoming one of the important means of environmental protection.

The main features of irradiation technology are:

(1) Good disinfection effect, thoroughly killing harmful microorganisms.
(2) It can avoid secondary pollution. Most of the irradiation process is carried out at room temperature, and some chemical pollutants can be physically degraded into harmless substances without the addition of other chemical reagents or catalysts, generally not causing secondary environmental pollution. Because the energy of the gamma rays, X-rays, and electron beams used for irradiation are all lower than the threshold energy for nuclear reactions, no nuclear reactions will occur, and therefore no induced radioactive substances will be produced to cause radioactive pollution.
(3) It can solve problems that are difficult to address using conventional methods. Since the mechanism of irradiation reactions is fundamentally different from that of typical chemical reactions, applying radiation technology to problems that are hard to solve with conventional techniques often yields satisfactory results. For instance, this applies to the treatment of polymers, the recycling of polytetrafluoroethylene, and the removal of nitrogen oxides from the exhaust gases of fuel and coal combustion.
(4) It is safe and reliable. The implementation of irradiation technology is straightforward and does not require any manual operations. Modern scientific and technological levels can fully ensure the safety of workers. In addition to the necessary protective screens, an automatic interlock system is usually installed between the radiation source and the personnel entrance. When the entrance is opened, the radiation source automatically descends into the depths of the pool. If the radioactive background value in the environment exceeds the allowable level, the radiation source will immediately shut down, and the dose monitoring and alarm system will automatically signal for personnel to leave the scene.
(5) It has strong adaptability and a wide range of applications. Since the intensity and energy of the radiation source are not affected by any external conditions, irradiation does not have any special requirements for external conditions. The parameters that need to be controlled in the irradiation process are mainly the use of time, distance, and orientation to control the irradiation dose, which is very simple.

The biggest drawback of irradiation technology is the high cost of the radiation source. To ensure absolute safety for personnel, the equipment must have a certain degree of automation, and the infrastructure investment is also higher than that of general facilities.

Radiation treatment mostly uses gamma rays or electron beams. Gamma rays mainly come from reactors or radioactive isotopes produced by reactors, while electron beams come from electron accelerators.

Radioactive isotopes produced by reactors can be used for the irradiation treatment of environmental waste. The most commonly used radioactive isotopes in industry are ^{60}Co and ^{137}Cs. The half-life of ^{60}Co is 5.26 years, and each decay emits two types of gamma rays with energies of 1.17 and 1.33 MeV. However, the half-life of ^{137}Cs is 30 years, so it does not need to be replaced frequently during use. Reactors can also serve as online irradiation sources, with the irradiated objects placed in the reactor cooling pool for online irradiation. When the dose level at the irradiation point is known, the dose level received by the irradiated object can be determined by controlling the irradiation time.

(1) Wastewater Treatment

Water pollution is a serious problem of global concern today. As production activities become increasingly diverse, the components of pollutants in wastewater are becoming more complex, and the allowable discharge standards for these pollutants are becoming stricter. Traditional water treatment methods such as activated sludge can no longer completely solve all wastewater purification problems. There is an urgent need to develop an economical, simple, effective, and capable of treating a variety of component pollutants wastewater treatment technology. Radiation technology is a promising water treatment technology with broad application prospects.

Radioactive isotopes produced by reactors can be used for the treatment of domestic sewage and industrial wastewater, usually in the form of radiation sources, such as ^{60}Co and ^{137}Cs sources. The basic principle is that water molecules will generate a series of highly active radiation decomposition products under radiation, such as OH, H, H_2O_2, etc. These products can decompose or modify organic matter in waste (polluted) water. The use of radiation source radiation treatment can significantly eliminate TOC (Total Organic Carbon), BOD (Biological Oxygen Demand), COD (Chemical Oxygen Demand) in urban sewage, and inactivate pathogens in sewage. For wastewater containing azo dyes and anthraquinone dyes, irradiation can completely decolorize them, with a TOC removal rate of 80–90% and a COD removal rate of 65–80%. Wastewater containing lignin is easily degraded when irradiated with gamma rays under oxygen-rich conditions. According to research reports, radiation sources can also effectively treat detergents, organic mercury pesticides, plasticizers, nitrosamines, chlorophenols, and other harmful organic substances. The combined use of radiation source irradiation technology and conventional wastewater treatment technology (such as coagulation, activated carbon adsorption, ozone activated sludge, etc.) has a synergistic effect and can improve the treatment effect. When used in conjunction with the activated carbon method, after the carbon has adsorbed organic matter, gamma-ray irradiation can regenerate the activated carbon, achieving recycling.

The Soviet Union once built a radiation source irradiation treatment pilot plant for the treatment of wastewater from an antibiotic factory, with a daily wastewater treatment capacity of 15,000 m^3. The treated wastewater indicators are all better than conventional treatment methods. Hungary, Canada, Japan, and other countries have similar pilot plants. The world's industrialized countries have also made positive progress in the development of sludge radiation treatment technology. Among them, the pilot plant in Munich, Germany was established the earliest and has been successfully operated for several decades. It is reported that the wastewater treatment capacity is 150 m^3 per day, with a treatment cost of approximately 4.4 Deutsche Marks per m^3. The irradiated sludge can be used as fertilizer.

(2) Solid Waste Treatment

Solid waste refers to solid and semi-solid waste materials that pollute the environment, produced in manufacturing, construction, daily life, and other activities. Solid waste includes household garbage, industrial solid waste, and agricultural waste. The production of solid waste in China is gradually increasing. The Ministry of Ecology and Environment pointed out that in 2018, the production of general industrial solid waste in 200 major cities reached 1.55 billion tons. According to an analysis by China Merchants Bank Research Institute, from 2018 to 2020, the production of general industrial solid waste in China will reach 2.73 billion tons, 2.87 billion tons, and 3 billion tons, respectively.

1) Treatment of Household and Industrial Waste

The Russian Institute of Physics and Power Engineering has meticulously designed a new technology for treating household and industrial waste using fast neutron reactors. This technology is realized by building a garbage incinerator in the nuclear reactor. Household and industrial waste is regularly loaded into the furnace, where it falls into an oxygen-blowing slag tank for combustion and gasification. To maintain high temperatures in the furnace, a small amount of power coal is continuously added. When the furnace temperature reaches about 1500 °C, all the waste will be gasified, with the mineral-containing part melting in the slag and the metal sinking to the bottom after melting. Then, the metal that has sunk to the bottom is regularly cast into castings and transported out for processing; the slag cleared from the furnace is used for processing building materials; the gas in the furnace, reaching up to 1750 °C, is drawn out by an exhaust fan for cooling and used for chemical production. The furnace and the gas in the furnace are cooled by liquid sodium. This liquid sodium circulates between the double walls of the waste reactor shell. At the same time, the liquid sodium, which can reach temperatures as high as 500 °C, can also heat the steam generator, with the resulting steam used to drive the steam turbine generator, and the condensed hot water flowing back into the boiler room. In this way, metal, building materials, chemical products, electricity, and heat can be obtained from waste treatment. According to tests, a set of such fast neutron reactor equipment can treat 2×10^4 tons of household and industrial waste per year, with a power generation capacity of 5 kW, which is self-sufficient.

2) Waste Plastic Treatment

In the treatment and disposal of solid waste, waste plastic is always a thorny issue due to its difficulty in degradation. For example, polytetrafluoroethylene (PTFE). Because it cannot be decomposed by biochemical methods, mechanical crushing is difficult, and a large amount of toxic fluorides are produced during high-temperature treatment, creating a difficult situation to handle.

Japan has used a combined method of γ-ray irradiation and heating, followed by mechanical crushing, to obtain polytetrafluoroethylene waxy powder of different molecular weights, which can be used as excellent lubricants and additives. Polyvinyl chloride releases a hundred times more vinyl chloride when used, and is therefore banned in some countries. However, after irradiation with a certain dose of γ-rays, it does not produce vinyl chloride, thereby expanding its use.

Waste plastic can also be induced to degrade by radiation methods. Early research on radiation-induced plastic degradation was completed in the 1950s and 1960s. Similar to rubber, the large molecules of plastic generally decompose by breaking the C–C bond, and radiation-induced degradation produces gaseous, liquid, and solid small molecule products, which can be used as raw materials for appropriate synthesis.

Currently, China has about thousands of tons of waste PTFE waiting to be recycled each year. PTFE is expensive and chemically very stable. It will not degrade in the environment for decades, which has a negative impact on the environment. Therefore, the recycling of waste PTFE has important economic value and environmental significance. γ-rays and electron beams can be used for the degradation of PTFE. By using γ-ray irradiation, a radiation degradation $G = 12.8$ is obtained (G value is a commonly used parameter in the field of radiation degradation, which means the number of molecules changed per 100 eV energy absorbed), and by using electron beam irradiation, a $G = 41.2$ is obtained. In order to obtain the required molecular size product fraction during irradiation, the radiation-induced process can be controlled by the dose, dose rate, and temperature used. All kinds of non-degradable products have a wide range of applications. For example, perfluoroalkene can be converted into fluorinated surfactants with special properties after irradiation, or it can be oxidized into perfluorocarboxylic acids needed for special purposes. The perfluoroalkanes that appear as by-products can be used as insulation materials for high-quality salts, solvents, and lubricants. And these conversion products can all be used as raw materials for synthesis.

(3) Sludge Treatment

Sludge is an inevitable byproduct of wastewater treatment, with the most common steps being screening, primary sedimentation, biological treatment, secondary sedimentation, and sterilization. Sludge is produced during the sedimentation processes before and after biological treatment. Sludge contains a large amount of energy and biological value, making it an excellent fertilizer and soil amendment for farmland. However, it cannot be used directly due to the presence of numerous pathogens. As early as 1979, the U.S. Environmental Protection Agency (EPA) issued CFR257 regulations requiring the control of pathogens in sludge. Later, the agency suggested in a supplemental environmental impact report that sludge from sewage systems is a useful resource. Conventional methods such as composting, pasteurization, or chemical treatment are not ideal and difficult to industrialize. Radiation technology can overcome the shortcomings of common treatment methods and is considered a promising sludge treatment method internationally. Both γ-ray sources and electron beam irradiation can be used for sludge treatment.

The advantages of radiation treatment of sludge are: it can kill bacteria and viruses in the sludge, and the disinfection effect is more reliable than heat treatment; it does not destroy organic nitrogen compounds in the sludge, so it does not reduce the fertility of the sludge or produce unpleasant odors (the pasteurization method results in significant nitrogen loss, reducing the fertility of the sludge); it can prevent weed seeds in the sludge from germinating, but it does not affect the germination of normal seeds; the treatment temperature is low (25–30 °C), reducing corrosion to factory equipment; irradiated sludge has good dewatering performance, eliminating the need for chemical flocculants and some corresponding equipment. After irradiation sterilization, sludge can be used directly as fertilizer in farmland.

Industrialized countries around the world have made positive progress in developing sludge radiation treatment technology. Germany built the world's first sludge irradiation treatment pilot plant in Munich in 1973, using ^{60}Co as the radiation source, and later added a ^{137}Cs radiation source. The plant uses strong γ-radiation to kill bacteria in the sludge. The irradiated sludge still contains the original nutrients and can be used as fertilizer. Its performance is far superior to sludge treated with composting and pasteurization, providing valuable experience for research in this field. The United States, India, Australia, and other countries have also established sludge irradiation facilities, but most of these facilities only operated for 2–4 years, and no full-scale sludge irradiation facilities have appeared since.

(4) Others

Cellulose is the main component of urban and agricultural waste. Japan has used irradiation to treat sawdust, waste paper, straw, etc., obtaining alcohol through saccharification and fermentation. The United States uses acidification followed by irradiation to obtain glucose from such cellulose, with a recovery rate of up to 56%. Spoiled food can be used as animal feed after irradiation treatment.

3.4.3 Application of Accelerators in the Environment

Various low-energy accelerators are widely used in various sectors of the national economy, including activation analysis, ion implantation machines for large accelerators, radiation modification, and non-destructive testing in industry. In agriculture, they are used for variety improvement, disease eradication, and food preservation. In medicine, they are used for cancer treatment, radiation sterilization, and production of short-lived isotopes. In defense, high-pulse power accelerators can be used as X-ray simulation sources. In environmental protection, strong beams generated by accelerators are used for desulfurization and denitrification treatment of flue gas from coal combustion.

High-energy electron accelerators can produce high-energy electrons with energies of 1–10 MeV, which have certain penetrating power. When the electron beam passes through a thin dispersed fluid, part of it is reflected back into the fluid for irradiation, increasing the dose level of the irradiated body. The accelerator can be shut down immediately when not in use. The use of electron beams generated by accelerators for irradiation treatment is becoming an important means of environmental protection. The technology of electron beams is widely applicable to the irradiation treatment of exhaust gas, wastewater, and solid waste.

The high-energy electron beam generated by an electron accelerator can induce physical, chemical, and biological effects in certain materials, effectively killing bacteria, viruses, and pests. This technology has been widely used in industrial production for material modification, new material production, environmental protection, processing production, sterilization and disinfection of medical and health products, and food sterilization and preservation. Like cobalt source irradiation, it has the characteristics of room temperature, non-damage, no residual toxicity, environmental protection, low energy consumption, easy operation, high degree of automation, and suitability for large-scale industrial production. Compared with cobalt sources (with a radiation efficiency of about 20%), the greatest advantage of high-energy electron beam irradiation is that the irradiation beam is concentrated and directed, the energy utilization is full, the irradiation efficiency is over 80%, and it does not produce radioactive waste. With the soaring price of cobalt sources and the rising cost of waste source treatment, electron accelerator irradiation devices have obvious economic advantages. The construction of a high-energy electron irradiation center with a high-power electron accelerator with an energy of 10 MeV has obvious social and economic benefits and immeasurable potential value while developing the irradiation processing industry and conducting research on irradiation technology and new fields of irradiation.

(1) Waste Gas Treatment

The application of accelerators in waste gas treatment is mainly used for the treatment of harmful flue gas. SO_2 and NO_x in the atmosphere are the main pollutants, which mainly come from flue gas emissions from chimneys. The usual flue gas desulfurization and denitrification technologies mainly include solid-phase adsorption and regeneration technology, wet simultaneous desulfurization and denitrification technology, absorbent injection method, etc., most of which will encounter difficulties of high cost or complex equipment, such as desulfurization with lime spray method, removal of NO_x with acid, alkali absorption or catalytic reduction method, etc. High-energy radiation chemistry is a new type of flue gas desulfurization and denitrification technology, mainly divided into electron beam irradiation method (EBA) and pulsed corona method (PPCP). Among them, the electron beam irradiation method is a better developed method at present. The method of electron beam irradiation can remove SO_2 and NO_x in flue gas, help purify the atmosphere, prevent the formation of acid rain, and also obtain by-products such as nitrocellulose and sulfur rubber, which can be used as fertilizers. It can also reduce the difficulty and cost of operation, and almost no secondary wastewater is produced under dry conditions.

In the 1990s, accelerators manufactured by the Institute of Nuclear Physics of the Siberian Branch of the Russian Academy of Sciences were supplied to Poland and Japan for smoke purification. Poland's poor-quality coal produces large amounts of toxic sulfur and nitrogen oxides when burned. When the smoke is irradiated, all oxides become solid precipitates that can be used as fertilizers. In Japan, accelerators produced in Russia have also been used to purify the smoke produced by burning garbage, avoiding the pollution of the environment by smoke.

1) Electron Beam Irradiation Method for Desulfurization and Denitrification

Advantages of the Electron Beam Irradiation Method

Since 1972, Japan has been widely conducting basic research and semi-industrial experiments on the removal of SO_2 and NO_x in flue gas by electron beam irradiation. After many years of research and development, it has gradually moved towards industrialization from small trials, pilot tests, and industrial demonstrations. Basic research and semi-industrial and industrial experiments that have begun in Japan, Germany, Italy, Poland, the United States, and other countries show that electron beam irradiation technology has the following advantages: it can simultaneously desulfurize and denitrify, and can achieve a desulfurization rate of over 90% and a denitrification rate of over 80%; since the electron beam method of desulfurization and denitrification is a dry treatment technology, it does not produce wastewater or waste residue; no catalyst is needed; the system is simple, easy to operate, and the process is easy to control; it has good adaptability and load tracking for flue gas with different sulfur contents and changes in the amount of flue gas, the reaction generates usable by-products, the by-products are a mixture of ammonium sulfate and ammonium nitrate, which can be used as fertilizers, and the cost of desulfurization and denitrification is more economical than traditional methods. Possible disadvantages include high electron beam dose requirements, high electricity demand, high operating costs, and the need to filter aerosols after irradiation.

In addition, the electron beam irradiation method can effectively purify other industrial waste gases, such as volatile organic compounds, automobile exhaust, toxic gases, and waste gases from incinerators. Of course, the technology of electron beam treatment of polluted waste gas involves many different physical and chemical mechanisms, such as energy absorption, gas phase reactions, particle formation, and the mutual influence of gas–solid phases.

Reaction Mechanism of the Electron Beam Irradiation Method

High-energy electron beams of 500 to 800 keV are generated using cathode emission and accelerated by an electric field. When these electron beams irradiate flue gas, they induce radiolytic chemical reactions, producing free radicals such as OH, O, and HO_2. These free radicals can oxidize SO_2 and NO_x, forming H_2SO_4 and HNO_3. Prior to irradiation, stoichiometric amounts of ammonia are added to the flue gas. The resulting aerosols of H_2SO_4 and HNO_3 interact with the NH_3 introduced into the reactor, forming byproducts such as $(NH_4)_2SO_4$-and NH_4NO_3. These byproducts are collected using methods such as electrostatic precipitation and are directly used as fertilizers. After the flue gas is irradiated by the electron beam, the main reaction processes are as follows:

(a) Generation of Free Radicals

Coal combustion flue gas generally consists of main components such as N_2, O_2, steam, CO_2, and components such as SO_2 and NO_x. When irradiated with an electron beam, most of the beam energy is absorbed by nitrogen, oxygen, and steam, generating a large number of highly reactive free radicals.

(b) Oxidation of SO_2 and NO_x

SO_2 and NO_x are oxidized by free radicals and other active species, forming sulfuric acid and nitric acid.

(c) Formation of Ammonium Sulfate and Ammonium Nitrate

Sulfuric acid and nitric acid react with the pre-injected ammonia to neutralize, forming aerosol particles of ammonium sulfate and ammonium nitrate. If there is still unreacted SO_2 and NH_3, thermal chemical reactions continue on the particle surface to form ammonium sulfate.

Process Flow of Electron Beam Irradiation Method

The experimental device consists of five main parts: flue gas parameter adjustment system, accelerator irradiation treatment system, ammonia addition device, byproduct collection device, and monitoring control system. The process flow of the electron beam irradiation flue gas desulfurization and denitrification industrial test device consists of flue gas cooling, ammonia addition, electron beam irradiation, and byproduct collection. The flue gas discharged from the boiler at about 130 °C enters the cooling tower after electrostatic dust removal. In the cooling tower, the flue gas is cooled to a temperature suitable for desulfurization and denitrification (65 °C) by spraying cooling water. The dew point of the flue gas is usually 50 °C, so the cooling water is completely vaporized in the tower and generally does not produce wastewater that needs further treatment. The cooled and humidified flue gas is sent to the reactor, and then stoichiometric ammonia gas is injected into the reactor according to the concentration of SO_2 and NO_x and the set removal rate. The flue gas in the reactor is irradiated by an electron beam, oxidizing SO_2 and NO_x to form sulfuric acid and nitric acid, which are neutralized with the injected ammonia to form ammonium sulfate and ammonium nitrate. The irradiated flue gas is sent to the byproduct collector, and the ammonium sulfate and ammonium nitrate in the flue gas are captured and recovered by a dry electrostatic precipitator. The purified flue gas is pressurized by a fan and discharged into the atmosphere from the chimney.

2) Pulsed Corona Plasma Method

The fundamental principle of the Pulsed Corona Plasma Process (PPCP) is similar to that of EBA. Both methods utilize high-energy electrons to activate, ionize, or dissociate gas molecules such as H_2O and O_2 in the flue gas, generating highly oxidative free radicals. These radicals then catalytically oxidize SO_2 and NO_x via plasma reactions, converting them into SO_2 and NO_x (or their corresponding acids). In the presence of additives, the corresponding salt is formed and precipitates. The difference between the two methods lies in the source of high-energy electrons: the EBA method is obtained through cathode electron emission and external field acceleration, while the PPCP method is generated by corona discharge itself. It uses a high-voltage power supply with a steep, narrow pulse front edge (rise time 10~100 ns, tail time 100~500 ns, peak voltage 100~200 kV, frequency 20~200 Hz) and a power load—corona electrode system (corona reactor) combination. This generates a streamer corona plasma in the gap between the corona and the corona reactor electrode, thereby oxidizing and removing SO_2 and NO_x. The advantage of the PPCP method is that it can also remove dust simultaneously. Studies have shown that dust in flue gas is beneficial to the desulfurization and denitrification efficiency of the PPCP method. Therefore, the PPCP method integrates the removal of three types of pollutants, and its energy consumption and cost are lower than the EBA method, making it an attractive flue gas treatment method.

(2) Wastewater Treatment

Electron beam radiation can treat water and wastewater. The irradiation effect produces active substances in the water, such as OH radicals that can gasify and decompose any organic pollutants in the water. Under certain economic conditions, the scale of wastewater treatment is determined. An appropriate electron beam irradiation device is selected to ensure that the flow rate of the treated wastewater receives a sufficient dose uniformly. Factors to consider include the energy distribution of the radiation, its penetration ability in water, the geometric shape of the volume of interaction between the radiation and water, and the thickness of the radiation dose usually considered. As early as the Soviet era, scientists conducted large-scale experiments using accelerators to improve the ecological environment. For example, in Voronezh, groundwater polluted by waste from rubber product production was purified. After 10 years of effort, the groundwater within a 30 km radius was finally purified.

Irradiation can also effectively kill microorganisms in the water. The method of chlorination treatment of secondary sewage can kill microorganisms, but it produces toxic chlorinated organic compounds, such as trichloromethane. Electron beam irradiation also effectively inactivates phages. With a relatively small dose (such as 0.25~1 kGy) of irradiation, 90% of common bacteria (such as *E. coli*, Salmonella, etc.) can be killed, and all bacteria are eliminated with a dose of 10 Gy. Using electron beam irradiation instead of chlorination treatment of secondary sewage will not produce toxic chlorinated organic compounds and can effectively sterilize. The sterilization effect depends on the thickness of the water layer in the irradiation area, the properties of the water, and the impact of different types of pathogens on radiation sensitivity. The sterilization efficiency of electron beam irradiation also depends on the penetration ability of electrons in water. In a flowing system, due to the intense mixing of the liquid, the survival curve of the irradiated bacteria is more effective than those irradiated under stable conditions. At the same time as killing microorganisms, the brown color usually present in liquid waste also disappears, and the water becomes odorless.

Due to the continuous expansion of the production and application scale of synthetic surfactants in industry and life, the problem of wastewater purification is becoming increasingly complex, and correspondingly, the pollution of reservoirs is intensifying. These substances are mainly used to produce detergents that are far superior to ester soaps. There are also some substances in the drainage ditches of industrial enterprises that can be used in the process of production. The characteristics of surfactants are high stability when chemically reacting with alkali solutions and salt solutions, slow biochemical oxidation, and high foaming performance. Although the desired effect can be obtained by processing water with activated carbon or ion exchange resin, it is not economical. Radiation can cause surfactants to break down into lighter, relatively easy-to-eliminate substances. Unlike biological purification, radiation purification acts on all compounds. In the presence of oxygen, it oxidizes them. When radiation and ozone act together, the purification efficiency is significantly improved. This can be explained as the conversion of HO_2 free radicals into OH free radicals, the latter of which is a strong oxidant for most organic compounds.

Wastewater and sludge contain usable organic and inorganic substances that can be used as agricultural fertilizers or as nutrients added to feed. In addition to irradiating to degrade toxic synthetic substances in wastewater, special treatment is also needed to kill pathogenic microorganisms. The general disinfection method is to heat the wastewater at about 70 °C for several tens of minutes. At the same time, the efficiency of treating wastewater with a radiation dose of 30 kGy is also being studied. From the perspective of sterilization effect, the two methods are almost the same. Studies on the sediment of irradiated wastewater show that these sediments are actually very similar to the fertilizer obtained by the thermal treatment of wastewater. The volume of sludge formed by various original water samples during sedimentation is 1 to 2 orders of magnitude smaller than the volume of sewage transported to the disinfection plant, so a very low radiation power is needed. From the perspective of application prospects, radiation disinfection of wastewater may become a very cheap method, because this method can also simultaneously remove polymer impurities.

Another approach to purifying wastewater using electron beam irradiation technology is the regeneration of used activated carbon. Activated carbon has strong adsorption properties, making it useful for eliminating pollutants in wastewater. However, the cost of regenerating used activated carbon using current methods is high. Since used activated carbon has organic matter attached to its surface, electron beam irradiation technology can effectively regenerate it. Under a nitrogen environment, the recovery rate of the adsorption capacity of activated carbon is the highest, and there is almost no loss of activated carbon after irradiation. During the irradiation process, the higher the temperature of the activated carbon and the larger the current of the electron beam, the higher the recovery rate of the adsorption capacity. By comparing the isotherm adsorption curves of regenerated activated carbon and new activated carbon in a sodium dodecyl sulfate solution, it was found that the isotherm adsorption curves of the two are almost the same.

(3) Solid Waste Treatment

The solid waste treated by electron beam irradiation can be divided into two categories: waste that needs to be irradiated for disinfection, such as urban sewage sludge, biomedical waste, and garbage from international airports and seaports; waste rubber and plastic that are regenerated through irradiation treatment.

1) Sludge Treatment

Sludge, a resource containing a large number of pathogens, can be used directly as fertilizer in fields after irradiation sterilization. In addition to gamma rays, electron beams can also be used for sludge treatment. In Japan, sludge disposal is a challenging problem. Approximately 60% of the sludge in inland and coastal areas needs to be disposed of. The Japan Atomic Energy Research Institute has begun to study an effective sludge treatment process: electron beam sterilization to produce compost. Traditional composting methods must use the heat generated during composting to sterilize the sludge. In traditional methods, composting is done by microorganisms, but the heat generated by composting can both sterilize and kill microorganisms, and it takes a long time to produce. In the process studied at the Takasaki Radiation Chemistry Research Center of the Japan Atomic Energy Research Institute, sterilization is done first, followed by composting, and the best conditions for composting can be selected to achieve better results. The composting rate is greatly affected by temperature, with the optimal temperature being 40–50 °C. The optimal pH is 7–8. To allow aerobic bacteria to ferment, oxygen needs to be supplemented in granular sludge with a diameter of about 5 mm. With the irradiation method, the time to emit carbon dioxide is only 2–3 days, while the traditional method requires more than 10 days. To kill pathogenic bacteria, the fermentation temperature needs to be above 65 °C. In the new process, sterilization and composting are separate, so the best conditions for composting can be selected, shortening the production cycle to the minimum.

2) Biomedical Waste Treatment

Waste from hospitals, research and diagnostic laboratories, etc., is considered a potential source of pollution. Therefore, it poses a threat to public health. This category also includes waste from dissections, animal tissues, and body parts. According to researchers, about 85% of hospital waste is non-infectious waste, with the rest considered biomedical waste. Most or all of a hospital's waste is treated as biomedical waste.

In North America, it is not allowed to dispose of biomedical waste in landfills without proper treatment (such as incineration). However, about 20% of hospital waste may be plastic (such as polyethylene, polypropylene, polyvinyl chloride, etc.). If incinerated, toxic gases will be produced, which must be removed. Additionally, to prevent the spread of infectious diseases during waste treatment, biomedical waste can be disinfected through irradiation.

3) Port Waste Treatment

International airport and seaport waste (such as food scraps, plastic, cellulose, etc.) may contain potential infectious pathogens of animal filterable viruses (such as foot-and-mouth disease, swine vesicular disease, African swine fever, etc.). Therefore, in most countries, there are regulations specifically for the treatment of this international waste. In Canada, the above-mentioned waste is heated to above 100 °C for at least 30 min, or incinerated to achieve the purpose of disinfection and sterilization. According to regulations, containers used for storing and transporting waste should be cleaned and disinfected before reuse. A possible alternative method for international waste disinfection is irradiation (gamma rays or electron beams). The irradiation device can be set up in the airport area and can work automatically and continuously. This concept design is reliable and economical, and can meet modern environmental cleanliness requirements. Of course, if appropriate, biomedical waste and waste from international airports and seaports can be disinfected by irradiation in a single irradiation facility.

4) Rubber and Plastic Treatment

Butyl rubber is a raw material used in the manufacture of automobile inner tubes. From a safety perspective, tires typically have a short lifespan, resulting in a large number of discarded tires each year. Using radiation methods to recycle the rubber from old tires is a well-established practice. The recycled material gains valuable processing characteristics, and by incorporating it into the components of new tires, their durability can be improved. When the radiation dose reaches 70 kGy, the plastic properties of the original rubber resin and the recycled rubber mixture are significantly enhanced, while other physical properties such as tensile strength are only slightly weakened at this dose. In commercial products, 25% of the rubber resin is based on recycled rubber. The process of recycling rubber includes: first cutting old tires into small pieces, then irradiating them with high-energy electron beams or gamma rays to break down the large molecular network structure, while also inducing cross-linking. These processes alter the mechanical properties of the rubber. Therefore, by grinding the raw materials with less energy and separating the broken fibers, the raw materials can be mixed with other components to produce new tires. The entire recycling process is environmentally friendly and produces almost no pollution. It should be noted that some further irradiation effects have also been observed, such as an increase in modulus values, hardness, and strut quality with increasing dose, but a decrease in corrosion resistance.

5) Other Applications

Accelerators also have other environmental applications, such as high-current continuous beam proton accelerators that can be used for nuclear waste treatment, nuclear fuel production, and clean nuclear energy generation. The strong beam produced by the accelerator can bombard nuclear waste, transforming long-lived radioactive elements into useful or short-lived elements. A subcritical nuclear reactor driven by a high-current proton beam of 1 GeV, tens of milliamperes, produced by an accelerator, can generate electricity in a safe and clean manner. The beam produced by the accelerator and the target material undergoing nuclear reactions is an effective way to produce nuclear fuel.

Discussion Questions

(1) What is radionuclide tracing technology? Please briefly describe.
(2) What is radiation processing? What are its important applications?
(3) What is the principle of radioactive isotope logging?
(4) Briefly explain the principle of tracer method application in agriculture.
(5) What are the important applications of radioactive isotope tracing technology in agriculture?
(6) What are the applications of radiation technology in pest sterilization and biological growth stimulation?
(7) What are SPECT (Single Photon Emission Computed Tomography) and PET (Positron Emission Tomography), and what are their important applications?
(8) What are the advantages of using nuclear technology for qualitative and quantitative analysis of environmental samples compared to conventional non-nuclear technology?
(9) Briefly describe the main features of radiation technology and its applications in environmental science and environmental protection.

References

China Association for Science and Technology. (2018). 2016–2017 Discipline Development Report: Nuclear Technology Applications. Beijing, China Science and Technology Publishing House.

Li GZ, Chen G. (2005). Applications of Nuclear Technology in Biological Sciences and Agriculture. Beijing, China Forestry Publishing House.

Li S, Zha LF, Zhao WY. (1994). Nuclear Energy and Nuclear Technology. Shanghai, Shanghai Scientific and Technical Publishers.

Luo SZ. (2015). Applications of Nuclear Technology. Harbin, Harbin Engineering University Press.

Wang CX. (2002). Applications of Nuclear Energy and Nuclear Technology. Beijing, Atomic Energy Press.

Wei YX, Jia BS. (2011). Introduction to Nuclear Energy and Nuclear Technology. Harbin, Harbin Engineering University Press.

Chapter 4
Nuclear Power Technology and Its Applications

4.1 Reactor Technology

4.1.1 Fission Reactors

Fission reactors typically consist of an active core where a chain fission reaction is maintained, and most of the fission energy is also generated there (Wu et al. 2014). The core contains nuclear fuel, which consists of fissile nuclei and usually also contains fissionable nuclei. For thermal neutron reactors, fission reactions are mainly induced by thermal neutrons, so a moderator is needed to slow down the high-energy neutrons produced by the fission reaction to thermal neutrons. The best moderators are substances composed of elements with low atomic mass, large neutron scattering cross-sections, and low neutron capture, such as ordinary water (containing hydrogen), heavy water (containing deuterium), beryllium, beryllium oxide, and carbon (graphite). In addition, a neutron reflector is arranged around the core to scatter back many neutrons that have already escaped from the core, reducing neutron leakage and thus lowering the critical mass of fissile isotopes. The reflector can use the same or other specific substances as the moderator. In many power reactors, water is used as a moderator and reflector (and also as a coolant). But in other reactors, fission is caused by high-energy neutrons, as in fast reactors, where the reflector must avoid containing moderating substances.

The fission energy generated by fission reactions in the core of a nuclear reactor is converted into kinetic energy on the spot, which is then conducted out of the fuel elements and carried away by an appropriate coolant. Coolants that have been used include liquid water, liquid sodium (or sodium–potassium alloy), certain organic compounds, air, carbon dioxide, and nitrogen gases. If the energy released in the reactor is to be converted into electrical energy, the heat must be transferred from the coolant to other working substances to generate steam or hot gas, which is then used in conventional turbine generator systems. The higher the temperature of the steam or other working medium, the higher the efficiency of the output power. Therefore,

L. Yu et al., *Nuclear Technology and Safety*, Nuclear Science and Technology,
https://doi.org/10.1007/978-981-95-2856-1_4

it is generally required to increase the operating temperature of the power reactor under safe and reliable conditions.

From an economic perspective, the specific power of the reactor, i.e., the heat release rate per unit mass (or volume) of fissionable material, should be as high as possible. The heat release rate is proportional to the fission rate, which in a given reactor core is determined by the neutron density, i.e., the number of neutrons per unit volume. By changing the neutron density in the core, control of the nuclear reactor can be achieved, including startup, shutdown, and operation at any power level. These processes are generally achieved through the movement of control rods or the adjustment of boron-containing liquids. These control rods and the materials in the liquids easily absorb neutrons and are known as neutron poisons. The introduction of nuclear poisons reduces the reactivity (or neutron multiplication ability) of the core, thereby reducing the neutron density and the power level of the reactor, and vice versa.

Nuclear reactors are classified into different types according to the kinetic energy (or speed) of the neutrons that cause the main fission. Most of the neutrons released in fission are high-energy neutrons. If fission is caused by high-energy fast neutrons, this type of reactor is called a fast neutron reactor (fast reactor). The fuel in this type of reactor must contain a sufficient proportion (more than about 15%) of fissionable isotopes and should contain convertible substances. In fast reactors, both the core fuel and the reflector layer, known as the conversion zone, contain convertible substances, which are transformed into fissionable nuclei by neutron capture. Fast neutrons have a small capture cross-section, so if neutron loss due to escape can be kept to a minimum, each fission may have more than one neutron available to convert the convertible nuclei in the core and conversion zone into fissionable nuclei. In this case, the fissionable material produced by neutron capture may exceed the fissionable material consumed. This type of nuclear reactor is called a breeder reactor, which achieves the breeding of fissionable material. A power breeder reactor is one that can produce more fissionable material than it consumes while producing power. In a fast reactor with plutonium-239 as the fissionable isotope and uranium-238 as the convertible isotope, the same fissionable isotope is produced and consumed. Another similar fast reactor uses uranium-233 and thorium-232 as the fissionable and convertible isotopes, respectively, and the same fissionable isotope is produced and consumed; theoretically, the breeding of uranium-233 can also be achieved in some reactors that do not rely on fast neutron fission. A fast neutron reactor that uses more plutonium-239 produced from uranium-238 by consuming uranium-235 is also a breeder reactor. Uranium-235 is the only fissionable material found in nature, and its reserves are small. So, one day all available uranium-235 will be exhausted. At that time, the further use of the remaining uranium-238 and thorium-232 will depend on the application of plutonium-239 or uranium-233 to maintain the fission chain.

Thermal neutron reactors have more choices for moderators, coolants, and nuclear fuel materials, and are more flexible than fast neutron reactors. In most commercial thermal neutron reactors, either natural uranium (containing 0.7% uranium-235) is used as fuel, with heavy water or graphite as the moderator, or fissionable isotope uranium containing 2% to 5% uranium-235 is used as fuel, with ordinary water as the moderator. This way, the fuel contains a considerable proportion of convertible material uranium-238 (in a few cases, thorium-232). During the operation of the reactor, some of the convertible isotopes are transformed into fissionable isotopes plutonium-239 (or uranium-233). However, because the probability of thermal neutrons in a thermal reactor being captured in non-fission reactions with fuel and structural materials is greater, plus some neutron loss due to escape, less than one neutron per fission is available to convert convertible nuclei into fissionable nuclei. Thus, the amount of plutonium-239 in a thermal neutron reactor is usually less than the amount of uranium-235 consumed to maintain the fission chain. Therefore, it is difficult to design a thermal neutron breeder reactor using uranium-235 (or plutonium-239) as fuel. But when uranium-233 is used as the fissionable material, the situation is different, and a thermal neutron breeder reactor with uranium-233 as the fissionable isotope and thorium-232 as the convertible isotope is possible.

4.1.2 Brief Introduction to Types of Nuclear Reactors

There are many types of nuclear reactors, which can usually be classified in the following ways (Chen et al. 2019).

(1) Classification by Purpose

Research and experimental reactors: Used to study neutron characteristics, and then carry out experimental research and teaching in nuclear physics, reactor physics, thermal hydraulics, structure, operation and control, radiation protection, as well as biology, material characteristics, etc.

Power reactors: Reactors that use the power generated by nuclear energy to propel ships, generate electricity and provide heating.

Production Reactors: Primarily used for producing new fission materials such as Uranium-233, Plutonium-239, and various isotopes with different uses.

(2) Classification by Neutron Energy Causing Fission

Thermal Neutron Reactors: Fission is mainly caused by thermal neutrons. The reactor contains a large amount of light material as a moderator to slow down fission neutrons to thermal neutrons.

Intermediate Neutron Reactors: Fission is mainly caused by intermediate neutrons. The reactor contains only a small amount of moderator, insufficient to thermalize fission neutrons.

Fast Neutron Reactors: Fission is mainly caused by fast neutrons. The reactor contains no moderator. The prominent advantage of fast reactors is their ability to breed nuclear fuel.

(3) Classification by Moderator

Light Water Reactors: Use light water (H_2O) as a moderator, which can be further divided into pressurized water reactors and boiling water reactors.

Heavy Water Reactors: Use heavy water (D_2O) as a moderator.

Graphite Reactors: Use graphite (C) as a moderator.

(4) Classification by Coolant

Water-Cooled Reactors: Use light water and heavy water as coolants.

Gas-Cooled Reactors: Use carbon dioxide or helium as coolants.

Liquid Metal-Cooled Reactors: Use sodium and sodium–potassium alloys as coolants, or lead or lead–bismuth alloys as coolants.

(5) Classification by Fuel Used

Reactors Using Natural Uranium as Fuel: Such as natural uranium heavy water reactors, natural uranium graphite reactors.

Reactors Using Enriched Uranium as Fuel: Such as low-enriched uranium light water reactors, high-enriched uranium fast neutron reactors.

Reactors Using Thorium as Fuel: Such as thorium high-temperature reactors.

Reactors Using Plutonium as Fuel: Such as plutonium breeder reactors.

(6) Classification by Fuel Arrangement

Uniform Reactors: Nuclear fuel is uniformly distributed in the reactor core.

Non-Uniform Reactors: The fuel is made into "fuel elements", which are then distributed in the reactor core in a certain way. In summary, the forms and structural materials of various types of nuclear reactors are listed in Table 4.1 for comparison.

Looking at the types of reactors, light water reactors account for about 80% of the nuclear power plants that have been built, are under construction, and will be built. In light water reactors, the majority are pressurized water reactors. The fuel elements of pressurized water reactors are usually rod-shaped. The elements are made by encapsulating core blocks made of low-enriched UO_2 ceramic fuel in a zirconium alloy metal casing. The enrichment level of uranium is generally 3% to 5%. Water serves as a moderator and also as a coolant. The cooling water carries heat out of the reactor core, generating steam in the steam generator (secondary side), which is then directed to the turbine generator for power generation or direct work by the secondary circuit. The moderating and thermal conductivity properties of water are quite good, but the probability of neutron absorption is relatively high, so light water reactors must use uranium with a certain degree of enrichment as fuel. To raise the outlet water temperature of the reactor to about 300 °C without boiling, the pressure must be increased to about 14 to 16 MPa, and a pressure vessel capable of withstanding high pressure is needed to house the reactor core. Hence, this type of reactor is called a pressurized water moderated and cooled reactor, or simply a pressurized water reactor (PWR).

4.1.3 Trends in Reactor Technology Development

(1) Development History

Table 4.1 Forms and structural materials of nuclear reactors

Type of nuclear reactor		Fuel	Moderator	Coolant
Light water reactor (LWR)	Pressurized water reactor (PWR)	Enriched uranium	Light water	Light water
	Boiling water reactor (BWR)	Enriched uranium	Light water	Light water
Graphite reactor	Carbon-cooled reactor (CCR)	Natural uranium	Graphite	Carbon dioxide
	Advanced carbon-cooled reactor (ACR)	Enriched uranium	Graphite	Carbon dioxide
	High-temperature carbon-cooled reactor (HTCR)	Enriched Uranium thorium	Graphite	Helium
	Light water-cooled reactor (LWCR)	Natural uranium Enriched uranium	Graphite	Light water
Heavy water reactor (HWR)	Light water-cooled type	Natural uranium Enriched uranium Plutonium	Heavy water	Light water
	Heavy water cooled type	Natural uranium	Heavy water	Heavy water
Fast breeder reactor (FBR)		Enriched uranium Plutonium	None	sodium Sodium–potassium alloy

From the 1950s to the present, nuclear power reactor technology has evolved from the first to the third generation, and research on the fourth generation of nuclear reactors is currently accelerating. The main purpose of the first generation of nuclear reactors was to verify the technical feasibility and reliability of nuclear power generation. The reactors represented by the Shippingport Atomic Power Station in the United States, the Indian Point Energy Center Unit 1, the Obrigheim Nuclear Power Plant in Germany, the Chinon Nuclear Power Plant in France, and the Mihama Nuclear Power Plant Unit 1 in Japan, the Soviet icebreaker "Lenin", the American nuclear merchant ship "Savannah", etc. The second generation of nuclear reactors, based on the first generation, achieved commercialization, standardization, serialization, and mass production, greatly improving the economic performance of nuclear power plants and the power performance of ships. From the late 1960s, pressurized water reactors, boiling water reactors, and heavy water reactors with electrical power above 30 MW were built internationally, and nuclear power began to enter the era of commercial application, as well as the rapid development stage of nuclear-powered ships, making nuclear power economically competitive with traditional power generation methods, and greatly improving the endurance of ships. The oil crisis of the 1970s promoted the development of second-generation nuclear reactor technology.

Compared with the early stage, the late second-generation reactor type has greatly improved in all aspects. Some nuclear power countries are still promoting the upgrade and life extension research and plans of the second-generation reactors to maximize their economic value. The second-generation nuclear power plants include pressurized water reactors (PWR), boiling water reactors (BWR), heavy water reactors (PHWR), and graphite-moderated water-cooled reactors (RBMK), which have become relatively mature after many years of development and have not had major accidents for a long period. However, the Chernobyl nuclear accident in 1986 led to a new understanding of nuclear power safety by the international community, raised higher requirements, and began to have systematic quantitative assessment standards and norms for the safety and economy of nuclear power plants. For example, the United States launched the "Advanced Light Water Reactor User Requirements" document (i.e., URD document) in the 1990s, Europe provided the "European User Requirements for Light Water Reactor Nuclear Power Plants" (i.e., EUR document), and the International Atomic Energy Agency formulated the Nuclear Safety Regulations (NUSS) series, etc.

Internationally, nuclear power plants that meet the URD or EUR evaluation standards are referred to as third-generation nuclear power plants (Yan et al. 2023). Third-generation nuclear power plants have adopted passive safety systems that use principles such as gravity, convection, and diffusion, and equipped the nuclear power plant reactors with passive safety systems that do not require external power sources for operation. Compared with the second generation, its equipment has been simplified, the probability of accidents has been further reduced, and significant breakthroughs have been made in safety and economy. The representative third-generation nuclear power plants in China are shown in Table 4.2.

Although the third-generation reactors have improved in terms of safety, their advantages in uranium resource utilization and fuel breeding are not obvious compared to the second-generation reactors. Passive systems cannot completely eliminate inherent safety issues within the reactor, and theoretically, there is still the possibility of severe accidents. Therefore, to achieve sustainable development of nuclear energy and consider the inherent safety features of reactors, the concept of fourth-generation reactors has been proposed and related research is being conducted. This includes six representative types of reactors: ultra-high temperature gas-cooled reactors, gas-cooled fast reactors, molten salt reactors, sodium-cooled fast reactors, lead-cooled fast reactors, and supercritical water reactors.

Currently, China is transitioning towards third-generation nuclear reactor technology in its newly constructed nuclear power plants, and even towards fourth-generation reactor technology. In the early stages of the development of new nuclear energy technologies, the inevitable upgrades will lead to increased design and construction costs, making them less competitive compared to other forms of energy. The economic viability of the new generation of nuclear power will face significant pressure. Analysis of the current construction of third-generation nuclear power plants shows that the technical requirements and difficulties of structural design, equipment manufacturing, and civil installation have generally increased, leading to a significant increase in construction costs. In addition, innovative research in other forms of energy has greatly improved their economic viability. For example, the emergence of shale gas technology in the United States has reduced the cost of

Table 4.2 Mainstream third-generation nuclear power plant reactors

Name	Unit	Reactor type	Power/MW	Technology
ABWR	General electric	Boiling water reactor	1,350	Advanced boiling water reactor
AP600	Westinghouse	Pressurized water reactor	610	With passive safety systems
AP1000	Westinghouse	Pressurized water reactor	1,090	With Passive Safety Systems
EPR	FramatomANP	Pressurized water reactor	1,600	High power and enhanced dedicated safety systems
SWR1000	FramatomANP	Boiling water reactor	1,013	Meets EUR requirements
System80+	Former ABB-CE	Pressurized water reactor	1,350	Meets NRC document requirements
ESBWR	General electric	Boiling water reactor	1,380	No recirculation pump, natural circulation, passive safety systems
GT-MHR	U.S./Russia	High-temperature gas-cooled reactor	288	Uses weapon-grade plutonium as nuclear fuel, modular reactor
IRIS	Westinghouse	Pressurized water reactor	300	Modular, integrated reactor
ACRI000	Atomic energy of Canada	Heavy water reactor	1,200	Simple design, high efficiency, reduced cost, introduction of passive systems
CAP1400	State power investment corporation	Pressurized water reactor	1,400	Based on AP1000 technology, domestic nuclear power unit
Hualong One	CGN/CNNC	Pressurized water reactor	1,000+	Fully independent intellectual property Rights, meets Both URD and EUR requirements

natural gas power generation; the slowdown in global economic development has led to a significant drop in the prices of traditional fossil fuels (coal, oil, etc.), gradually reducing the cost of thermal power generation. The large-scale development of clean energy sources such as wind and solar power poses further challenges to the development of nuclear energy. Therefore, the development of the new generation of nuclear power technology will focus more on the following main aspects.

(1) Pursuing Better Safety

The probability of a nuclear power plant experiencing a core meltdown and a large release of radioactivity has been reduced from 10^{-4} and 10^{-5} per reactor-year to 10^{-5} and 10^{-6} per reactor-year (one in a hundred thousand to one in a million), extending the concept of inherent safety from the nuclear power unit to include the natural safety concept of the entire nuclear fuel cycle system.

(2) Continuously Improving the Economy of Nuclear Power

In the face of the gradual reduction in the cost of thermal and hydroelectric power, as well as the large-scale development of clean energy sources such as wind, solar, geothermal, and tidal power, it is necessary to develop more economical nuclear energy technologies to reduce their construction and power generation costs.

(3) Meeting the Requirements for Sustainable Environmental and Ecological Development

Nuclear energy does not emit pollutants such as sulfur dioxide and carbon dioxide, which is something conventional energy sources cannot achieve. However, the generation of long-lived radioactive nuclides and their continuous accumulation must be properly managed to meet the requirements for sustainable environmental and ecological development.

(4) Meeting the Requirements for Sustainable Resource Utilization

The uranium-235 currently used as nuclear fuel also has a certain lifespan, so it is necessary to develop nuclear fuels other than uranium-235 and new nuclear power technologies, such as breeder reactors and fusion reactors. If fusion nuclear power technology can be realized, it will be a permanent solution in terms of nuclear fuel, meeting the long-term sustainable development of resource utilization.

(5) Meeting the Requirements for Nuclear Non-Proliferation

Strict control over the production of separated plutonium, research on new fuel cycle processes, no separation of uranium and plutonium in the fuel of fast neutron breeder reactors, and the integration of nuclear power plants with post-processing are necessary to prevent the proliferation of fuel bred in reactors.

(2) Fourth-Generation Reactors

In 2001, the Generation IV International Forum (GIF) was established by the United States, South Korea, the United Kingdom, Switzerland, Canada, Brazil, Japan, France, and other countries, along with the European Atomic Energy Community. The research and development of the fourth-generation nuclear energy system began to enter the international cooperation model, and the "Technology Roadmap for the Fourth Generation Nuclear Energy System" was released in 2002. According to the development roadmap for the fourth-generation reactors published by the GIF organization in 2014, the Lead-cooled Fast Reactor (LFR) is progressing rapidly, followed by the Sodium-cooled Fast Reactor (SFR). Other technologies such as the Gas-cooled Fast Reactor (GFR), Supercritical Water Reactor (SCWR), Molten Salt Reactor (MSR), and Very High Temperature Reactor (VHTR) are also under development (Zhang 2023). Below is a brief introduction.

(1) Gas-cooled Fast Reactor (GFR)

Research on gas-cooled reactors began in the 1950s, and they are among the earliest types of reactors developed internationally. High-temperature gas-cooled reactors were developed based on early gas-cooled reactors and improved gas-cooled reactors, starting in the 1960s. The United Kingdom built the experimental high-temperature gas-cooled Dragon reactor in 1960, and the United States completed the Peach Bottom high-temperature gas-cooled reactor in 1967. Both the gas-cooled fast reactor and the very high-temperature gas-cooled reactor discussed below were developed based on this foundation.

The gas-cooled fast reactor is a fast neutron spectrum, helium-cooled reactor that uses a closed fuel cycle (see Fig. 4.1). The helium coolant outlet is at a high temperature, which can be used for power generation, hydrogen production, or high-efficiency heat treatment. The reference reactor is a helium-cooled system with an electrical power of 288,000 kilowatts, an outlet temperature of 850 °C, and a high thermal efficiency can be achieved using a direct-cycle gas turbine. The reactor core uses a mixture of actinide elements as particle fuel, made into prismatic or plate fuel assemblies.

The main advantages of gas-cooled fast reactors are the chemical stability of the coolant at high temperatures and the absence of a need for neutron moderators. Key technologies that need to be addressed in gas-cooled fast reactors mainly include the discharge of core decay heat, nuclear fuel cycle technology, and the manufacturing technology of fuel elements for gas-cooled fast reactors. Gas-cooled fast reactors are

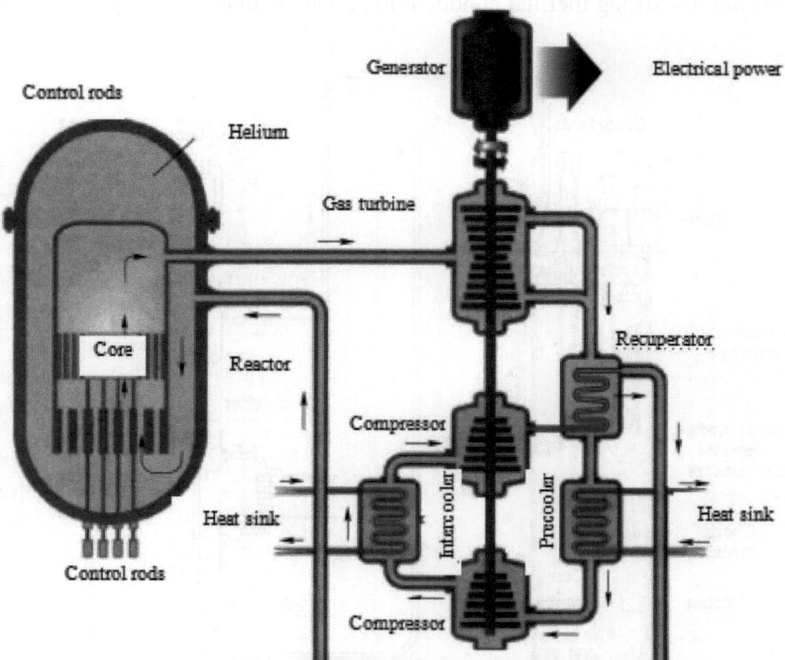

Fig. 4.1 Schematic of a gas-cooled fast reactor

quite similar to the very high-temperature gas-cooled reactors that will be introduced later.

From the 1990s to the present, the main international research on high-temperature gas-cooled reactors has been the 10 MW high-temperature gas-cooled experimental reactor (HTR) built by China and the 30 MW high-temperature gas-cooled experimental reactor (HTTR) built by Japan. Both of these reactors are currently in operation. China is a world leader in high-temperature gas-cooled reactors. In addition, the United States also has corresponding reactors for conducting related research.

(2) Lead-cooled Fast Reactor (LFR)

The lead-cooled fast reactor uses lead or lead–bismuth eutectic liquid metal as the coolant for the fast neutron reactor. Its research began in the Soviet Union in the 1950s, and in 1963, the Soviet Union built the world's first nuclear power device using lead–bismuth alloy as the reactor coolant. A typical lead-cooled fast reactor (see Fig. 4.2) usually includes: a lead-based reactor core, a steam generator, a coolant pump, a refueling device, and an external wall of the reactor. Similarly, a neutron moderator is not needed in the lead-cooled fast reactor, the layout is compact, and miniaturization is easy to achieve.

The lead-cooled fast reactor has several safety advantages: the coolant has a low melting point and a high boiling point, the reactor can operate under low pressure, the coolant has strong thermal conductivity, good chemical stability, weak reaction

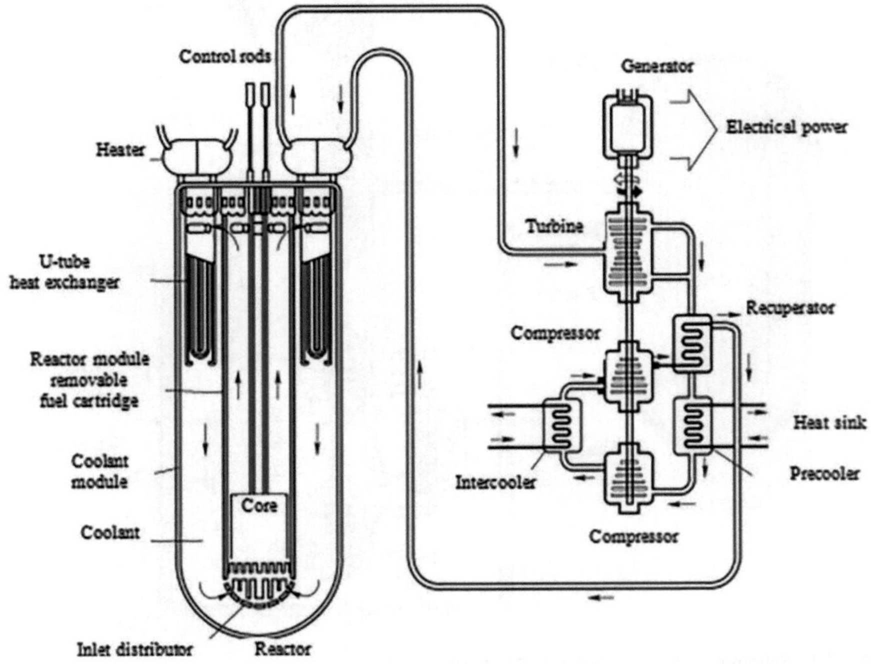

Fig. 4.2 Schematic of a lead-cooled fast reactor

with air and water, reducing the risk of coolant pipe rupture, explosion, etc.; strong natural circulation ability, the reactor outlet coolant temperature can reach 550 °C, or even up to 800 °C, and can achieve passive residual heat discharge under accident conditions; the lead-based reactor can achieve a closed fuel cycle, effectively convert uranium-238 and effectively manage actinide elements, greatly reducing the harm of radioactive waste.

Although lead and its alloys have good chemical stability, key issues such as compatibility with structural materials, coolant flow heat transfer characteristics, and chemical process control need further research. Because the neutron absorption and moderation cross-sections of lead are very small, and its neutron scattering ability is strong, it can be combined with accelerator devices to form an accelerator-driven subcritical system, which can transmute long-lived nuclear waste, which is of great significance for the sustainable development of nuclear energy.

(3) Molten Salt Reactor (MSR)

The liquid molten salt reactor (see Fig. 4.3) uses thorium or uranium dissolved in fluoride salts (such as lithium fluoride, sodium fluoride, etc.) to form a fusion material as fuel. There are mainly two types of molten salt reactors, one is a liquid fuel molten salt reactor, which no longer uses solid fuel core blocks, but dissolves fissionable breeding materials in fluoride salts. This method eliminates the possibility of solid fuel element damage and core melting. The other is a solid fuel molten salt reactor, similar to the encapsulated fuel pebbles of a high-temperature gas-cooled reactor.

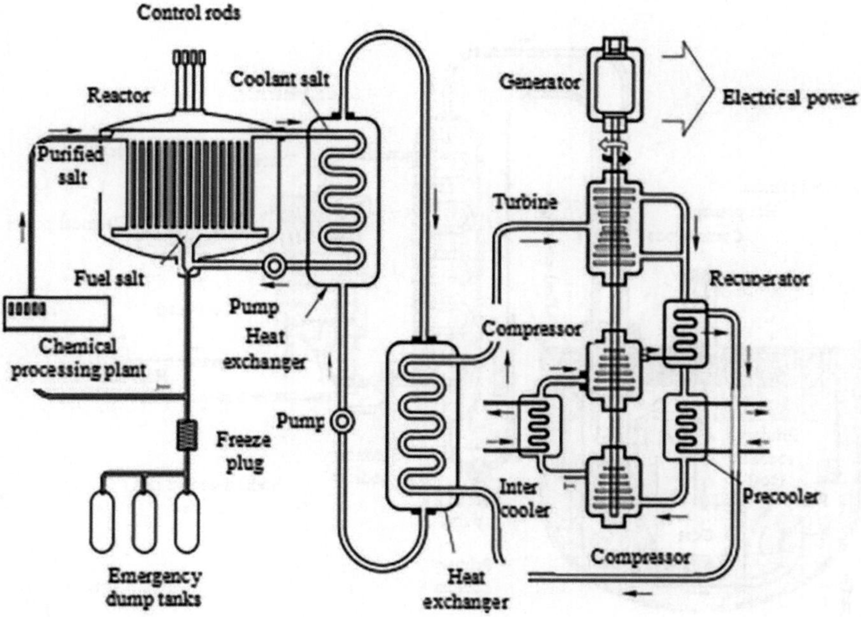

Fig. 4.3 Schematic of a molten salt reactor

Liquid fuel molten salt undergoes fission reactions in the core to release heat, which is absorbed and carried away by itself, without the need for additional coolants. Liquid fuel molten salt is both a heat carrier and a heat source for releasing fission energy. The main features of the molten salt reactor are: good heat transfer performance of molten salt fluorides, no radiation, operation at normal pressure, no violent reaction with air or water, negative temperature coefficient and void coefficient, no serious accident risks such as heat exchanger rupture and coolant loss, and unique advantages in nuclear safety. In addition, the molten salt reactor can also use thorium-based fuel, achieve nuclear fuel recycling, and improve the utilization rate of nuclear fuel.

(4) Sodium-cooled Fast Reactor (SFR)

The Liquid Sodium-cooled Fast Reactor (see Fig. 4.4) is a fast neutron spectrum, sodium-cooled reactor that can utilize actinide elements and convertible uranium in a closed fuel cycle. It can effectively manage the recycling of actinide elements and uranium-238. There are two schemes for this cycle: a sodium-cooled reactor with a power output of 150,000 to 500,000 kilowatts, using uranium–plutonium-actinide–zirconium alloy fuel, based on a high-temperature metallurgical treatment fuel cycle; a large-scale sodium-cooled reactor with a power output of 500,000 to 1,500,000 kilowatts, using uranium–plutonium oxide fuel, based on an advanced water treatment fuel cycle. The outlet temperature of sodium in both schemes is approximately 550 °C, generating high-temperature steam for power generation through sodium-sodium and sodium-water intermediate heat exchangers.

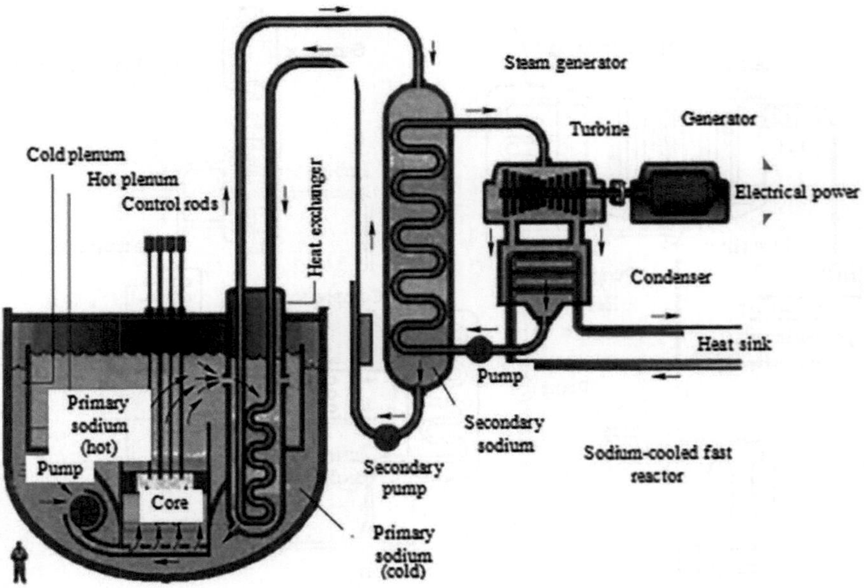

Fig. 4.4 Schematic diagram of sodium-cooled fast reactor

Internationally, France, Russia, India, China, South Korea, and Japan have carried out the design and development of commercial demonstration fast reactors. The commercial demonstration fast reactor BN800 built by Russia was connected to the grid for power generation in December 2015. The cumulative operating years of sodium-cooled fast reactors worldwide exceed 430 reactor-years, and the technology of sodium-cooled fast reactors has been engineering-verified. The largest sodium-cooled fast reactor in the world is the Super Phoenix in France (1200 MW). China began researching sodium-cooled fast reactor technology in the mid-1960s. After decades of technological accumulation, China completed its first sodium-cooled fast reactor, the China Experimental Fast Reactor, in 2010, achieving its first criticality.

(5) Supercritical Water Reactor (SCWR)

In the 1960s, the idea of using supercritical fluids in nuclear reactors began to emerge in the United States and the Soviet Union, but it did not attract attention until the 1990s. Real research and development activities started in Japan in 1989. Since the late 1990s, Europe, the United States, Canada, and South Korea have begun to develop supercritical water-cooled reactors (see Fig. 4.5) under the framework of the fourth-generation nuclear energy system. The supercritical water-cooled reactor system operates above the thermodynamic critical point of water (374 °C, 22.1 MPa) using high-temperature, high-pressure water. The supercritical water coolant can make the thermal efficiency about 1/3 higher than the current light water reactor. The coolant does not change state in the reactor and is directly connected to the energy conversion equipment. The supercritical water directly vaporizes to drive the turbine blades when it reaches the turbine, simplifying the power plant auxiliary facilities. The power of the reference system of the supercritical water reactor is 1,700,000 kilowatts, the operating pressure is 25 MPa, the outlet temperature of the reactor is 510 °C, and the fuel is uranium oxide, with passive safety facilities similar to those in boiling water reactors.

The supercritical water-cooled reactor has two types of reactors: thermal neutron spectrum and fast neutron spectrum. The thermal reactor adopts an open fuel cycle scheme; the fast reactor adopts a closed fuel cycle.

Currently, there are more than 10 conceptual designs of supercritical water-cooled reactors proposed by various countries, which differ in neutron energy spectrum, moderator, and fuel assembly structure.

(6) Very High Temperature Gas-cooled Reactor (VHTR)

The Very High Temperature Gas-cooled Reactor (see Fig. 4.6) is a reactor that uses a once-through uranium fuel cycle, graphite moderator, and helium gas cooling. The reactor generates heat, which can raise the outlet temperature of the core helium gas to 1000 °C. It can produce hydrogen or process heat for the petrochemical industry or other industries. The reference reactor uses a core with a thermal power of 600,000 kilowatts, connected to an intermediate heat exchanger to transfer process heat. The reactor core is prismatic block or spherical (like China's HTR-10). In terms of hydrogen production, this system provides heat that can be effectively used by the thermochemical iodine–sulfur process.

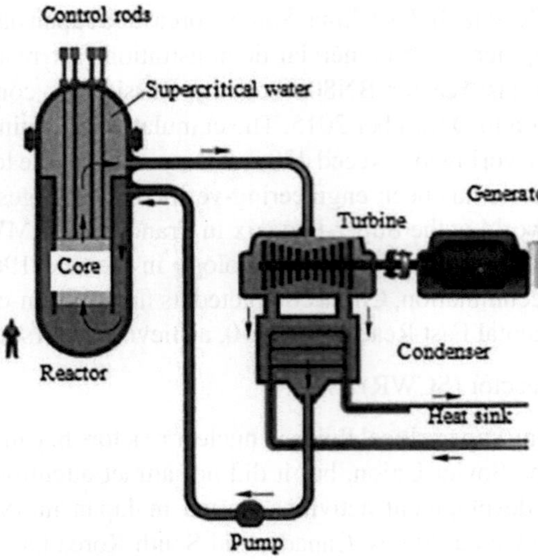

Fig. 4.5 Schematic diagram of supercritical water-cooled reactor

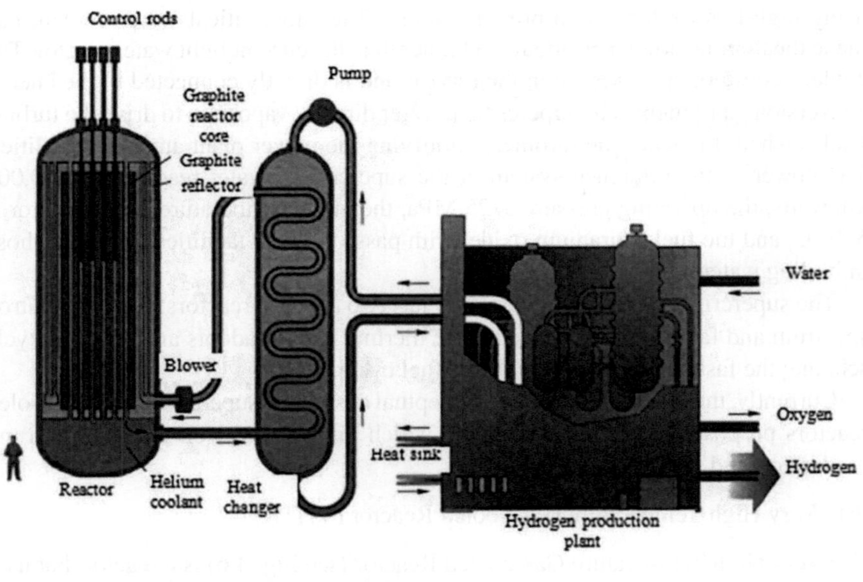

Fig. 4.6 Schematic diagram of very high temperature gas-cooled reactor

The Very High Temperature Gas-cooled Reactor system can provide process heat for high-temperature, energy-intensive non-electric processes. If power generation equipment is added to this system, it can meet the needs of combined heat and power. In addition, this system uses a uranium/plutonium fuel cycle, which minimizes waste.

It also retains the inherent safety features of modular high-temperature gas-cooled reactors.

(3) Nuclear Power for Icebreakers

Currently, the world's active polar icebreakers are mainly concentrated in Russia, the United States, Canada, Finland, Sweden, Denmark, and other countries. Among them, the United States owns the world's most powerful conventionally powered icebreaker, while Russia has the most advanced technology, the most polar icebreakers, and the strongest polar operation capabilities. Russia is also the only country in the world with nuclear-powered polar icebreakers. Russia has a leading advantage in icebreaker technology, equipment, and operational application.

The Arctic, often referred to as the "second Middle East," is rich in oil and gas resources. The U.S. Geological Survey estimates that the Arctic region holds about 30% of the world's undiscovered natural gas resources and about 13% of the world's undiscovered oil resources. At the fifth International Arctic Forum in April 2019, President Putin stated, "Investments in the Arctic region now account for over 10% of Russia's total investments." Putin indicated that Russia would increase its investment in nuclear-powered icebreakers, predicting that by 2035, the Russian Arctic fleet would have at least 13 heavy icebreakers, nine of which would be nuclear-powered. Russia hopes to achieve year-round navigation of the Northern Sea Route with its powerful heavy nuclear-powered icebreakers and strengthen its strategic presence in the Arctic region. Russia is vying for dominance in this region with traditional rivals Canada, the United States, and Norway.

So far, Russia leads in nuclear-powered icebreaker technology and remains the only country in the world with nuclear-powered icebreakers. All nuclear-powered icebreakers are managed and operated by the Rosatom State Atomic Energy Corporation. Four generations and six types of nuclear-powered icebreakers have been developed to date.

The first generation of nuclear-powered icebreakers was the "Lenin," followed by the second generation "Arktika" and "Taymyr" classes, the third generation "LK-60" class, and the fourth generation "Leader" and LK-40 classes, which are currently under development.

To date, Russia has built 10 nuclear-powered icebreakers, but only four are in service, one completed its first trial voyage on December 14, 2019, and the other five have been retired. These 10 icebreakers can be divided into three generations and four types.

The "Lenin" was the world's first nuclear-powered icebreaker, built in 1956, commissioned in 1959, and retired in 1989. The ship is 134 m long, 27.6 m wide, 16.1 m high, with a draft of 10.5 m, icebreaking capability of 1.7 m, a displacement of 19,240 tons, and a maximum speed of 19.6 knots. The main power unit has a power of 44,000 horsepower. It was originally equipped with three 90 MW OK-150 pressurized water reactors, but after an accident during refueling, the reactor core was severely damaged and replaced in 1970 with two 159 MW OK-900 pressurized water reactors, providing 32 MW of propulsion for the propellers. The ship mainly carried out exploration and rescue activities in the Arctic Ocean and performed icebreaking and guiding transport ships on the Northern Sea Route.

The second generation of nuclear-powered icebreakers includes eight ships, two "Taymyr" class and six "Arktika" class. Of the "Arktika" class, only the "Yamal" and "50 Years of Victory" are still in service, while the other four have been retired. They are all 148.5 m long, 30.5 m wide, 17.25 m high, with a displacement of 23,000 tons, icebreaking capability of 2.5 m, a displacement of 19,240 tons, and a maximum speed of 20.8 knots. The main power unit has a power of 75,000 horsepower. They are equipped with two 171 MW OK-900A pressurized water reactors, providing 54 MW of propulsion for the propellers. The "50 Years of Victory" was launched for trial navigation in 2006 and officially delivered in 2007. The ship is 159 m long, 30 m wide, 17.2 m high, with a draft of 11 m, a full-load displacement of 25,000 tons, a maximum speed of 20.8 knots, a maximum icebreaking thickness of 2.8 m at a speed of 18 knots, a total power of 75,000 horsepower, and is equipped with two nuclear reactors.

The "Taymyr" class is equipped with a 171 MW KLT-40 M pressurized water reactor, providing 35.5 MW of propulsion for three propellers, mainly used for icebreaking operations in shallow areas such as coasts, harbors, and even river mouths. The "Taymyr" class includes the "Taymyr" and "Vaygach." They are both 150 m long, 28 m wide, with a draft of 8.1 m, a displacement of 18,500 tons, an icebreaking thickness of 1.7 m, and a speed of 18.5 knots. Both icebreakers are currently in service.

The third generation of nuclear-powered icebreakers is the LK-60YA class, currently the largest and most powerful icebreakers in the world, including the "Arktika," "Sibir," and "Ural," mainly used to replace the "Arktika" and "Taymyr" classes. The lead ship "Arktika" completed a two-day navigation test on December 14, 2019, but used diesel fuel power. The "Arktika" is 173.3 m long, 34 m wide, with a displacement of 33,500 tons, and a maximum speed of 22 knots. It has a service life of 40 years and costs about 36.9 billion rubles (about $640 million). It is equipped with two 175 MW RITM-200 nuclear reactors, providing 60 MW of propulsion for the propellers, has a 9-level icebreaking capability, and can break through 3 m thick ice. The second ship "Sibir" has a displacement of 33,540 tons, is 173.3 m long, 34 m wide, with a draft of 10.5 m, equipped with two 175 MW NPS-200 nuclear reactors, three propellers for propulsion, and a speed of 22 knots. It was launched on September 22, 2017. The third ship "Ural" was launched in Saint Petersburg on May 25, 2019. In addition, in August 2019, Rosatom and the Baltic Shipyard signed a contract to build two more of this type of nuclear-powered icebreaker. It is evident that this type of nuclear-powered icebreaker has three main features: the largest ship size, stronger power, and higher icebreaking efficiency.

The 4th generation "Leader" class nuclear-powered icebreaker is currently under development in Russia, designed for year-round navigation of large-capacity transport ships with a load capacity exceeding 100,000 tons and a width of over 50 m in the Northern Sea Route. It will provide year-round escort for natural gas transport ships and oil tankers, including those transporting hydrocarbon fuels to South Korea, Japan, and China. The "Leader" class will have a total displacement of 71,380 tons, a length of 209 m, a width of 47.7 m, and a height of 20.3 m. The ship will be equipped with two RITM-400 nuclear reactors, each with a propulsive power of 120 MW and a thermal power of 315 MW. The propulsion system of the ship includes four propellers. The expected speed of this type of icebreaker in water will reach 24 knots (44 km per hour), and 12 knots in 2-m-thick ice, with an ice-breaking thickness of nearly 4 m, significantly improving on the parameters of the LK-60 class. The designed service life of the ship is 40 years. Russia plans to build at least three "Leader" class nuclear-powered icebreakers, each estimated to cost 1.6 billion dollars. All three are expected to be delivered by 2033.

4.2 Space Nuclear Power

Space nuclear power includes space nuclear power sources and nuclear thermal propulsion systems, which provide electrical energy and propulsion power for spacecraft. Space nuclear power can meet the needs for electrical energy and propulsion power of communication satellites, space stations, space transport, space weapons, space combat platforms, deep space exploration, extraterrestrial bases, etc.

The energy source of space nuclear power is nuclear energy. Currently, space nuclear power mainly utilizes the fission energy generated by nuclear reactors and the decay energy of radioactive isotopes. Depending on different mission requirements, space nuclear power devices can convert nuclear energy into electrical energy and propulsion power in various ways (Su et al. 2020).

4.2.1 Space Reactors

(1) Introduction to Space Reactors

Artificial satellites, spacecraft, space stations, and other spacecraft require long-term, highly reliable power sources for their operation and equipment. However, conventional chemical batteries are heavy and have a short lifespan, making them unsuitable for satellites and spacecraft. Therefore, most satellites and spacecraft use a combination of solar cells and storage batteries for power. However, solar cells cannot operate in shadowy or deep space environments and cannot be used too close or too far from the sun. Even within the solar system, it is estimated that the solar energy received by Neptune per unit area is less than 1/900 of that on Earth, making it unusable. Additionally, large solar cell arrays create significant drag for maneuvering flight and low-orbit flight, requiring a large amount of fuel for orbit maintenance, and also present technical difficulties in installation and deployment: large deployment areas, complex structures, difficulty in achieving high precision and high stability attitude control; large exposure to space debris, meteorites, and external impacts, and susceptibility to radiation. As humans continue to explore and utilize space, the power and lifespan requirements for space energy sources are increasing, making it difficult for these conventional energy sources to meet the long-term operational needs of space stations. Therefore, humans have turned to nuclear energy, using space reactors to meet the requirements of manned spacecraft and large space stations for high specific power, long lifespan, high reliability, and resistance to electromagnetic interference.

Space reactors can use highly enriched uranium as fuel, zirconium hydride as a moderator, and beryllium as a reflector in thermal or fast neutron reactors; they can also use liquid metal lithium or sodium as a coolant in fast neutron reactors, directly converting fission energy into electrical energy through thermoelectric converters, or converting thermal energy into electrical energy through steam turbines or gas turbines. The technology of space nuclear reactor power sources is challenging and requires a long development cycle. However, space nuclear reactor power sources have significant advantages: they can easily provide high power supply, providing thousands to millions of watts of electrical energy for spacecraft; they have high energy density, and under high power, their weight-to-power ratio is superior to the combined power supply system of solar cell arrays and storage batteries, with lighter weight, smaller volume, smaller area, less drag, smaller impact surface, and better concealment; they have a large power adjustment range, with the ability to quickly increase power, and high maneuverability; they do not depend on solar radiation, do not need sun orientation, and can work continuously all day and in all weather; they have good environmental adaptability, with strong resistance to space debris impact, and can work in harsh conditions such as dust, high temperature, and radiation. Therefore, space nuclear reactor power sources are ideal power sources for space exploration and are irreplaceable power sources for deep space exploration.

As early as the 1950s, the Soviet Union/Russia and the United States officially began to develop space nuclear power technology, investing a large amount of manpower, material resources, and financial resources, and achieved significant results. Figure 4.7 is the TOPAZ-II space reactor developed by Russia. This is a fast neutron reactor moderated by zirconium hydride. The reactor core is surrounded by a beryllium reflector, and the reactor power is controlled by the rotation of the beryllium reflector drum. Each reflector drum has a 116-degree sector of boron carbide as a neutron absorber, and the reactor is controlled by changing the position of the sector absorber by controlling the rotation of the drum. The reactor core is 37.5 cm high and 26 cm in diameter, composed of 37 fuel elements of 96% enriched uranium dioxide, with a fuel load of 27 kg. The reactor is cooled with a liquid alloy composed of 22% sodium and 78% potassium. The thermal power of the reactor is 115 kW, with an electrical output power of 5.5 kW. The core reactor outlet temperature is 843 K, the inlet temperature is 743 K, and the coolant mass flow rate is 1.3 kg/s.

In the field of space nuclear reactor technology, there is a type called a space thermionic nuclear reactor, as shown in Fig. 4.8. It is made by fabricating uranium fuel elements like a diode, using nuclear fuel and tungsten metal sheets to make the cathode, and niobium metal to make the anode. After vacuuming, cesium vapor is added to facilitate electron emission from the cathode. This device is also known as a thermionic emission conversion reactor. The reactor generates heat through fission,

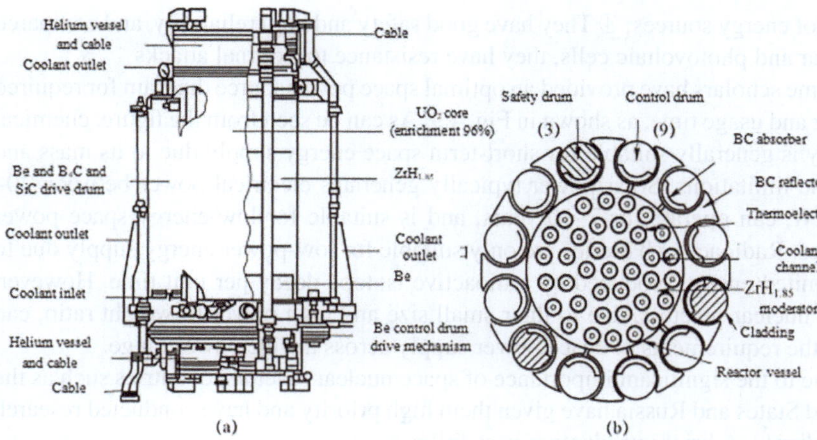

Fig. 4.7 TOPAZ-II Reactor

heating the tungsten to high temperatures of 1300–2000 °C, causing the tungsten to emit a large number of electrons, which form a current through the diode and external circuit. This type of reactor does not require moving parts for power generation, and the entire device can be made small and light. Currently, the power can reach more than 10 kW, and with further improvements, it can reach 1000 kW. It will be used as the main power source for large communication satellites and interstellar spacecraft.

In general, the main advantages of space nuclear reactors are: ① They can meet all the requirements of space flight, providing both long-term low-power supply and short-term high-power supply; ② They can utilize existing nuclear power technology, reducing risk and cost while saving development time; ③ They can be made small and lightweight, especially with a power-to-weight ratio much smaller than other

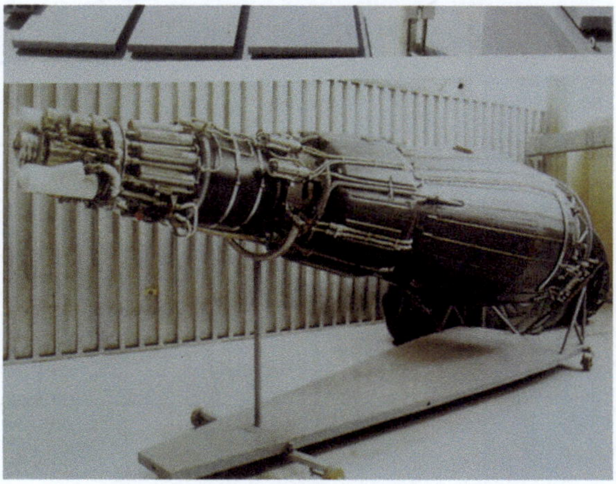

Fig. 4.8 Schematic diagram of space thermionic nuclear reactor

types of energy sources; ④ They have good safety and high reliability, and compared to solar and photovoltaic cells, they have resistance to external attacks.

Some scholars have provided an optimal space power source diagram for required power and usage time, as shown in Fig. 4.9. As can be seen from the figure, chemical energy is generally suitable for short-term space energy supply due to its mass and volume limitations. Solar power typically generates electrical power between 20–100 kW, can operate for 7–10 years, and is suitable for low-energy space power sources. Radioactive batteries are only suitable for low-power energy supply due to the limited energy produced by radioactive isotope decay per unit time. However, space nuclear reactors, due to their small size and high power-to-weight ratio, can meet the requirements of space power supply across a wide power range.

Due to the significant importance of space nuclear reactors, countries such as the United States and Russia have given them high priority and have conducted research accordingly. A brief introduction is as follows.

(1) United States

The United States began research on space power reactors in the 1950s and 1960s. On April 3, 1965, the United States successfully launched the 500W SNAP-10A space reactor into space. This was the first and only space reactor sent into space by the United States, but it was abandoned after only 43 days of operation due to equipment failure. In 1983, the United States began research on the SP-100 type space nuclear power reactor, a fast neutron reactor cooled by liquid lithium, which can operate at temperatures above 1000 °C and has a full-power lifespan of 7 years. The reactor

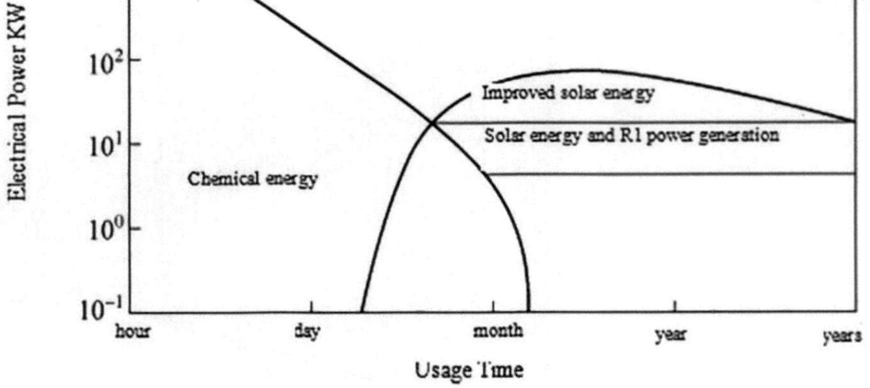

Fig. 4.9 Optimal space power source for required power and usage time

core is equipped with 878 uranium nitride (UN) fuel rods with an enrichment level of 87%-97%, capable of producing 2.4 MW of thermal power. UN fuel expands significantly at temperatures above 1300 K, so a large gap is maintained between the fuel rod and the cladding to minimize cladding deformation. To prevent the UN from melting when heated, the maximum fuel temperature is limited to 2000 K. In addition to fuel rods, the reactor core also has 7 safety rods.

In 2012, Los Alamos National Laboratory conducted a space reactor technology solution verification test on the Flattop device in Nevada. In this test, a high-enriched uranium core, a single heat pipe (located in the core center hole), and two Stirling generators were used, ultimately generating 24W of power, proving the feasibility of this technology solution. The test lasted less than six months and cost less than one million dollars. On May 2, 2018, NASA held a press conference at the Glenn Research Center in Cleveland, announcing the completion of the Kilopower kilowatt-level space reactor test. The test was conducted at the Nevada National Security Test Site from November 2017 to March 2018.

Kilopower is a kilowatt-level space nuclear reactor with simple structure, low cost, and uses mature technology. It can serve as a power system for deep space exploration and an energy supply system for extraterrestrial outposts. Like traditional nuclear reactors, Kilopower uses the energy produced by uranium-235 fission. Unlike traditional reactors that use water as a coolant and uranium dioxide as fuel, Kilopower uses sodium heat pipes to passively conduct heat from the core to the Stirling generator and uses a uranium–molybdenum alloy as fuel.

Kilopower uses a uranium–molybdenum alloy core surrounded by a beryllium oxide reflector layer, with a power output of 1 to 10 kW, capable of continuous operation for 10 years without refueling. Its operation is controlled by inserting and removing a single boron carbide rod from one end of the reactor. The beryllium oxide reflector layer can bounce neutrons back into the core, improving the efficiency of the self-regulating fission reaction. The heat pipes closely surrounding the core transfer the core's heat to the Stirling generator, converting thermal energy into electrical energy. An umbrella-shaped cooling array is set up to dissipate excess heat (see Fig. 4.10).

Kilopower uses a modular design, combining a self-supporting reactor device with a Stirling generator to provide energy for deep space probes or extraterrestrial outposts like Mars. Four Kilopower reactors can meet the power needs of a Mars outpost. Moreover, the capacity of Kilopower is expected to expand to 10 to 100 kW, or even to the megawatt level in the future. In the future, astronauts only need to install a few such devices on the surface of the moon or Mars to ensure the energy supply of the outpost for 10 years or even longer.

(2) Soviet Union/Russia

Although the Soviet Union successfully developed and applied Polonium-210 radioactive isotope batteries early on, the focus of development was on space nuclear reactor power sources and nuclear thermal propulsion. Starting in 1961, the Soviet Union developed four types of space nuclear reactor power systems: ROMASHKA converter-reactor, BUK type space nuclear reactor power source, TOPAZ-1 type

Fig. 4.10 Structure diagram of the Kilopower space heap

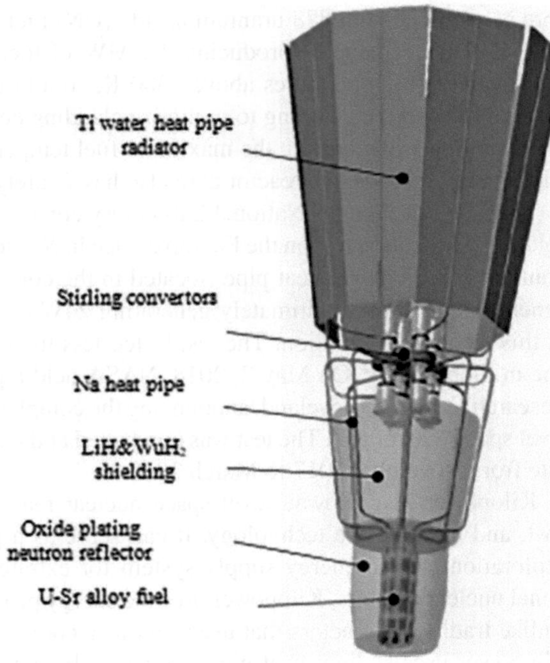

space nuclear reactor power source, and TOPAZ-2 type space nuclear reactor power source. The first two are small fast reactors with direct thermoelectric conversion. The latter two are superthermal neutron reactors with direct thermionic conversion. The Soviet Union launched the first "Cosmos" satellite powered by a nuclear reactor in 1967, and to date, dozens of Cosmos series satellites powered by reactors have been operating in space. The nuclear reactors of the Cosmos series satellites have two types: one is called Romashka, a fast neutron reactor, using uranium-235 ceramic fuel with an enrichment level greater than 90%, converting thermal energy into electrical energy through the principle of thermoelectric conversion; the other is called Topaz, a superthermal neutron reactor, using zirconium hydride as a moderator, using uranium with an enrichment level greater than 90% as fuel, converting thermal energy into electrical energy through the principle of thermionic emission.

In addition, the Soviet Union specifically established the "IGR" high-flux graphite pulse reactor, "IVG-I" experimental reactor, and "IRGIT" experimental prototype reactor to provide test conditions consistent with the actual operating conditions of the nuclear thermal propulsion system. The dynamic test of the fuel elements of the nuclear thermal propulsion system was completed on the "IGR" reactor, the life test of the fuel assembly was completed on the "IVG-1" reactor, and the "IRGIT" experimental prototype reactor was operated to a power level of 90 MW. The fuel elements and fuel assemblies of the nuclear rocket engine were successfully developed, the RD-0410 type nuclear rocket engine test model was built, and several test series of the full-size nuclear rocket engine reactor were completed on the famous Baikal test stand.

After the disintegration of the Soviet Union, the Russian government issued the "Russian Aerospace Nuclear Power Development Concept" in 1998, emphasizing the need to continue to maintain the international leading position in the field of space nuclear power. In October 2009, Russia announced a plan to develop a megawatt-level nuclear spacecraft. The investment is $600 million, the research and development period is 9 years, and it was launched in 2018.

Russia has accumulated more than 400 reactor-years of operating experience in the research of space reactors, and is still researching space reactor types with higher efficiency, longer life, and electrical power up to 400 kW than previous Cosmos series reactors.

(2) Characteristics of Space Reactors

Due to the significant differences between the operating environments of space reactors and terrestrial reactors, the requirements for various aspects of the reactor are also very different, mainly manifested as:

(1) During the launch of the satellite, the reactor must withstand an acceleration of about 10 g, and at the same time, it will generate a lot of vibration. When entering orbit, it is in a weightless state, so the state of the liquid in the reactor and its related structures will change;
(2) In space, the temperature of local space is very low, even zero degrees Celsius, and the nuclear reactor device must maintain an appropriate operating temperature;
(3) Space is a high vacuum state. Without filling gas, organic electrical insulators may vaporize, causing the reactor to malfunction;
(4) To prevent the impact of cosmic dust on the reactor, it is required that the reactor must be equipped with a protective container.

Due to the above special operating environment and characteristics of space reactors, the following requirements are put forward for space reactors:

(1) To adapt to the limited space of space satellites or spacecraft, the reactor must be small in size and light in weight to reduce the load during launch;
(2) To accommodate long-distance space travel and work, the reactor must have high self-control and reliability, capable of operating on ground commands without the need for manual maintenance and adjustment.
(3) To ensure safety on the ground and during the launch process, as well as the safe operation of the reactor in space, it is necessary to ensure that the reactor does not leak or contaminate radioactivity.

Due to these requirements, the technical requirements for space nuclear reactors are higher and the conditions are more stringent. Therefore, it is necessary to develop and improve on the basis of existing terrestrial nuclear reactor technology. At the same time, some new technologies, such as new control technologies and thermoelectric conversion technologies, need to be developed.

4.2.2 Space Nuclear Power

Currently, there are two forms of nuclear power as a space power source: one is a space nuclear reactor power generation system; the other is a radioactive isotope battery. Since the United States began launching spacecraft equipped with nuclear power sources into space in 1961, more than 20 spacecraft have been equipped with over 40 nuclear power sources, most of which are Radioisotope Thermoelectric Generators (RTGs), and a few are nuclear reactor power sources (such as SNAP-10A). Since the Soviet Union used the first space nuclear reactor as a radar monitoring ocean satellite power source in 1967, a large number of space nuclear reactor power sources and a small number of radioactive isotope power sources have been applied.

(1) Reactor Power Source

The nuclear reactor power generation system has high power, long life, compact structure, light weight, high specific power, high efficiency, and can work under strong radiation conditions. It has high maneuverability and survivability in space, so it can be used as a nuclear power propulsion for high-power radars, satellites, and rockets. Space nuclear reactor power sources can also be used for deep space exploration, global communication, resource exploration, material processing, and the establishment of space bases. Therefore, space nuclear power is an ideal space power source.

In terms of space nuclear power, the United States formulated a "System for Nuclear Auxiliary Power (SNAP)" plan as early as 1956. This plan named radioactive isotope thermoelectric generators with odd numbers, and nuclear reactors with even numbers, such as SNAP-10A. From the mid-twentieth century, the nuclear auxiliary power system plan and the nuclear rocket plan implemented by the United States both included the content of space nuclear reactors, aiming to achieve the power supply of spacecraft and nuclear power propulsion of rockets.

In April 1965, based on ground tests, the United States sent a space nuclear reactor SNAP-10A into an orbit 1300 km high. The designed power of the reactor was 540W, the actual power reached 630W, and it was shut down after 43 days due to a power regulator failure, and it still remains in a long-life high orbit.

In 1978, the Soviet "Cosmos-954" satellite fell in Canada, exposing for the first time that Soviet satellites used nuclear reactors as space power sources. In 1979, the United States began to develop a new generation of nuclear reactors to meet the power needs of future space systems. This nuclear device is an evolution of the Space Power Advanced Reactor (SPAR) plan, later named SP-100. In 1983, based on the summary of past work, the United States found that it was a historical mistake to stop developing space nuclear reactor power sources, so it proposed to focus on developing the SP-100 space nuclear reactor system. The SPAR/SP-100 nuclear device is the first stage of the SP-100 plan. After three years of comparison from 1983 to 1986, the United States finally established a fast neutron breeding reactor that is compatible with static thermoelectric energy conversion devices, using uranium nitride as fuel elements, niobium-based alloys as cladding, and liquid metal lithium as both the coolant of the main loop of the device and the working fluid of the heat pipe radiation embedded plate cooling cycle.

After 1986, the SP-100 space nuclear reactor system plan mainly focused on the design, research, and manufacturing of the reactor and energy conversion system, as well as nuclear device testing, and selected four power generation systems, focusing on the reactor thermoelectric power generation system. The goal is to develop a space nuclear reactor power source with a power of 100 kW, a life of 7 to 10 years, a weight of 3t, and direct thermoelectric conversion, with applications in space weapons and nuclear power propulsion.

The Soviet Union's space nuclear power program took a completely different technical approach from the United States. From the outset, they focused on developing space nuclear reactor power sources. The Soviet Union launched nearly 2,000 satellites into space, 35 of which used nuclear reactor power sources and a small number of radioactive isotope power sources. In 1964, the Kurchatov Institute of Atomic Energy developed the Romaska space nuclear reactor ground device, which was installed in Cosmos-198 and launched in December 1967. Romaska uses a fast neutron reactor, thermoelectric conversion, and has an electrical power of 0.5 to 0.8 kW. It is simple in structure, compact in size, and reliable in use. In 1970, the Obninsk Institute of Physics and Power Engineering in the Soviet Union developed a new type of TOPAZ-1 space nuclear reactor power device, which was first installed in Cosmos-1818 satellite and launched in February 1987, and for the second time in Cosmos-1867 satellite and launched in June 1987. This space nuclear reactor is a thermal neutron reactor, using zirconium hydride as a moderator, uranium-235 with an enrichment degree greater than 90% as fuel, uranium-potassium liquid metal forced circulation heat transfer, thermal ion energy conversion, and an electrical power of 7 kW. Experts believe that TOPAZ was the most promising space nuclear power system at the time. Since the United States began cooperating with Russia in 1989, it has leased six TOPAZ-2-2 s from Russia (with a maximum output power of 5.5 kW, DC current, thermal power of 135 kW, maximum coolant temperature at the outlet of 550 °C, maximum launch pole temperature of 1650 °C, working life of 1.5 years, device length of 3900 mm, maximum diameter of 1400 mm, weight of 1000 kg, and a reactor margin ensuring a lifespan of 3 years). This is the second generation product of the TOPAZ space reactor. The fourth generation TOPAZ-2-4 space reactor has an increased electrical power of 30 kW. If the TOPAZ system is changed from a thermal neutron reactor to a fast neutron reactor, the electrical power can reach 100 kW. Therefore, the TOPAZ type thermal ion space nuclear reactor power source is also considered to be the most advanced space nuclear power source in the world so far.

(2) Isotope Power Source

Isotope power sources generate electricity from the heat produced by the decay of radioactive isotopes. The first isotope used in isotope power sources was Polonium-210, which has a heat output of 141 W/g and a half-life of 4.5 months. Such a working cycle obviously cannot meet the needs of long-term space flight, so later researchers chose Plutonium-238. Plutonium-238 has a half-life of 89 years and a heat output of

0.55 W/g, but this isotope fuel is relatively expensive. A cheaper isotope is Strontium-90, which has a half-life of 18 years and a heat output of 0.93W/g. However, the radiation hazard of Strontium-90 is somewhat greater than that of other isotopes. Using plutonium as the heat source for isotope batteries is relatively safe, as these isotopes only emit alpha particles during their decay process and do not emit gamma radiation. When using these isotopes as heat sources, they are generally well sealed in metal tubes in the form of alloys, oxides, or carbides. For example, in the isotope batteries developed by Russia, the isotope fuel tube is placed in a graphite layer, and the exterior of the graphite cylinder is tightly connected to the thermoelectric conversion component. The main principle of thermoelectric conversion is that when two different materials form a closed circuit, a current of a certain direction and size will be generated in the circuit when the temperatures at the two ends of the metal are different. The current and voltage generated by this thermoelectric converter are proportional to the temperature difference at the two ends of the converter. Depending on the isotope used, the temperature at the hot end is generally between 350 and 650 °C, and the temperature at the cold end is between 100 and 250 °C. The current thermoelectric conversion efficiency of isotope power sources can reach 4% to 6%, with a power-to-weight ratio of up to 200 kW/kg.

The power of isotope power sources is generally not large, because radioactive isotopes continuously release heat according to the decay law, and their heat dissipation problem must be considered when not in use. When the satellite is installed on the launch rack and is in a pre-launch state, the isotopes should be pre-installed in the power source. However, at this time, the satellite does not need power, or only needs a small amount of power, so a large amount of excess heat must be dissipated. If the power of the isotope power source is large, a large heat dissipation area is required, so its power must be limited. In contrast, nuclear reactor power sources can achieve controlled heating and do not need to solve the problem of excess heat discharge when not in use. However, nuclear reactor power sources cannot be made very small, mainly due to the limitation of the critical size of the reactor, so isotope power sources have superiority in cases where the power is relatively small. Since 2003, the United States has begun to implement the so-called Prometheus plan. On a technical level, this plan includes the development of a new generation of radioactive isotope power source systems.

4.2.3 Nuclear Electric Rockets and Nuclear Rockets

Today's development of space technology owes much to chemical rockets. However, for a given mass of fuel, chemical rockets emit lower power, thus imposing strict limitations on spacecraft. Technically, the maximum speed that chemical rockets can achieve is relatively low, i.e., their exhaust speed is not high. Rockets using hydrogen and oxygen as propellants can only give probes leaving Earth's orbit a maximum speed of about 10 km/s, while nuclear rockets can reach a maximum speed of 22 km/s. At such high speeds, we can send probes directly to Mars and can shorten the journey from about seven years to three years. Probes using nuclear thrusters can be launched from the ground with conventional rockets and the nuclear thrusters can be started after reaching 800 km above the ground.

Nuclear propulsion methods for spacecraft can be divided into two types: nuclear electric propulsion (nuclear electric rockets) and nuclear thermal propulsion (nuclear rockets).

(1) The nuclear electric propulsion system is a product of the combination of a space nuclear reactor and an electric rocket. The nuclear electric propulsion system, which evolved from general electric propulsion to nuclear electric propulsion, belongs to the second generation of special rocket engines, and achieved interstellar travel and space exploration in the mid-twentieth century. The nuclear electric propulsion device provides electrical energy converted from nuclear energy to the electric rocket, causing the propellant (such as mercury or neon) to ionize and accelerate, becoming a plasma state propellant that is expelled from the nozzle at high speed to generate thrust. The earliest considered energy conversion method for the nuclear electric propulsion system was to use a closed gas turbine cycle. This method can make more use of existing mature technologies, using the gas generated by high-temperature gas-cooled reactors to directly drive turbines to do work, making the turbines generate electricity. There are two major factors affecting efficiency in this energy conversion method. One is the material problem of the turbine blades at high temperatures. To achieve higher efficiency, the turbine blades need to work at high temperatures, but the material and lifespan of the blades limit their operating temperature. Another factor affecting efficiency is the temperature of the exhaust. The lower the exhaust temperature, the higher the efficiency, but the problem brought about by the low exhaust temperature is the need for larger radiators, which will increase the mass and volume of the radiator.

To overcome the shortcomings of the energy conversion of the turbine engine, later scientists began to study the use of magnetohydrodynamic energy converters on spacecraft propellers. Figure 4.11 is a structural diagram of a nuclear electric propulsion device with a magnetohydrodynamic energy converter. This propulsion device uses a magnetohydrodynamic energy converter to convert the thermal energy generated by the reactor into electrical energy. The electric rocket of the nuclear electric propulsion device has the characteristics of high specific impulse, small thrust, long life, high precision, and high reliability. It is suitable for various spacecraft for position maintenance, attitude control, orbit correction (auxiliary propulsion) as well as interstellar travel, interstellar exploration, orbit transfer (main propulsion), etc. Using an electric rocket to transfer a spacecraft from a low Earth orbit to a high orbit or geosynchronous orbit can increase the mass of the effective payload by several times to tens of times compared to direct launch using a launch vehicle. According to analysis, to achieve satellite orbit change, a large power space power source of tens of kilowatts or more is required, which requires a nuclear electric propulsion device with a large specific power (>0.05 kW/kg). Assuming that the thrust of a single electric rocket is 0.15N and the required power is 3.6 kW, the combined thrust of 30 electric rockets used for orbit transfer can reach 4.5N, and the required power is about 100 kW.

The development of nuclear electric propulsion devices is difficult and requires simultaneous research, design, and testing of space nuclear reactors, energy converters, and electric rockets. Available nuclear reactors include thermal ion reactors, pulse reactors, etc.

(2) The nuclear thermal propulsion device (i.e., nuclear rocket engine) replaces the combustion chamber of a liquid fuel rocket engine with a nuclear reactor, replaces chemical energy with nuclear energy, and heats the propellant (hydrogen, helium, or nitrogen) to extremely high temperatures, which is then expelled at high speed through the exhaust nozzle, generating a large thrust (several kN, capable of sending tens of tons of payload into Earth orbit). An American company has designed a nuclear rocket engine. The fuel of this engine is housed in a porous metal plate of a hollow coil, and a lithium hydride sheath is wrapped around the fuel coil as a moderator, which can slow down the neutrons

Fig. 4.11 Structural diagram of the nuclear power propulsion unit

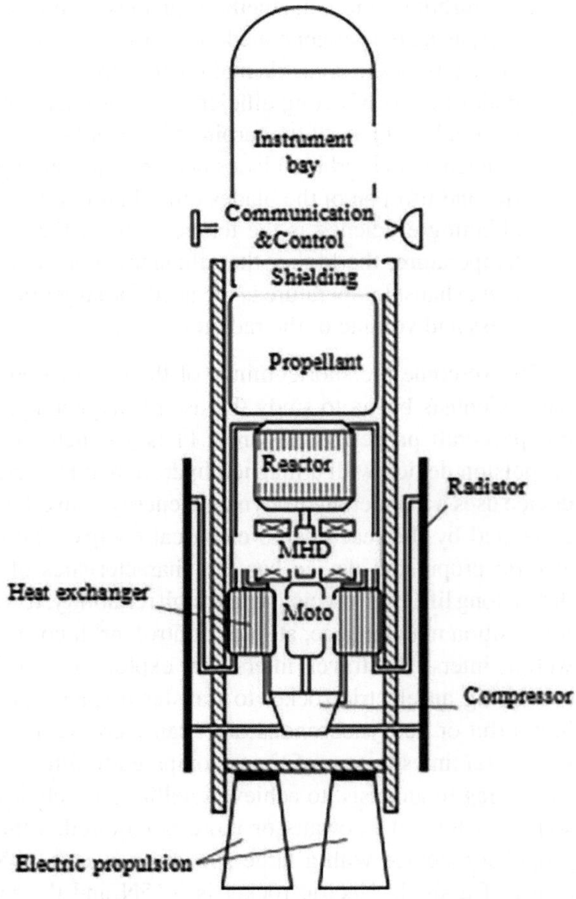

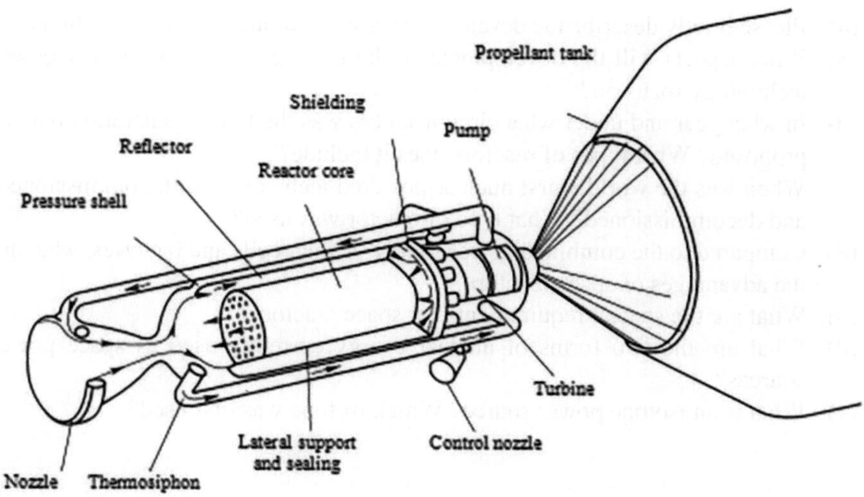

Fig. 4.12 Schematic diagram of a nuclear rocket engine device

produced by fission in the fuel. Liquid hydrogen, which serves both as a coolant and a propellant, flows from the outside of the coil inward, and quickly turns into a gas during the heating and flow process towards the center. The high-temperature hydrogen gas at 3,000 K will flow at high speed along a channel in the center of the fuel coil and then flow out through a small nozzle. One advantage of this nuclear propulsion device is that its propellant (hydrogen) can not only be supplied on Earth, but also has extensive sources on giant planets in the outer solar system, as well as in the water ice of distant satellites and planets. Figure 4.12 is a schematic diagram of a nuclear rocket engine device.

The schematic diagram of the engine device mainly consists of a nuclear reactor energy source, a propellant storage tank, and a turbopump, etc. This form of propulsion device system is simple, but because the temperature inside the reactor is very high, it has high requirements for reactor technology and materials. The specific impulse of the nuclear rocket engine is very high (>900 s), it can generate an electric power of more than 200 kW, has a lifespan of 600 days, and can greatly reduce the amount of propellant used. Nuclear rockets can also shorten the time required for flight in achieving spacecraft orbit transfer, which is beneficial to reduce the radiation hazard of cosmic rays to astronauts, so it is suitable for interstellar travel, such as manned Mars exploration, etc.

Discussion Questions.

(1) Please briefly describe the working principle of the fission reactor. What is the difference between a thermal neutron reactor and a fast neutron reactor?
(2) What types of reactors are there, classified by the coolant and fuel used?
(3) What are pressurized water reactors and boiling water reactors? What is their main difference?

(4) Please briefly describe the development history of nuclear reactor technology.
(5) What aspects will the development of the new generation of nuclear power technology focus on?
(6) In what year and under what circumstances was the fourth-generation reactor proposed? What types of reactors does it include?
(7) When was the world's first nuclear-powered icebreaker built, commissioned, and decommissioned? What type of reactor was used?
(8) Compared to the combined power supply of solar cells and batteries, what are the advantages of space reactors?
(9) What are the special requirements for space reactors?
(10) What are the two forms of nuclear energy currently used as space power sources?
(11) What is an isotope power source? Which isotope was first used?

References

Chen WZ, Zhang LM, Xiao HG. (2019) Principles of Nuclear Reactor Engineering (Volume I). Beijing, China Atomic Energy Press.
Su ZT, Yang JC, Ke GT. (2020). Space Nuclear Power. Shanghai, Shanghai Jiao tong University Press.
Wu HC, Cao LZ, et al. (2014). Nuclear Reactor Physics. Beijing, China Atomic Energy Press.
Yan CQ, Ding M, Bian HZ. (2023), Advanced Nuclear Reactors. Beijing, Science Press.
Zhang DH. (2023). Advanced Fast Neutron Reactor Technology. Shanghai, Shanghai Jiao tong University Press.

Chapter 5
Nuclear Reactor Safety Issues and Safety Design

5.1 Reactor Safety Characteristics

Since the construction of the first reactor, scientists have recognized the unique safety characteristics of reactors (Luo 2015). These special safety issues mainly come from the uncontrollable release of radioactive nuclides under accident conditions (Editorial Board of Registered Nuclear Safety Engineer Training Series 2004). The inherent hazard manifested in:

5.1.1 The Dramatic Increase of Nuclear Power Caused by Nuclear Fission

Due to rapid fission and energy release, there is a possibility of a dramatic increase in fission power in the reactor (Zhu 2010). The reason why humans can use nuclear energy is mainly based on the following facts: In the process of nuclear fission, a unit mass of nuclear fuel will release a huge amount of energy. The fission reaction triggered by neutrons also releases additional neutrons. It is through the combination of these two conditions that we can design a reactor that can both form a self-sustaining chain reaction and continuously release energy.

In order to obtain a stable energy output, a stable nuclear fission chain reaction must be maintained. That is, the number of neutrons in the reactor should be balanced. In other words, only when the number of neutrons triggering the fission of fissile nuclides uranium or plutonium in the reactor core is stable, the reactor is in a stable state. We use the effective multiplication factor k to describe the degree of this neutron balance. If $k = 1$, the number of neutrons is constant, the reactor reaches a critical state critical, and the power generated by the reactor core is stable. If k is less than 1, the reactor is subcritical, and the power of the reactor core decreases continuously,

L. Yu et al., *Nuclear Technology and Safety*, Nuclear Science and Technology,
https://doi.org/10.1007/978-981-95-2856-1_5

finally reaching a minimal level. If k is greater than 1, the reactor is supercritical, and the power of the reactor core increases rapidly.

In practical applications, we use reactivity ρ to describe the degree to which the reactor deviates from criticality (see Chap. 2). Correspondingly, when $\rho = 0$, the reactor is in a critical state, when ρ is less than 0 (negative reactivity), the reactor is in a subcritical state, and when ρ is greater than 0 (positive reactivity), the reactor is in a supercritical state.

In an uncontrolled supercritical state, the reactor power may increase rapidly, the fuel generates huge thermal stress, and if the cooling capacity cannot keep up with the demand, the reactor core will be damaged. Therefore, to ensure nuclear safety, one of the basic safety functions of the reactor is introduced: control of reactivity. However, in some reactivity accident conditions, the nuclear fission reaction of the nuclear reactor is out of control. For example, introducing a positive reactivity exceeding β (delayed neutron fraction β, about 0.0065), the reactor is instantaneously supercritical and will produce much higher power than the design power. Even for a reactor with a thermal power of only 100 MW, when a control rod ejection accident occurs in the reactor, the instantaneous maximum power of the reactor can reach more than 10,000 MW within 1 s! The excessively high power causes a serious imbalance in the heating, heat transfer, and heat transport of the fuel elements. The fuel fragments into hot particles, which cause the cooling water to evaporate dramatically, leading to steam explosions and even hydrogen explosions. The reactor can be completely destroyed. The Chernobyl nuclear accident is typical cases. Therefore, to ensure nuclear safety, the reactivity introduced by the reactor must always be controlled within the delayed neutron fraction β (called delayed supercritical). In fact, nuclear power plants or nuclear power reactors always operate under very small reactivity disturbances, rarely exceeding 0.001.

In fact, modern reactor designs inherently ensure negative reactivity feedback coefficients, which fundamentally differ from the core design of the Chernobyl nuclear power plant that experienced an accident. Under abnormal conditions, the negative reactivity feedback coefficient will automatically educe the reactivity, thus slowing down the fission chain reactions. Simultaneously, during operation, multiple redundant control systems (such as control rods, chemical compensation control, etc.) strictly control the rate and amount of reactivity introduced. In abnormal conditions, the reactivity control system automatically initiates an emergency, achieving an emergency shutdown of the reactor. Moreover, the explosions that occur within the reactor and the containment vessel are fundamentally different from atomic bomb explosions. Due to the relatively low concentration of fissionable fuel, it is impossible to reach the critical volume required for a nuclear explosion, so a nuclear explosion will not occur. Whether it's the Chernobyl nuclear power plant or the Fukushima nuclear power plant, explosions are mainly caused by the combination of overheated steam or high concentrations of hydrogen and oxygen. High concentrations of hydrogen are mainly produced by the chemical reaction of zirconium alloy fuel cladding material with cooling water under high temperature conditions, so the explosions in accidents are chemical and physical explosions, completely different from the nuclear explosions of atomic bombs, and the two should not be confused.

5.1.2 Nuclear Fission and Decay in Reactors Release Strong Radioactive Radiation

Compared to general industrial devices, the inherent hazard of reactors lies in the fact that in addition to releasing a tremendous amount of energy during the nuclear fission process, a large amount of radioactive material is also generated. That is, the atomic nuclei within the reactor will produce new atomic nuclei and a large number of rays such as alpha, beta, gamma, and neutrons during the fission process. Statistical analysis shows that the fragments produced during the nuclear fission process include about 40 elements and 250 different isotopes. After the reactor has been operating for a while, the isotopes present in the core will reach thousands (Wu et al. 2014). Analysis of spent fuel removed from the reactor reveals that isotopes with mass numbers between 85–105 and 130–150 make up the majority. Moreover, most fission products are unstable and will continuously decay, emitting rays such as alpha, beta, and gamma rays, which, if directly irradiated onto the human body, may cause ionizing radiation damage. We refer to these newly produced unstable nuclei as radioactive fission products. The quantity of radioactive fission products will decay over time, and the time it takes for their quantity to decay to half of the initial value is called the half-life. The half-life of fission products varies, from seconds to hundreds of years, so nuclear power devices remain radioactive for a long time. Among the fission products, those that easily escape from fuel pellets mainly include gaseous krypton (Kr), xenon (Xe), and halogen isotopes that are easily soluble in water. A 10 MW reactor can accumulate a total amount of radioactive isotopes reaching 10^{18} Bq over its entire operating life. Under normal circumstances, most of these fission products and the rays they produce are contained by multiple layers of barriers such as fuel element cladding, the primary coolant pressure boundary, and the reactor vessel wall. For example, more than 98% of radioactive fission products can be retained within uranium dioxide ceramic cores, with less than 2% of krypton (Kr), xenon (Xe), and iodine (I) and other gaseous radioactive substances diffusing into the gap between the fuel core and the element cladding. Therefore, the fuel elements of general pressurized water reactors use zirconium alloy cladding for high-quality encapsulation, and the radioactive isotopes produced during operation are mostly retained within the elements. The necessary condition for radioactive release is element damage. Only when the fuel elements are severely overheated, melted, or disintegrated, or when cracks are produced in the fuel element cladding during the processing and manufacturing process, can a large amount of radioactive material be released. Once the barriers that contain these radioactive materials, such as the fuel element cladding, the primary circuit pressure boundary, and the containment shell (or reactor vessel) are damaged, radioactive fission products may diffuse into the environment or other compartments, causing serious harm to workers, the public, and the environment.

5.1.3 Decay Heat Continues to Be Released After Reactor Shutdown

Unlike conventional fossil fuels, reactors continue to generate heat for a considerable length of time after shutdown. This heat comes from the kinetic energy of various radiations released during the decay of fission products, a form of energy known as decay heat. The magnitude of decay heat depends on the quantity of radioactive fission products. As long as radioactive fission products exist in the fuel elements, decay heat exists, so reactors have decay heat whether they are operating at power or after shutdown. Fission power can be regulated and controlled by controlling the chain reaction of neutrons with fissionable nuclei, but the generation of fission products is currently uncontrollable, making the generation of decay heat difficult to control. Some fission products have half-lives of several years or even longer, so reactor decay heat can persist for a long time, but it gradually decreases over time. In a typical nuclear fission reaction, the energy released by a single fission of uranium-235 is about 200 MeV (compared to fossil fuels, each carbon atom produces only 4 eV of heat when burned). Of this, 187 MeV of energy is released instantaneously during the fission process, including 165 MeV of kinetic energy of fission products, 5 MeV of kinetic energy of fission neutrons, 7 MeV of prompt gamma-ray energy, and 10 MeV of gamma-ray energy when neutrons are captured; an additional 23 MeV of energy is released over time, including 7 MeV of kinetic energy when fission products undergo beta decay, 6 MeV of gamma-ray energy produced with beta decay, and 10 MeV of neutrino radiation energy accompanying beta decay. Because neutrinos interact very weakly with matter, the above 10 MeV of energy is not retained in the core and is not included in the energy released by nuclear fission. Therefore, after a nuclear fission reaction, about 13 MeV of energy is retained in the core due to the delayed beta decay of fission products. That is, after the nuclear fission chain reaction stops, there is still 6.5% of the residual heat that is gradually released later.

After the reactor is safely shut down, the main source of heat in the core becomes the beta decay heat of the fission products, which rapidly decreases over time. If the reactor is shut down after a long period of stable operation, the decay heat will quickly drop to 1.5% of the power before shutdown after 1 h, to 0.6% after 1 day, to about 0.2% after 1 year, and still 0.02% after 10 years. For example, in a million-kilowatt pressurized water reactor nuclear power plant, the decay heat produced 1 h after shutdown after 100 days of full-power operation is about 48 MW, a considerable amount of heat. This is one of the important differences between nuclear power plants and thermal power plants. Therefore, whether it is the decay heat of the fuel after the reactor is shut down or the decay heat of the spent fuel, it must be ensured that it can be removed in time. If the fuel elements lose effective cooling, it is likely to cause a nuclear accident. The accidents at the Three Mile Island nuclear power plant in the United States and the Fukushima nuclear power plant in Japan were both caused by the inability to timely discharge the decay heat of the reactor, causing the fuel elements to overheat and even melt (Wu 2017).

The operating power and operating history of the reactor determine the quantity of fission products and the magnitude of decay heat. When the reactor is operating at power, the decay heat accounts for a small portion of the total, so the focus is mainly on fission power; after shutdown, the fission power quickly decreases to negligible within a few minutes, and at this time decay heat plays a dominant role. Nuclear power plants generally operate at full power for a long time, so the decay heat is significant. For example, a nuclear power plant reactor with a thermal power of 3000 MW operates at full power for a long time, and the decay heat at the beginning of shutdown reaches more than 100 MW. The characteristics of low power operation and intermittent operation of ship nuclear power devices determine that their decay heat is relatively small. For a reactor with a thermal power of 100 MW, the decay heat about 1 h after shutdown is about 1 MW or even lower. Therefore, for the same type of accident, considering only decay heat, the response time for a ship nuclear power device accident is relatively relaxed, and the accident process is relatively slow. To deal with this less than 5% of decay heat and its derived consequences, nuclear power devices need to be equipped with a large number of systems, equipment, instruments, and personnel.

5.1.4 High Temperature and High-Pressure Water and Steam

In order to keep the high-temperature coolant from boiling as a whole, pressurized water reactors must maintain a high working pressure. The working pressure of the primary loop of a nuclear power device is generally between 13 and 16 MPa. Once the pressure boundary is broken, high-temperature and high-pressure water will spray out. For ship nuclear power devices, this not only threatens cabin structural components, junction boxes, cables, and other equipment, but also directly leads to an increase in cabin pressure, endangering the integrity of the cabin. For ship nuclear power devices, once the secondary loop pressure boundary is broken, high-temperature and high-pressure steam will spray into the main engine room or auxiliary machine room, directly threatening the on-site post operators and also affecting the ship's power and power supply guarantee.

Like any other power device, nuclear power devices may also have design errors, manufacturing defects, construction and installation errors, operation and maintenance errors, equipment failures, and various operator errors. These errors or mistakes are also the main sources of risk for nuclear power devices.

The above indicates that to ensure the safe operation of the reactor, it is necessary to prevent the release of a large amount of radioactive fission products accumulated in the fuel elements, and to prevent their potential release into the surrounding environment. The usual design is to provide multiple physical barriers to isolate radioactive materials from the environment. This forms the three basic safety functions that are well-known in the industry today: under normal operation and accident conditions of the reactor, all safety facilities should satisfy the three major safety functions of controlling reactivity, ensuring core cooling, and containing radioactive products.

5.2 Safety Issues of Nuclear Power Devices

The basic safety issues of nuclear power devices can be summarized into three aspects: critical safety, thermal safety, and radiation safety.

(1) Critical Safety

The nuclear fission reaction in the nuclear reactor is achieved by the chain fission reaction of neutrons. If the total number of neutrons in the reactor does not change over time, the reactor is in a critical state; if the number of neutrons decreases over time, the reactor is in a subcritical state; if the number of neutrons increases over time, the reactor is in a supercritical state. Unlike atomic bombs, nuclear power devices must be able to achieve critical control of the reactor. If the fission reaction continues to be out of control, the reactor's nuclear power will increase sharply, eventually leading to the nuclear fuel elements burnout. Therefore, under any circumstances, the nuclear fission process in the reactor must be kept under control, allowing the energy produced by nuclear fission to be released according to external demand.

On ship nuclear reactors, the control of reactor criticality is mainly achieved by raising or lowering the control rods in the reactor. There are generally two types of initiating events that can cause the reactor's fission reaction to lose control: the control rod is uncontrollably pulled out, and the control rod is ejected.

The criticality of the reactor is unrelated to the power level of the reactor, meaning that the reactor can operate critically at very high power levels; at very low power levels, the reactor can also maintain criticality. When a ship's nuclear power device is operating, even if the load remains constant, it is difficult to achieve true physical criticality. In fact, the reactor power fluctuates slightly around criticality, but due to the negative temperature feedback effect of the pressurized water reactor, the reactor can achieve power self-stability.

Furthermore, reactor supercriticality only reflects that the number of neutrons produced in the reactor at a certain time period is more than the number of neutrons that disappear. The reactor power will gradually increase, but it does not indicate that the reactor will become uncontrollable. In fact, the operation of the reactor from zero power to full power is achieved through the supercriticality of the reactor. Supercriticality can be divided into delayed supercriticality (reactivity is less than the delayed neutron fraction β, β is usually about 0.0065) and prompt supercriticality (reactivity is greater than the delayed neutron fraction β). In layman's terms, during delayed supercriticality, the speed of neutron growth is still controllable, and the timely action of personnel or the protection system can quickly terminate the supercritical state of the reactor; during prompt supercriticality, the speed of reactor power increase is very fast, and when personnel or the protection system has not had time to act, the reactor may already be on the verge of being burned out. In fact, the control rods of ship nuclear power devices are all set with the highest lifting speed. Even if the operator uses the highest rod speed and continuously raises the control rods, as long as the protection system can work normally, the reactor will not undergo prompt supercriticality. Only when one or more control rods are quickly ejected, the reactor may undergo prompt supercriticality.

(2) Thermal Safety

In order to maintain the integrity of multiple barriers, thermal safety limits have been proposed in the design of nuclear power devices, mainly including the following parts.

1) Nuclear Fuel Pellet Temperature Limit

The main mechanisms of nuclear fuel pellet degradation are cracking and melting. Especially after the fuel pellet melts, not only the integrity of the first barrier is destroyed, but the integrity of the fuel cladding will also be seriously threatened. Therefore, it is necessary to set a limit for the temperature of the fuel pellet. The highest temperature of the fuel is generally in the center of the nuclear fuel pellet, which is difficult to measure. Since the highest temperature of the fuel mainly depends on the linear power density of the fuel (proportional to power), the fuel's linear power density is limited by limiting the reactor power. There are different reactor power limits for single main pump operation or low-speed operation of the main pump. If the reactor power exceeds the limit during operation, the control system will automatically act to limit the reactor power. During the entire life of a ship's nuclear power device, when operating at full power steady state, the highest temperature of the fuel pellet is generally controlled between 1500 and 1700 °C; it will be higher during over-power transients, but the safety margin is sufficient.

The primary cause of fuel pellet temperatures exceeding their limit is the severe mismatch in heat generation, heat transfer, and heat transport in the reactor. If a large positive reactivity is rapidly introduced, the reactor power will surge. Due to the transient heat generation of the fuel elements being far greater than the heat transfer and transport, the fuel pellet temperature will rapidly rise. If the protection system can act in time to suppress the increase in reactor power and bring the reactor to an acceptable state, the fuel pellet temperature will not reach its melting point; otherwise, the fuel pellet will melt at high temperatures. These types of reactivity accidents during power operation occur quickly, with the accident time measured in seconds. For incidents like the Three Mile Island nuclear accident and the Fukushima nuclear accident, the fuel elements lost effective cooling, and under the effect of decay heat, the fuel pellet temperature continued to rise until it exceeded the melting point. These accidents generally take several tens of minutes or a few hours. For naval nuclear power systems, due to the relatively low operating power and decay heat, such accidents develop more slowly.

2) Cladding Thermal Limits

The main failure mechanisms of cladding include damage caused by material manufacturing defects, damage under the expansion force generated by fuel pellet irradiation, instability caused by the pressure difference inside and outside the cladding, foreign object wear, damage caused by excessive temperature stress between the inner and outer walls of the cladding, material corrosion damage, cladding hydrogen embrittlement, and burning or overheating melting caused by boiling crises, etc.

The thermal limits of the cladding mainly include two aspects:

Firstly, to prevent the fuel elements from "burnout". During reactor power operation, the fuel elements are restricted from experiencing a boiling crisis. A boiling crisis is caused by the deterioration of the heat transfer conditions of the fuel elements. Under certain conditions, if the heat flux density on the surface of the fuel element cladding (proportional to reactor power) reaches a certain level, a vapor film will form on the outer surface of the fuel element cladding. Since the thermal conductivity of the vapor film is far lower than the convective heat transfer of the fluid, the heat cannot be transferred to the coolant in time. All this heat acts on the cladding, causing the cladding temperature to rapidly rise to a very high value in a very short time, possibly exceeding the melting point of the cladding, causing the cladding to burn through, commonly referred to as burnout. Burnout generally occurs during high power operation. The heat flux density at burnout is usually defined as the critical heat flux density; the ratio of the actual heat flux density to the critical heat flux density is the burnout ratio. The burnout ratio varies for fuel elements in different locations and at different lifetimes. The reactor design should ensure that the minimum burnout ratio in the fuel elements meets safety criteria to ensure reactor safety. Generally, for naval nuclear power systems, the steady-state condition ensures a minimum burnout ratio $DNBR_{min} = 2.0\sim2.3$; normal dynamic conditions: $DNBR_{min} = 1.3\sim1.4$. The thermal limits of the cladding can be achieved by limiting reactor power, reactor operating pressure, reactor outlet temperature, and optimizing reactor power distribution.

Secondly, to prevent the fuel elements from "dryout". When an accident such as a total power loss occurs in the reactor, although the reactor can shut down in time, the decay heat will "dryout" the coolant around the fuel elements, i.e., heat it into steam. When the steam's heat transfer performance is poor and cannot dissipate the residual heat, the temperature of the fuel element cladding and combustible poisons will continue to rise:

568 ~ 623 K: Normal operating temperature of the cladding;
970 ~ 1020 K: Combustible poisons begin to soften;
1020 ~ 1070 K: Cladding begins to swell;
1223 K: Cladding begins to perforate;
1273 ~ 1373 K: Zirconium-water reaction is evident;
1477 K (1204 °C): Acceptable standard for emergency core cooling system;
1523 K: Heat release from zirconium-water reaction equals decay heat;
2030 K: Zirconium-4 melts.

The safety limit is usually set so that the maximum surface temperature of the fuel element cladding does not exceed 1204°C, mainly to prevent the zirconium-water reaction from becoming intense and the cladding becoming extremely brittle. Since the melting point of the cladding material is lower than that of the uranium alloy core, during accidents such as total power loss, coolant loss of flow, or coolant loss, the cladding will fail before the fuel pellet.

3) Coolant Temperature Limits

The primary purpose of coolant temperature limits is to prevent the coolant from boiling over, control the rate of temperature rise and fall of the coolant, otherwise, the thermal stress, thermal shock, and flow instability generated will pose a serious threat to the safety of the reactor. This criterion is generally achieved by limiting the outlet temperature of the reactor coolant, the temperature difference between the inlet and outlet, the temperature difference between the regulator and the hot end coolant of the reactor, and the rate of temperature rise and fall of the coolant.

(3) Radiation Safety

Radiation safety focuses on the integrity of multiple barriers, ensuring that personnel and the environment are free from the harm of radioactive radiation. Any nuclear activity involving personnel must limit the radiation dose that personnel may be exposed to and assess the impact on the environment. To minimize the release of radioactive substances into the environment, radioactive substances are confined within a certain range through multiple barriers (Zhu et al. 2019). This is also an important manifestation of the principle of in-depth defense. The number of barriers required mainly depends on the quantity of radioactive nuclides and the types of isotopes, the effectiveness of a single barrier, internal and external hazards, and potential consequences of failure. For a typical pressurized water reactor, multiple barriers mainly include the following parts.

The first and second barriers are the fuel core and the cladding. The nuclear fuel of a pressurized water reactor uses low-enriched uranium dioxide, which is sintered into cores, stacked in zirconium alloy cladding tubes, and welded at both ends with end plugs. Most of the fission products are contained in the uranium dioxide cores, and only a small part of the gaseous fission products will diffuse into the gap between the core and the cladding. Considering various deteriorations that may occur during normal operation and anticipated operational events, the fuel elements must be able to withstand the expected in-reactor irradiation and environmental conditions, and ensure that their performance will not further significantly deteriorate.

The third barrier is the primary circuit pressure boundary that contains the coolant. The primary circuit pressure boundary of a pressurized water reactor consists of the reactor vessel and the external coolant circuit, which includes the steam generator heat transfer tube, the regulator, the pump, and the corresponding connecting pipes. In order to ensure the tightness and integrity of the second barrier, there should be enough design margin in terms of structural strength, and attention should also be paid to the material selection, manufacturing, and operation of the barrier.

The fourth barrier is the containment system. The containment system is used to ensure that the radioactive substances released into the environment under design basis accidents are below the specified limits. It mainly includes sealed structures, related systems for controlling pressure and temperature, and facilities for isolating, managing, and eliminating fission products and other substances that may be released into the containment atmosphere. In the event of an accident, it can prevent radioactive substances from leaking into the external environment and is the last line of defense to ensure the safety of nuclear power plants. The containment can also protect important systems and equipment of the nuclear power plant from being damaged by various external attacks. According to design requirements, all identified design basis accidents must be considered in the design of the containment system. In addition, facilities for mitigating the consequences of certain selected severe accidents must also be considered. There are strict requirements for the sealing of the containment, and there should be enough margin in the design of structural strength to cope with the changes in pressure and temperature caused by the large rupture of the coolant pipe and prevent a large amount of radioactive nuclides from leaking out. In addition, regular leak checks should be carried out for the containment to ensure its sealing.

During normal operation, minor cracks may appear on the surface of the cladding of a few fuel elements, and a small amount of fission products and activation products will enter some auxiliary systems in the nuclear auxiliary plant, such as the chemical and volume control system and the spent fuel pool. These radioactive products mainly exist in liquid or gaseous form and are controlled by the following methods:

1) Maintain a relative negative pressure on-site to prevent radioactive substances from spreading to other areas. Areas with radioactive iodine should also maintain a negative pressure compared to surrounding areas.
2) Collect radioactive gases through the radioactive waste treatment system, and after their radioactivity decays to an acceptable level, send them to a chimney equipped with filters and iodine adsorption devices for monitored discharge. Low-radioactivity waste gas can be directly discharged through the chimney after filtration.
3) After the radioactive waste liquid is collected, it is sent to the boron recovery system or waste liquid treatment system for filtration, desalination, degassing, evaporation, and storage monitoring, and then stored in the waste liquid treatment system storage tank. After sampling and analysis meet the discharge standards required by the environmental protection department, it is discharged to the environment under monitoring.

The radiation sources of a ship's nuclear power device mainly include alpha rays, beta rays, gamma rays, and neutrons. Due to these rays being blocked by four barriers, their impact on the health of workers is very minimal. For a ship's nuclear power device, under normal operating conditions, apart from the reactor compartment, the radioactivity levels in the compartments where the crew are located are very low, and the radiation levels to the human body are far below the levels stipulated for occupational radiation workers in national standards. When the reactor is shut down, no neutrons are generated, and due to the sufficient shielding layers set up in the reactor, alpha and beta rays cannot escape from the reactor, and only a small part of the gamma rays can penetrate the shielding layer. Radioactive materials can only enter other compartments when multiple barriers fail, and this probability is extremely low.

5.3 Safety Design Principles and Requirements

The safety design of the reactor mainly includes the basic principles of nuclear safety design, safety management requirements, technical requirements, and design requirements, as shown in Table 5.1. The first level in the table can be regarded as the

Table 5.1 Reactor design safety principles and requirements

Name	Content
Basic principles	Defense in depth
	Multiple barriers
Safety management requirements	Management responsibilities
	Design management
	Verified engineering practices
	Operational experience and safety research
	Safety evaluation
	Independent verification of safety evaluation
	Quality assurance
Main technical requirements	Requirements for defense in depth
	Safety functions
	Accident prevention and safety characteristics
	Radiation protection and acceptance standards
Design requirements	Radiation protection and safety grading of acceptance
	Overall design basis
	Reliability design
	Human factors engineering principles
	Measures for in-service testing, maintenance, repair, inspection, and monitoring
	Quality requirements
	Safety analysis
	Aging
	Decommissioning

basic safety concept, and the latter three levels can be regarded as the implementation of the basic safety concept. This section mainly introduces the basic principles and main design requirements.

5.3.1 *Basic Principles*

(1) Defense in Depth

Defense in depth refers to the establishment of multiple layers of defense in all nuclear safety activities, so that individual failures can be compensated or corrected, without causing harm to workers, society, and the environment. The concept of defense in depth is the basis of nuclear safety technology and should be reflected in all stages of the reactor's life cycle (including design, construction, operation, maintenance, and decommissioning). Nuclear power devices generally set up five levels of defense:

prevention, monitoring, protection, mitigation, and emergency response. The five levels of defense have overlapping parts.

The first line of defense is prevention. Its main purpose is to prevent deviations from normal operation and system failures. Measures taken include: considering redundancy, independence, and diversity; designing, constructing, operating, and maintaining nuclear power devices in a comprehensive and conservative manner; selecting appropriate design standards and specifications; controlling the manufacturing and construction process of equipment; considering human factors engineering; managing the safety of equipment aging and wear; utilizing operational experience; conducting safety analysis under various operating conditions; independent verification of safety evaluations; detailed analysis of in-service inspections and decommissioning, etc.

The second line of defense is monitoring. Its main purpose is to monitor and correct deviations from normal operating conditions to prevent anticipated operational events from escalating into accident conditions. Measures taken include: setting up control and protection systems determined in safety analysis and formulating good operational procedures.

The third line of defense is protection. Its main purpose is to guide the nuclear power device to a safe and controllable state in the event of an accident by setting up special safety facilities for design basis accidents and formulating accident handling procedures. The main measures taken include: designing inherent safety features; designing for fault safety; setting up special safety facilities for design basis accidents; formulating accident handling procedures for design basis accidents.

The fourth line of defense is mitigation. Its main purpose is to limit radioactive materials within the barrier as much as possible when core damage is inevitable, by setting up physical barriers and means of protecting physical barriers (including special safety facilities), and formulating management guidelines for severe accidents beyond the design basis. The main measures taken include: formulating supplementary measures and accident management guidelines to prevent accident expansion for leading sequences and representative accident conditions that may lead to severe accidents; setting up measures that can mitigate the consequences of severe accidents.

The fifth line of defense is nuclear emergency response. Nuclear emergency response refers to the state of having to take certain actions beyond normal work procedures immediately to control the occurrence of nuclear and radiation accidents or to mitigate the consequences of accidents. Nuclear emergency response is the last line of defense in the nuclear safety defense-in-depth system, and it is also the bottom line. Its main purpose is to mitigate the radioactive consequences caused by the release of radioactive materials by taking protective measures such as shielding, evacuation, and iodine tablet intake, and emergency response measures such as engineering rescue, pollution control, and elimination, through appropriate emergency preparedness. The main measures taken include: considering nuclear accident emergency issues within a certain range, such as establishing emergency channels, emergency control centers, etc.; formulating off-site emergency response action plans aimed at protecting the public and the environment, and preparing technical data and documents required for emergency plans, such as source terms, initiating events, accident sequences, etc.

The principle of defense-in-depth is the foundation of nuclear safety work, and other safety principles are derived from it. Its goals are: first, to compensate or correct equipment failures or human errors; second, to maintain the effectiveness of the barrier itself and prevent fault propagation; third, to protect workers, the public, and the environment from radiation damage when the effectiveness of the barrier itself cannot be fully maintained. Generally speaking, severe reactor accidents are caused by multiple failures (including human failures and equipment failures), and multiple failures are due to unacceptable defects in some defense components. The core of defense-in-depth is to place all safety-related activities, including all aspects related to organization, personnel behavior, or design, under overlapping measures, so that even if one type of failure occurs, there will be appropriate measures to detect, compensate, or correct. That is to say, throughout the entire lifespan of the reactor, all layers of defense or measures taken to ensure safety must be guaranteed to be effective. We cannot tolerate defects at any level on the grounds that there are other levels of defense. Each level must be considered as the last line of defense. All levels must be balanced, and we cannot say that if one link is strong enough, other links can be simpler. The occurrence of a nuclear accident depends on the weakest link in prevention. At the same time, preventing accidents is a priority. On the one hand, it is relatively easy and effective to prevent deviations from normal operating conditions of the reactor and degradation of performance, capability, and status. On the other hand, the overall cost of preventing accidents is relatively smaller than the cost of mitigating accidents. Among the five levels of defense, the reactor has greater difficulty in dealing with severe accidents and nuclear emergencies.

(2) Multiple Barriers

To prevent the release of radioactive materials, nuclear power devices are equipped with four barriers to contain radioactivity. The first barrier is the fuel pellet, which retains over 98% of the radioactive fission products. The remaining less than 2% of radioactive materials are sealed by the second barrier, the fuel element cladding. The first and second barriers can also be collectively referred to as the first barrier, i.e., the fuel element. If the cladding is damaged, the high-strength pressure vessel and the closed primary circuit system, serving as the third barrier, can completely seal the radioactive materials leaking from the cladding. If the third barrier leaks, the fourth barrier, i.e., the reactor containment (or reactor compartment), can shield the radioactive materials, preventing their release to other plant locations (or compartments) to ensure personnel safety.

Each physical barrier is a critical protective measure. Therefore, the reliability of each barrier must be improved to achieve its design function and prevent subsequent barrier failure due to damage or failure of a barrier.

5.3.2 Design Requirements

(1) Safety Grading

Similar to nuclear power plants, the compartments, systems, and equipment (including software) related to ship nuclear power devices are classified into safety grades based on their safety functions. Their design, construction, and maintenance must ensure that their quality and reliability are commensurate with this grading. Mechanical equipment in ship nuclear power devices is classified into safety grade one, safety grade two, safety grade three, and non-safety grade (NS); electrical equipment (including power equipment, instrument control, and their power supply equipment) is classified into safety grade (1E grade) and non-safety grade (N1E grade).

Safety grade one applies to equipment and its supports that constitute the reactor coolant pressure boundary, the failure of which would cause the loss of reactor coolant exceeding the replenishment provided by the normal coolant supply. Examples include the reactor pressure vessel, the primary side of the steam generator, reactor coolant piping, control rod drive mechanisms, etc.

Safety grade two applies to pressure components and supports within the reactor coolant pressure boundary that are not classified as safety grade one, as well as equipment used to prevent anticipated operational events from leading to accident conditions and mitigating the consequences of accident conditions. Examples include small diameter pipes that are part of the coolant system pressure boundary, residual heat removal system equipment, piping and valves, purification system equipment, piping and valves, etc.

Safety grade three applies to other safety–critical equipment other than safety grade one and safety grade two. Examples include equipment, piping, and valves of the chemical shutdown system, piping and valves of the equipment cooling water system, etc.

Non-safety grade (NS) applies to equipment that does not belong to safety grade one, safety grade two, and safety grade three.

Safety grade (1E grade) electrical equipment applies to electrical equipment required to perform or support basic safety functions, prevent accidents, and mitigate the consequences of accidents. Examples include emergency power supply systems, instrument nuclear control equipment required for emergency reactor shutdown, dedicated safety facility required instruments, control and power supply equipment, etc.

Non-safety grade (N1E grade) electrical equipment applies to all electrical equipment other than safety grade (1E grade) electrical equipment. When a piece of equipment or part of it can provide two or more different safety functions, the equipment should be classified into the higher safety grade.

(2) Overall Design Basis

The design basis must specify the essential capabilities of the nuclear power device to adapt to the operating conditions and design basis accidents determined within the specified radiation protection requirements. The design basis includes normal operation technical specifications, reactor operating conditions, safety grading, accident analysis methods, and important assumptions, etc. In the design of normal operation, anticipated operational events, and design basis accidents, conservative design measures and good engineering practices must be adopted to ensure the safety of the reactor core; radiation doses are kept within the prescribed limits, as low as reasonably achievable. In addition to design basis accidents, the design must also consider beyond design basis accidents, including severe accident behavior. This section introduces some factors that need special consideration in the design of ship nuclear power devices.

1) Environmental Characteristics

In the design of ship nuclear power devices, special usage conditions such as tilt, sway, vibration, and impact, as well as environmental factors such as irradiation, radioactive contamination, salt spray, oil mist, mold, special waterways, sea conditions, weather, and hydrology should be considered. The design must consider all external facilities necessary to ensure the safety of ship personnel and the public during an accident, such as power supply and fire protection facilities.

2) Protection Against Natural Events

The design of safety-related compartments, systems, and components must ensure that the ship can safely shut down the reactor in the event of natural incidents such as tornadoes, typhoons, tsunamis, and special navigation channels that have been considered, and maintain it in a subcritical state for a long period of time thereafter. It should be able to dissipate the reactor heat and limit the escape of radioactive materials to acceptable levels. If a safety-related compartment, system, or component is damaged in an irresistible natural event due to a well-thought-out design, such a natural event may not be included in the design basis accidents of the ship's nuclear power device, but it must be ensured that the reactor is in a subcritical state.

3) Protection Against Human Events

The design of safety-related compartments, systems, and components must ensure that in the event of human events such as ship grounding, collision, fire, underwater close-range non-contact explosion, and accidents near the port that endanger the safety of the ship, the reactor can be safely shut down and maintained in a subcritical state for a long period of time thereafter. It should be able to dissipate the residual heat and limit the escape of radioactive materials to acceptable levels. If a safety-related compartment, system, or component is damaged in an irresistible human event due to a well-thought-out design, such a human event may not be included in the design basis accidents of the ship's nuclear power device, but it must be ensured that the reactor is in a subcritical state.

4) Combination of Events

The design basis of safety-related compartments, systems, and components must reflect the various combinations of human events, natural events, equipment failures, and operational errors that are likely to occur simultaneously or separately, and their resulting radiological consequences.

(3) Reliability Design

1) Principle of Inherent Safety

To date, nuclear power plants around the world have accumulated more than 10,000 reactor-years of operational experience, and the operational record of nuclear power plants (especially pressurized water reactor nuclear power plants) is good. However, the occurrence of the Three Mile Island nuclear accident, the Chernobyl nuclear accident, and the Fukushima nuclear accident shows that due to the extreme complexity of the nuclear power plant system, the safety of the nuclear power plant depends on the effectiveness of the measures taken. In nuclear power safety design, it is necessary to use inherent safety characteristics as much as possible, such as ensuring that the reactor has negative Doppler and moderator temperature feedback characteristics in pressurized water reactor design, and relying on gravity, stored potential energy, and other passive safety features, which can terminate the chain reaction in the reactor under abnormal conditions, dissipate the residual heat of the reactor, and bring the nuclear power device into a safe state.

2) Principle of Fail-Safe

That is, when a nuclear power device, system, or component fails, the nuclear power device should be able to enter a safe state without any triggering action. The design of extremely important systems and components for nuclear power device safety should implement the fail-safe principle as much as possible.

3) Single Failure Criterion

Equipment combinations that meet the single failure criterion can still maintain their assigned functions when a single random failure occurs at any part of them. All kinds of secondary failures caused by a single random event are considered as part of a single failure. General requirements for single failure:

(a) For some systems and components that require high reliability and must perform important safety functions, in order to achieve high reliability, in addition to requiring high quality, redundancy must also be used in equipment combinations. When necessary, diversity design should also be used.
(b) The minimum requirement for redundancy is: in all kinds of assumed initiating events, even if any component that affects important safety functions has a credible random failure, the design must still meet safety requirements.

According to the safety design requirements of ship nuclear power devices, the single failure criterion must be met in the following two types of equipment combinations.

(a) In each safety combination in the design of ship nuclear power devices;
(b) Each safety system specified in the design standards to meet the single failure criterion.

The following situations may be allowed not to comply with the single failure criterion:

(a) Extremely rare assumed initiating events;
(b) Consequences that are extremely unlikely to be produced by assumed initiating evebnts;
(c) For maintenance, repair, or regular testing, some equipment is not used for a limited time.

4) Principle of Diversity

Diversity should be applied to redundant systems or components performing the same function, that is, different attributes are introduced into redundant systems or components to improve system reliability. Ways to obtain different attributes include: adopting different working principles, different physical variables, different operating conditions, and using products from different manufacturers, etc. The use of the diversity principle can reduce common cause failures or common mode failures, thereby improving the reliability of some systems.

5) Principle of Independence

In order to improve system reliability and prevent common cause failures or common mode failures, system design should achieve system layout and design independence through functional isolation or physical separation.

(4) Principles of Human Factors

Engineering From a safety perspective, the workplace and working environment of reactor operators must be designed according to the principles of ergonomics. Human factors have two aspects of influence: on one hand, in abnormal conditions, if operators can take correct actions, they can make significant contributions to reactor safety; on the other hand, if operators fail to make correct judgments or use safety facilities or adopt incorrect response measures, it poses a significant threat to nuclear safety. One of the most important lessons from the Chernobyl and Three Mile Island nuclear accidents (from minor incidents to severe accidents) is that human errors can dramatically exacerbate accidents. Therefore, the safety design of reactors must facilitate operators to take successful actions within a limited time, in the expected surrounding environment, and under psychological pressure, and minimize the necessity for operators to intervene in the short term.

(5) Measures for Regular Testing, Maintenance, and Inspection

To maintain the ability of important structures, systems, and components related to reactor safety to perform their functions, they should be calibrated, tested, maintained, repaired, inspected, or monitored throughout the life of the reactor.

(6) Quality Requirements

The design, manufacture, installation, inspection, and testing of safety-related compartments, systems, and components must reach a quality level commensurate with the importance of their safety functions. The specifications and standards used must be listed and evaluated to determine their usability, suitability, and completeness, to ensure that the design can achieve the quality required by its safety functions.

To ensure that safety-related compartments, systems, and components can reliably perform their functions throughout the life of the reactor, a quality assurance outline must be developed and implemented. The relevant records of the design, manufacture, installation, inspection, testing, operation, and maintenance of safety-related compartments, systems, and components must be kept by the competent department or under its control.

(7) Safety Analysis

A safety analysis must be conducted on the reactor design, using deterministic and probabilistic analysis methods. Based on this analysis, the design basis for safety-important items must be established and confirmed. It must be demonstrated that the designed reactor can meet all the prescribed limits for radioactive release under various conditions and the acceptable limits for potential radiation doses, and that the defense-in-depth has played its role. The computer programs and analysis methods used in the safety analysis must be verified and confirmed, and all kinds of uncertainties must be fully considered.

(8) Aging

The design must provide appropriate margins for all safety-important structures, systems, and components to consider relevant aging and wear mechanisms and possible performance degradation related to service life, thereby ensuring that these structures, systems, or components can perform the necessary safety functions throughout their design life. Aging and wear effects under all normal operating conditions, tests, maintenance, decommissioning, and assumed initiating events and subsequent nuclear power plant states must be considered. Monitoring, testing, sampling, and inspection measures must be taken to evaluate the aging mechanisms anticipated at the design stage and identify unexpected situations or performance degradation that may occur during use.

(9) Decommissioning

To achieve reliable performance, the decommissioning of equipment and the expected maintenance, testing, and repair work on each individual safety system must be considered when designing the reactor and its safety systems. If the overall reliability of the system no longer meets the criteria used in design and operation after decommissioning, the reactor must be shut down; or if the decommissioned equipment cannot be replaced or repaired on time, the reactor must be put into a safe state.

5.4 Basic Strategy for Nuclear Accident Prevention

The basic strategy for nuclear accident prevention can be summarized as one center and two basic points. One center: prevent the massive release of radioactive materials. Two basic points: maintain the balance of heat generation and export; maintain the integrity of the radiation protection boundary.

Nuclear accident prevention requires the establishment of a concept of full-life prevention for nuclear facilities, that is, the strategy of defense-in-depth is implemented throughout the design, construction, operation, maintenance, and decommissioning stages of nuclear facilities. Focusing on nuclear safety goals, we must address issues in four areas: regulations and standards, equipment and technology, human resources, and organizational management.

5.4.1 Nuclear Safety in the Design Stage

High-level, high-quality design is a crucial foundation for ensuring the safety of nuclear equipment, and it is inherently essential. On the management level, it is necessary to implement a qualification licensing system for design units and designers, as well as a quality assurance system. In the work of a nuclear safety regulatory mechanism based on the entire lifespan, the independence, comprehensiveness, and timeliness of regulation are the most fundamental requirements. Personnel of nuclear facility user units have the responsibility to strengthen coordination with the design department according to national and local conditions. They also have the responsibility to feed back the operating experience of previous equipment to the design process of new equipment to reduce the phenomenon of incomplete consideration in design due to the designer's lack of understanding of the use of equipment. The drafting and formulation of laws, regulations, standards, and other documents involved in the design phase of nuclear equipment should be complete and standardized, progressive, mature, and forward-looking. They should be principled and flexible, but the bottom line of nuclear safety cannot be breached. Once laws, regulations, standards, etc., are established, their seriousness and authority must be maintained.

On the technical level, the full use of advanced, mature, and fully verified technical means to maximize the reliability of equipment is key to preventing accidents and reducing the probability of accidents. Good human–machine-environment design is conducive to the safe operation of the device. High-level automated monitoring and alarm device design (visual, auditory, olfactory, tactile) helps in the discovery and handling of the beginnings of accidents. Redundant design of important safety systems, fault-tolerant design, inherent safety design (i.e., natural safety and passive safety) can minimize the "innate deficiencies" of equipment and enhance the ability of reactor safety protection under accidents. Damage mitigation design (physical isolation, protective equipment) can alleviate the consequences of accidents. Nuclear emergency equipment and nuclear emergency technology must be fully considered at the time of nuclear equipment design.

On the personnel level, the state should formulate relevant policies with a developmental perspective and create conditions to continuously retain one or more high-level nuclear equipment design teams. A high-level team is important as it guarantees technological innovation; the continuity of the team is equally important as it guarantees the inheritance and development of technology. Without continuity, there will be a gap in talent, a break in equipment technology, and a lack of new equipment research and development capabilities.

5.4.2 Nuclear Safety in the Construction Phase

High-level construction quality is the lifeline of nuclear safety. Only by improving construction quality can high-level design be implemented, hidden dangers of accidents be eliminated, and the probability of accidents be reduced. The responsible units should establish a quality assurance system for regulatory representatives, design responsible units, general assembly factories, and equipment manufacturing factories according to national regulations and standards, ensuring that everything is rule-based. They should control the construction quality of the design responsible units, general assembly factories, and equipment manufacturing factories, ensuring that everyone is responsible. They should supervise the entire process from the construction phase to the debugging phase, ensuring that everything is supervised. They should attach great importance to the archiving and management of various information and documents, ensuring that everything is traceable.

For nuclear safety-grade equipment, it is necessary to organize identification tests conducted by regulatory representatives, design responsible units, general assembly factories, and equipment manufacturing factories to ensure that these devices can still perform their functions under various disturbances during normal operation or accident conditions. This identification test can be conducted through analysis or experiment. Identification tests are usually conducted on prototypes or model machines through specific test benches or laboratories. Equipment that has not been fully tested in principle cannot be installed, especially some special safety facilities, which are not used during normal operation and are only used at critical times during accident handling. If they are not fully verified and installed, their effectiveness under accident conditions cannot be guaranteed. During the installation and debugging of nuclear equipment, regulatory personnel should increase supervision, strictly accept according to standards, and promptly discover and rectify safety hazards.

5.4.3 Nuclear Safety in the Operation Phase

Most nuclear accidents in nuclear power devices occur during the operation phase. The guarantee of nuclear safety management during the operation of nuclear power devices mainly depends on the high-level nuclear safety management of the operator, advanced operation support technology, perfect operation manipulation procedures, good personnel quality, and meticulous nuclear safety supervision.

On the management level, the operator has direct responsibility for the safe operation of the nuclear power device, and the nuclear safety management department exercises supervisory power over the operation of the nuclear power device. Management personnel should be proficient in nuclear safety regulations and technical conditions, exercise their supervisory mission through regular tests and availability judgments, nuclear safety regular assessment systems, supervision and inspection mechanisms, and perfecting the detailed rules of nuclear safety supervision at all levels. The equipment use management department should formulate various policies conducive to the selection, training, assessment, appointment, retention, and motivation of nuclear-related talents. They should advocate a nuclear safety culture that emphasizes "attitude, responsibility, behavior". Only by immersing this culture into the bones of every employee can employees be elevated from behavioral constraints to conscious action.

On the technical level, to ensure the safe operation of nuclear power devices, a complete support system for the operation of nuclear power devices needs to be established. Good work should be done in tracking and analyzing the core fuel consumption status, conducting physical startup tests of reactors, tracking and analyzing the operating status of nuclear power devices, analyzing the risk of changes in the status of key equipment, calculating and evaluating accident safety margins and consequences, and formulating and improving various operation manipulation procedures.

Operating procedures and severe accident management guidelines are the legal basis for managing nuclear power facilities and ensuring operational safety. The development and refinement of these procedures and guidelines are crucial tasks during the operation phase of nuclear power facilities. Operating procedures can be divided into normal operating procedures and accident handling procedures, with the latter serving as the basis for operators to judge the occurrence of accidents, their nature, and to take emergency measures to handle them. When an accident occurs, operators can maintain the reactor in a safe state or transition it to a safe state according to the accident handling procedures. The content of the accident handling procedures consists of accident judgment procedures and accident handling guidelines.

Severe accident management is a strategy and action that includes preventing incidents from escalating into severe accidents, mitigating the consequences of severe accidents, and achieving a long-term stable safety state. Severe accident handling measures and management guidelines are the main support for severe accident management, providing reasonable handling strategies, intervention timing, corresponding handling measures, and implementation procedures for major hazards during the course of severe accidents.

At the personnel level, it is important to recognize that human behavior is a dual-impact phenomenon. The operation of nuclear power facilities has shown that a considerable number of accidents are caused by operator errors. In the factors that trigger accidents in the operation of nuclear power facilities, human errors account for more than 50%. On the other hand, correct, timely, and effective actions by personnel can play a crucial role in accident prevention, handling, and mitigation. Therefore, focusing on the training of nuclear-related personnel, especially improving the capabilities and qualities of nuclear power operation personnel and security personnel, is a fundamental measure for accident prevention and handling.

Operational experience feedback is a systematic method that prevents the recurrence of similar events and continuously improves the operational efficiency and safety level of equipment by discovering, analyzing, and summarizing abnormal events and human errors in the development, operation, and maintenance of equipment. This is achieved through reporting, screening, grading, analyzing, formulating corrective actions, implementing corrective actions, and evaluating their effectiveness. Currently, international organizations such as the International Atomic Energy Agency and the World Association of Nuclear Operators have established complete experience feedback systems.

5.4.4 Nuclear Safety During Maintenance

Nuclear safety during maintenance refers to measures taken during the maintenance process to ensure the reactor safety, the radiation safety of maintenance personnel and the surrounding environment, and to prevent nuclear accidents caused by maintenance quality issues. Nuclear activities during maintenance mainly include reactor vessel opening, fuel uploading and replacement, physical startup tests, and cold and hot state functional tests.

At the management level, nuclear safety management during maintenance includes maintenance qualifications, items, quality, and system management related to nuclear safety. Its management objectives are: first, to ensure nuclear safety during the maintenance process of the device and items; second, to maintain and restore the technical performance and safety functions of the nuclear power device; third, to protect the environment and personnel. The nuclear safety department should supervise maintenance activities: first, to verify whether the maintenance management department and maintenance units comply with nuclear safety standards and fulfill their commitments to maintenance safety; second, to urge the correction of behaviors that affect nuclear safety; third, to verify and confirm the set safety control points and their release conditions on-site.

5.4.5 Safety During Decommissioning

The decommissioning of nuclear facilities is the final and critical phase of a nuclear facility's life cycle. It is internationally acknowledged as a lage-scale, complex engineering prject with substantial technical challenges, major risks, and policy sensitives. It is necessary to follow national nuclear facility management policies and regulations, use special technologies and countermeasures, implement the dismantling of spent fuel, reactor systems, and other radioactive items, as well as the treatment and disposal of exhaust gases and waste liquids, to ensure nuclear safety, radiation safety, and environmental safety.

The nuclear safety issues during the decommissioning process mainly involve nuclear activities such as the unloading of nuclear fuel, the dismantling of reactors and radioactive items, the storage and cooling of spent fuel, and the transportation and management of spent fuel and radioactive items. In the implementation and operation stages of decommissioning, nuclear and radiation accidents may occur due to human factors, improper program procedures, and unsuitable process technologies, which can cause significant casualties, economic losses, ecological environmental pollution, and serious social impacts.

At the management level, to ensure nuclear safety during the entire decommissioning implementation process, according to engineering control theory, quality management during the decommissioning engineering practice process can be combined with specific nodes closely related to the realization of nuclear safety goals, setting nuclear safety monitoring points and formulating release conditions. Implementing nuclear safety monitoring point management is one of the important contents of implementing nuclear safety supervision and inspection. The main purpose is to check the implementation of nuclear safety regulations and standards before the start of some important stages and links in nuclear activities, confirm the consistency of the documents and actual situation submitted by the supervised units, and urge the correction of matters that do not meet the requirements of nuclear safety management in a timely manner.

On the technical equipment level, the decommissioning of nuclear facilities involves a wide range of aspects. The main technologies and equipment required include nuclear fuel unloading technology and equipment, radioactive decontamination technology and equipment, radioactive item dismantling technology and equipment, radioactive item quantity technology and equipment, radioactive item storage technology and equipment, and radioactive item transportation technology and equipment. The adoption and use of these technologies and equipment must ensure nuclear and radiation safety for personnel and the environment, while also minimizing the generation of secondary radioactive pollution and striving to minimize radioactive waste throughout the process.

Discussion Questions:

(1) What are the main manifestations of risk in a nuclear reactor?

(2) Under what circumstances can a nuclear reactor explode? What causes the explosion? Is the mechanism of an explosion within a reactor the same as that of an fission atomic bomb?
(3) How dose a reactor contain its highly radioactive substances?
(4) Is there decay heat when the reactor is operating at power? For a nuclear power reactor operating at 3000 MW for a long time, what is the initial decay heat after shutdown?
(5) What accidents can cause serious critical safety issues in nuclear power systems? Provide the conditions under which these accidents occur.
(6) To prevent the melting of fuel elements and the burning of cladding, what protective parameters are set by the protection system during reactor power operation, and how is protection carried out?
(7) What are the damage mechanisms of the cladding? What are the differences in thermal limits of the cladding during power operation and after reactor shutdown?
(8) What is nuclear safety function? What does multi-barrier protection include?
(9) Please explain the essence of the basic principle of in-depth defense.
(10) How is nuclear safety classification carried out for electromechanical equipment in nuclear power systems?
(11) Please explain the main principles of reliability design for nuclear power systems.
(12) What is the single failure criterion? What are its basic requirements?
(13) What is the basic strategy for nuclear accident prevention?
(14) What are the key points of nuclear safety at each stage of the full life cycle of a nuclear power system?

References

Editorial Board of Registered Nuclear Safety Engineer Training Series. (2004). Case Studies in Nuclear Safety. Beijing, China Environmental Science Press.
Luo SZ. (2015). Applications of Nuclear Technology. Harbin, Harbin Engineering University Press.
Wu HC, Cao LZ, et al. (2014). Nuclear Reactor Physics. Beijing, China Atomic Energy Press.
Wu YC. (2017). Introduction to Nuclear Safety. Hefei: University of Science and Technology of China Press.
Zhu JZ, Shan JQ. (2010). Nuclear Power Plant Safety. Beijing, China Electric Power Press.
Zhu JZ, Shan JQ. (2019), Nuclear Reactor Safety Analysis. Xi'an, Xi'an Jiaotong University Press.

Chapter 6
Safety Evaluation of Nuclear Reactors

6.1 Deterministic Safety Analysis Method

6.1.1 Overview

Deterministic safety analysis is the design basis for nuclear power systems. It is mainly used to assess whether the safety of nuclear power systems under normal operating conditions, transient operating conditions, and accident conditions meets various design standards. The safety analysis of nuclear power systems under normal operating conditions mainly ensures that the radiation doses received by workers and the public are within acceptable ranges, the leakage of radioactive materials in nuclear power systems is within acceptable ranges, and the emission of radioactive materials in the surrounding environment is within acceptable ranges. The transient analysis of nuclear power systems mainly assesses the reasonableness of the threshold settings of various alarm systems, automatic adjustment systems, protection systems, etc., to avoid transient events triggering accidents that destroy nuclear power systems. The analysis of design basis accidents in nuclear power systems is used to assess whether nuclear power systems (including specially designed safety facilities) can effectively respond to these accidents and ensure that the safety of nuclear reactors and the release of radioactive materials meet certain safety limit criteria.

Usually, the deterministic method classifies accidents into "credible" and "incredible" categories based on past experience and socially acceptable levels, and further classifies credible accidents as the initial events for deterministic analysis.

When using deterministic analysis methods to determine the type, capacity, and response speed of safety facilities, a hypothetical accident is needed as a design basis, and this accident is considered the maximum credible accident, i.e., it is believed that if the set safety facilities can prevent this accident, they can definitely prevent other accidents. Therefore, this maximum credible accident is often referred to as the design basis accident (DBA). For pressurized water reactor nuclear power plants, the double-ended guillotine break of the main coolant pipe cold section is usually

L. Yu et al., *Nuclear Technology and Safety*, Nuclear Science and Technology,
https://doi.org/10.1007/978-981-95-2856-1_6

considered the maximum credible accident, which is carefully considered in the design and strictly defended against. Even if such a serious initiating event occurs, serious consequences may not necessarily occur due to the strict defense of safety facilities such as the emergency core cooling system.

The safety analysis of accidents can be conducted through conservative methods or best-estimate methods. Due to the selection of design basis accidents and the significant uncertainty in the analysis models, safety review units require the use of conservative assumptions to ensure the envelope of the analysis results. Therefore, there are two basic assumptions in the analysis of nuclear reactor accidents: the invoked safety system meets the single failure assumption; operators do not intervene in the accident progression in the short term, and all actions rely on the automatic response of the device. The purpose of setting these two assumptions is to try to prove that the nuclear reactor still has the ability to maintain safety in the worst-case scenario. The advantage of the conservative method is that the analysis results have good inclusiveness, but this analysis method often leads to accident sequences that are completely unrealistic, and some important phenomena and beneficial operations are overlooked. Due to these shortcomings and the gradual maturity of the current best-estimate procedures, the best-estimate method is gradually accepted by safety reviewers.

The greatest advantage of the best-estimate method is that it can describe the actual process of accident development and make equally conservative evaluation conclusions based on some reasonable assumptions. The main disadvantage of the best-estimate method is that the theoretical analysis models for some accident phenomena are not perfect enough, and the empirical models are more numerous, resulting in insufficient universality and greater uncertainty of the model.

(1) Basic Logic of Deterministic Analysis

The basic idea of deterministic safety analysis is to assume that an accident has occurred, make reasonable assumptions about system response and personnel operations, then use mechanism analysis procedures to simulate the response of nuclear power devices, and derive the consequences of the assumed accident, mainly including multi-barrier integrity assessment and the size of radioactive release. Deterministic safety analysis must include:

1) Confirming that the operating limits and conditions of the nuclear power device meet the assumptions and requirements of normal operation design;
2) Assumptions of initiating events suitable for the design of nuclear power devices;
3) Analysis and evaluation of event sequences originating from assumed initiating events;
4) Comparison of various analysis results with radioactive acceptance criteria and design limits;
5) Establishment and confirmation of design basis;
6) Demonstrating that the expected transient events and design basis events can be managed through the automatic response of the safety system combined with the prescribed operator actions.

The corresponding steps mainly include:

First, determine the object of analysis, classify nuclear power device accidents, and determine the initiating events for accident analysis.

Second, determine the main design parameters and uncertainties for accident analysis; determine the initial conditions for calculation analysis; determine the assumption conditions for accident analysis (mainly including reactor power history, reactor power peaking factor magnitude and distribution, reactor feedback coefficient, various types of reactivity introduction, etc.); determine the safety facilities involved in accident analysis and possible mitigation measures; determine reactor protection settings and delay times; appropriately consider human intervention behavior, including human performance failure.

Third, determine the deterministic safety analysis program to be used. According to the research object, the accidents to be analyzed, and the analysis program used, divide the research object into control volumes and junctions; fill in the input card of the analysis program; perform steady-state calculations; according to different initiating events, perform accident calculation analysis one by one.

Fourth, according to the calculation results, draw dynamic curves of important parameters; compare with safety design criteria, analyze and draw conclusions.

(2) Specific Methods of Deterministic Analysis

Due to the severity of nuclear reactor accident consequences, people cannot fully verify the safety of devices under design basis accidents through experimental methods, but can only obtain the response characteristics of devices under accident conditions through simulation analysis, indirectly proving the safety of reactors. In order to accurately obtain the response characteristics of nuclear power devices under various accident conditions, it is necessary to establish dynamic characteristic analysis models of nuclear power devices and compile corresponding program systems based on these models.

The specific methods of deterministic analysis can be divided into three types: experimental measurement, numerical simulation, and operational experience feedback. The three types of methods can complement, guide, and cross-check each other.

1) Experimental Measurement

Experimental measurement is to obtain the dynamic response characteristics of certain parameters through experiments on actual operating devices or test benches. Specific experimental methods include frequency response method, pulse test method, random signal test method, and parameter tracking method, etc. By using one of these methods to measure the dynamic characteristics of the object, and then using computational mathematics to process the data, the dynamic model of the object can be obtained. The data obtained from experiments on actual devices can objectively reflect the dynamic characteristics of the device.

However, there are certain limitations to conducting dynamical studies of devices solely through experimental methods. The most significant limitation is the limited coverage of the experiments. Due to the unique nature of nuclear power devices, only some normal operating transients can be tested on actual operating devices. Full-range testing cannot be conducted for conditions that deviate from normal operation or for major accident conditions. Therefore, the dynamical models of some devices, systems, and equipment under these conditions cannot be fully obtained through experiments. Moreover, most experiments can only be conducted after the complete construction of the nuclear power device, which does not provide timely basis for the design of the device. When problems are found in the design, it is not easy to make modifications.

Principle test benches, verification test benches, land-based model reactors, and actual ship nuclear power devices can both verify the performance of devices, systems, and equipment through a large number of tests, and provide a verification platform for the development, modification, and improvement of deterministic safety analysis procedures. The deterministic safety analysis procedure is developed through steps such as principle experiments of components, establishment of mathematical models of components, development of component analysis procedures, experimental verification, development of system analysis procedures, experimental verification, trial use of procedures, modification and improvement, official issuance, continuous verification, and continuous improvement.

For specific devices and conditions of specific nuclear power devices, principle experiments of components and establishment of mathematical models of components are necessary. For ship nuclear power devices, this mainly includes the calculation formula for the critical heat flux density of fuel elements, specific equipment such as once-through steam generators, narrow gap channels, and the thermal–hydraulic calculation formula for equipment, flow instability analysis; simulation of thermal–hydraulic behavior of fluids under specific conditions such as natural circulation; simulation of operating characteristics of passive safety systems, etc.

2) Numerical Simulation

Numerical simulation is a method that establishes mathematical and physical equations to describe the dynamical characteristics of nuclear power device core physics, thermal-hydraulics, radiation dosage, and control logic, to obtain the response rules of various parameters over time, and the interrelationships between various parameters. The mathematical and physical equations that describe these devices, systems, and equipment are referred to as the dynamical models of the devices, systems, or equipment. Numerical simulation can compensate for some deficiencies of experimental measurement methods. By establishing a comprehensive dynamical model of the device, numerical simulation can not only simulate most operating transients, but more importantly, it can conduct anticipated safety evaluations for some major accidents and extreme accidents.

Many operating parameters of nuclear power devices are functions of both time and space coordinates, such as neutron flux density, coolant temperature, and pressure. If both the time function and the space function of these parameters are considered, and nonlinear partial differential equations are used to establish their dynamical models, this method is called the distributed parameter method, such as the three-dimensional space–time neutron dynamics model of the reactor. The distributed parameter method can more realistically reflect the actual response characteristics of parameters, but solving a large number of nonlinear partial differential equations is very complex, which brings inconvenience to the study of many problems.

If the microscopic distribution of the research quantity in space is not considered, and the spatial average value is studied, and the research quantity and other additive quantities are lumped d together and placed at the geometric center of the research body, this processing method is called the lumped parameter method. The lumped parameter method can simplify many partial differential equations describing dynamics into ordinary differential equations, greatly simplifying the solution. The point reactor neutron dynamics model often used in core physics is a kind of lumped parameter method. Although some simplifications used in the lumped parameter method facilitate modeling and calculation, some models cannot realistically reflect the actual response rules of parameters, and their application range is limited.

However, in actual applications, the distributed parameter method and the lumped parameter method can learn from each other and complement each other. However, the method of mathematical simulation is limited by the research status of many basic disciplines, such as two-phase flow and heat transfer. The current status of these basic researches directly determines the accuracy and reliability of the establishment and analysis calculation of mathematical models. Therefore, the dynamics models established under some accident conditions are not perfect, and there are still some differences with the experimental results. In addition, there are many difficulties in accurately solving a large number of dynamics equations. In order to meet the requirements of calculation accuracy and speed at the same time, it is necessary to use reasonable and effective numerical calculation methods.

Therefore, when modeling the dynamics of nuclear power devices, it is necessary to combine experimental measurement methods and mathematical simulation methods, and then develop corresponding accident analysis programs based on the established mathematical and physical models. Using these program systems to analyze and verify the dynamic conditions of nuclear power devices can verify whether the system and equipment meet the design requirements under various expected transient conditions; safety analysis of accident conditions can verify whether the design of the system and equipment meets the safety design criteria. The ultimate goal of numerical simulation is to achieve high-fidelity digital reactors and digital nuclear power plants.

3) Experience Feedback

In the long-term operation of nuclear power devices, some problems or events may occur due to misoperation or equipment failure, which are sometimes unavoidable. The important thing is how to deal with these failures or events. Practice has proved that summarizing experience and lessons in a timely manner, and learning from one example to prevent similar problems from recurring, is an important measure. Experience feedback is the best way to solve this problem and the most effective way to improve the level of safe operation.

Experience feedback is a systematic method that identifies, analyzes, and summarizes anomalies, incidents, and human errors in the development, operation, and maintenance of equipment. It involves reporting, screening, grading, analyzing, formulating corrective actions, implementing corrective actions, and evaluating their effectiveness. This process helps prevent the recurrence of similar incidents and continuously improves the operational efficiency and safety level of the equipment.

The technologies involved in the field of experience feedback mainly include operational information feedback system technology, high-reliability information collection and application technology, and information management technology. Operational information feedback refers to the technical system that collects, transmits, analyzes, processes, and summarizes relevant information in a timely manner when occasional incidents or anomalies related to safety and quality occur during operation, maintenance, and other service activities. Through the operational information feedback system, relevant parties can obtain the latest information in a timely manner, learn lessons from it, and prevent similar incidents from recurring. By establishing a standardized and institutionalized system for information feedback technology, the operational safety level of nuclear facilities can be continuously improved. With the rapid development of communication methods, the future development of this technology will be more closely integrated with high-tech communication methods, achieving the recording and transmission of images, sounds, pictures, etc. For some occasional phenomena that are difficult for personnel to capture, technologies such as video recording can achieve comprehensive recording and playback.

Currently, internationally, organizations such as the International Atomic Energy Agency (IAEA) and the World Association of Nuclear Operators (WANO) have established complete experience feedback systems. Countries such as the United States, France, Canada, Japan, and South Korea also have specialized agencies engaged in experience feedback work in their respective nuclear industries. Experience feedback work in China's nuclear power field has also been fully launched at the industry, group company, and nuclear power plant levels. At present, the industry level has established an experience feedback system that includes an organizational management system, a document system, and a network communication platform. Each nuclear power group company is gradually promoting the experience exchange, analysis, and feedback work of its affiliated nuclear power plants in various forms. Each nuclear power plant has also established an internal experience feedback system and has successively built computer management systems, greatly improving the effectiveness of nuclear power experience feedback.

6.1.2 Common Models in Deterministic Safety Analysis

(1) Nuclear Reactor Lumped Parameter Analysis Model

To ensure the safe operation of the reactor, nuclear power operators need to understand the response characteristics of important parameters under various conditions. Especially for accident conditions, the operator's analysis and prediction of the response of important parameters are extremely important for accident judgment and disposal. The composition and operating conditions of nuclear power devices are complex, and the theoretical research and safety analysis of their transient processes usually require the use of large-scale system analysis programs. During the operation of the ship reactor, due to the lack of an online operation support system, it is difficult for operators to use large-scale, detailed safety analysis system programs to analyze various conditions. One relatively simple and effective analysis model is the aggregate parameter analysis model, which helps operators theoretically analyze the macro response characteristics of important reactor parameters under various conditions.

1) Lumped Parameter Analysis Model

When performing transient analysis of a reactor, if the quantity under study changes little with space, that is, at any moment in the study area, the value of the quantity under study at any spatial location is very close to its average value in the study area, then for such transients, the microscopic distribution of the quantity under study in space can be ignored, and only its spatial average value can be studied (Chen et al., 2019). At the same time, the quantities under study and other additive quantities are lumped together and placed at the geometric center of the study body, like a "point mass". This method is called the lumped parameter method. For extensive quantities such as mass M and heat capacity, they can be concentrated at the geometric center, and the intensity quantities such as temperature and pressure are treated according to the spatial average value. The main purpose of introducing the lumped parameter method in reactor transient analysis is that the parameters of the actual dynamic process of the reactor change with time and space variables, and the solution is complex. The use of the lumped parameter method can get rid of the complexity brought by spatial coordinates and obtain the macro characteristics of related parameters in the dynamic process. The mathematical model established using the lumped parameter method is called the lumped parameter model.

2) Simplified Reactor Dynamics Model

The diagram of the simplified reactor dynamics model is shown in Fig. 6.1 (Zhu et al. 2019). This model mainly includes a point reactor neutron dynamics model, a core heat transfers and transport model, and various reactivity calculation models. The usual steps are: according to the initial reactivity size, calculate the reactor power through the point reactor neutron dynamics model, use the core heat transfer and transport model to calculate the heat transfer process of the reactor based on the input of reactor power, obtain the temperature and coolant temperature, density response characteristics of the fuel element, use the reactivity calculation model to calculate the reactivity feedback based on the changes in fuel temperature, coolant temperature, and density, and obtain the total reactivity size based on the reactivity introduced by the reactor control system, and then input it into the point reactor neutron dynamics equation for the calculation of the next cycle.

3) Point Reactor Neutron Dynamics Analysis Model

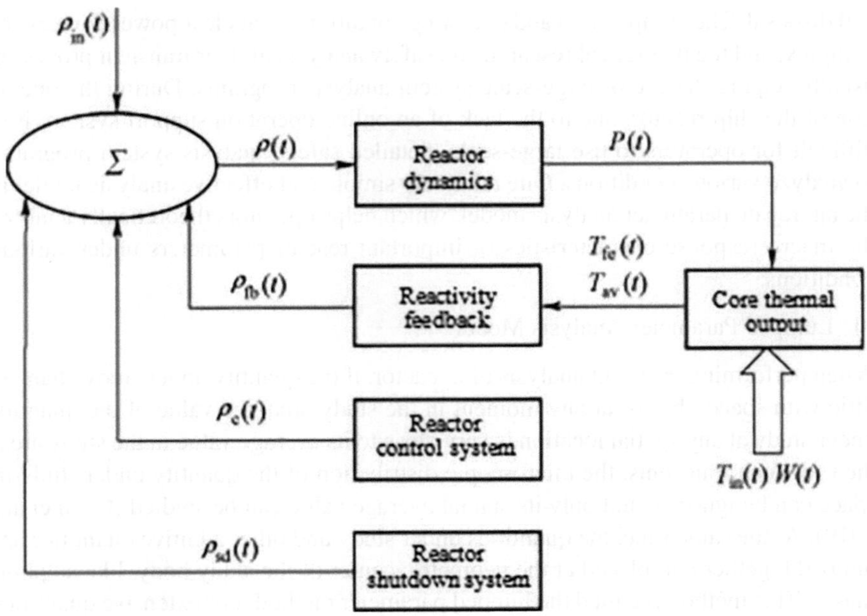

Fig. 6.1 Simplified reactor dynamics model

The so-called point reactor neutron kinetics analysis model refers to the assumption made when analyzing the neutron multiplication characteristics within the reactor core. It assumes that the neutron flux density and precursor concentration can be separated by space–time variables, and that the shape function does not change over time. Physically, this is equivalent to assuming that the spatial distribution of neutron flux density at different times remains unchanged. In other words, the changes in neutron flux density at various points within the reactor over time are synchronized, as if the reactor core has no linear dimensions. This can be handled using the lumped parameter method, hence the model is called the "point reactor model".

The point reactor model can meet engineering requirements when the reactor is not far from critical and the reactivity disturbance is not large. It is widely used in the fields of reactor transient safety analysis and reactor control. However, due to its crude single-group approximation assumption, it cannot fully describe the generation, absorption, leakage, and scattering processes of fast group and thermal group neutrons within the reactor. This model cannot describe the response of the spatial neutron flux density distribution in the reactor during transients, and therefore cannot obtain the axial, radial, and local power peaking factor needed for transient thermal–hydraulic calculations. It cannot perform calculations of the reactivity feedback effects in the three-dimensional space of the reactor.

The point reactor model includes the point reactor neutron kinetics equation group model of a single group and six groups of delayed neutrons. The former is relatively simple and often has analytical solutions, while the latter can only have analytical solutions in very limited special cases. When considering reactivity feedback effects, numerical methods must be used for solution. Due to the stiffness problem of the equation group, the selection of the calculation step size also has high requirements during numerical calculation.

Common numerical calculation methods for the point reactor neutron kinetics equation include the GRN method, Gear method, fully implicit difference method, generalized Runge–Kutta method, improved Euler method, weighted residual method, third-order Hermite method, high-order endpoint floating method, maximum eigenmethod, etc. (Chen et al. 2023a, b).

The GRN method converges quickly, and its convergence is not limited by the step size, but this method is not convenient for changing the step size or the order of the algorithm during calculation. The Gear method can flexibly change the step size and the order of the equation during calculation. The fully implicit difference method can achieve satisfactory accuracy as long as the time step size is appropriate, and it is unconditionally convergent. Its application is simple, and the development of computer computing capabilities has largely compensated for the slow calculation speed of this method.

Calculations show that under the condition of small reactivity introduction, various calculation methods are not sensitive to changes in step size, and the calculation results of each method are relatively close. Under the condition of large reactivity introduction, if the calculation step size increases, the difference in the calculation results of each method increases. If the step size is appropriate, the calculation results are relatively close. For most transient conditions, a calculation step size of about 0.001 s is sufficient to meet the calculation accuracy requirements.

4) Core Heat Transfer Lumped Parameter Model

The reactor heat transfer lumped parameter model usually models all fuel elements in the core as one zone and all coolants as one zone. The model that treats the fuelpellet, gas gap, and fuel rod cladding as one zone is called the core heat transfer single-zone lumped parameter model. If the fuel core block and fuel casing are each divided into one zone, it becomes a two-zone lumped parameter model.

(a) Core Heat Transfer Single-Zone Lumped Parameter Model

This model uses the lumped parameter method to establish the thermal balance equation of the fuel pellet zone, the energy conservation equation of the coolant zone, and the reactor heat transfer equation. Then it uses the established equations to obtain the core heat transfer single-zone lumped heat transfer equation, thereby solving the relevant thermal parameters. One important physical quantity is the core time constant τ (unit s), which physically represents the speed of heat transfer from the fuel element to the coolant.

(b) Core Heat Transfer Two-Zone Lumped Parameter Model

This model also uses the lumped parameter method, but establishes the energy conservation equations for the fuel pellet and fuel rod cladding separately. The establishment of the energy conservation equation for the coolant zone and the reactor heat transfer equation is similar to the single-zone lumped parameter model. It separately solves the heat transfer from the fuel pellet to the cladding, the heat transfer from the cladding to the coolant, and the heat transfer of the coolant, thereby solving the relevant thermal parameters, such as the temperature of the fuel and cladding, time constant, etc.

5) Simplified Reactor Hydraulics Model

When calculating the heat transfer of the reactor, it is necessary to obtain the coolant flow rate of the reactor. A simplified reactor hydraulics model is established to solve the coolant flow rate.

(a) Momentum Conservation Equation

For any hydraulic system, the system pressure drop equation can always be written as: Rate of change in flow in the channel = Pressure difference upstream and downstream + Pump head—Flow resistance—Gravitational pressure drop.

(b) Kirchhoff's Law

Kirchhoff's law is used to solve problems of pressure drop and flow rate calculation in fluid networks. The loop law states that the sum of pressure drops around any closed loop should be zero; the node law states that at each node in the fluid network, for incompressible fluids, the algebraic sum of flows at any instant is zero.

(c) Forced circulation flow equation under steady-state conditions

For a primary circuit system of a nuclear power device composed of N loops, the system flow equation under steady-state forced circulation conditions is:

Forced circulation driving pressure head (external force) + natural circulation driving pressure head (self) = total resistance along the path.

6) Reactivity feedback calculation model

The reactivity calculation expression can be represented as:

Total reactivity change = reactivity introduced externally + reactivity introduced by thermal–hydraulic feedback + reactivity introduced by the reactivity control system + reactivity introduced by the shutdown system.

The reactivity introduced by the reactivity control system can be obtained from the integral value curve and differential value curve of the control rod; the reactivity introduced by the shutdown system can be obtained from the reactivity curve introduced by the shutdown; the thermal–hydraulic feedback mainly includes fuel temperature feedback, coolant temperature feedback, coolant density feedback, coolant void fraction feedback, etc., and the reactivity feedback introduced by the change in coolant pressure can be obtained through the calculation of coolant density feedback. These feedback coefficients can take the average value, or they can be calculated using the polynomial fitting relationship of experimental data to calculate the reactivity feedback coefficient, usually merging the benefits of coolant temperature feedback and coolant density feedback into coolant temperature feedback.

(2) Space–time neutron dynamics analysis model of nuclear power devices

The transient analysis model of nuclear power devices is the basis for the application of deterministic safety analysis methods, mainly including the neutron dynamics model in the nuclear reactor and the system thermal–hydraulic model. In recent years, with the deepening of people's research on the response mechanism of the system after a nuclear reactor accident and the development of simulation calculation technology, the transient model of accident analysis has been continuously improved, and the accuracy of accident simulation analysis has been continuously improved.

In the past decade, with the development of computer hardware and software technology, international research on the numerical calculation methods of multi-dimensional multi-group space–time neutron dynamics equations has been widely carried out, and various advanced spatial variable discretization methods and time variable discretization methods have been widely used in the numerical calculation of space–time dynamics equation groups. Such as the modern coarse mesh method of spatial variable discretization (nodal expansion method, nodal Green's function method, analytical nodal method, nonlinear iterative nodal method), it can achieve satisfactory accuracy under a larger grid, greatly improving the calculation efficiency (Wu 2023). The three-dimensional space–time dynamics program system compiled according to these methods has been widely used in engineering, and now internationally renowned nuclear power device transient thermal–hydraulic analysis programs (RELAP5, CATHARE, TRACE) have a trend to replace simplified point pile models and one-dimensional models with multi-group three-dimensional space–time dynamics models.

1) Reactor Space–time neutron dynamics model and solution method
(a) Space–time neutron dynamics equation

The multi-dimensional multi-group neutron diffusion space–time dynamics equation group includes the neutron diffusion equation and the delayed neutron precursor concentration equation. In the Cartesian coordinate system, the multi-dimensional multi-group neutron diffusion space–time dynamics equation group of the reactor is a complex partial differential equation group, and the number of variables to be solved needs to be determined according to the specific number of delayed neutron groups and energy groups.

(b) Overview of solution methods

In the discrete solution of time variables, considering the convergence and stability requirements of the neutron diffusion dynamics equation group, the fully implicit backward Euler format can be used to discretize the time derivative term in the equation, and the time integration method is used for solution. The time derivative term of the neutron flux density in the equation is approximated by the fully implicit backward finite difference format.

Using the time integration method to solve the equation, the delayed neutron precursor concentration at time t_n can be obtained by integrating over the time interval $[t_{n-1}, t_n]$.

In order to solve the integral term in the equation, the specific expression of the neutron fission source term in the time interval $[t_{n-1}, t_n]$ with respect to time is needed. Consistent with the fully implicit backward finite difference approximation of the neutron flux density time derivative term, it is approximately assumed that the total fission source term in the time interval $[t_n$-$_1, t_n]$ is linearly changing. By solving these two equations simultaneously, we can obtain the precursor concentration of delayed neutrons at time t_n.

(c) Discretization of Spatial Variables

The numerical solution of the multi-group space–time neutron kinetics equation is very time-consuming, as a diffusion calculation must be performed for each time step. The diffusion calculation can be performed using the fine mesh finite difference method or the nodal method.

The finite difference method is used to replace the differential with a difference approximation when discretizing the system of differential equations. The finite difference method is simple to derive and has a solid theoretical foundation, but it has a significant drawback. To ensure computational accuracy, the spatial mesh size must be sufficiently small (approximately equal to the neutron diffusion length, 1–2 cm for pressurized water reactors). This increases the number of grid points after space discretization, resulting in a large number of unknowns. Especially when the dimension of the space variable is high, such as in two or three dimensions, the time consumed by directly numerically solving the multi-group space–time kinetics equations is astonishing. For example, for a pressurized water reactor with a thermal power of 140 MW, if the mesh size is equal to the diffusion length, there are approximately 74,200 spatial grid points in the entire reactor. Considering the time variable, the required storage capacity and computation time are staggering.

To improve computational efficiency, research on modern nodal methods began in the late 1970s. Representative nodal methods such as nodal expansion method, analytical nodal method, Green's function nodal method, and nonlinear iterative nodal method have been developed. These methods can achieve good accuracy under very coarse spatial meshes (generally one component per grid). The advent of the nodal method greatly improved computational efficiency, reducing the time to solve steady-state diffusion problems by an order of magnitude compared to the finite difference method, while still meeting the accuracy required for engineering. Therefore, this method has been widely used in steady-state reactor diffusion calculations. The success of the nodal method in steady-state problems naturally led to its application in solving time-dependent space–time kinetics problems. Currently, the nodal method is the best choice for solving multi-group space–time neutron kinetics equations.

2) Three-Dimensional Two-Group Space–Time Neutron Kinetics Equation for Reactors

(a) Equation

To accurately reflect the dynamic process of neutrons in time and space in a reactor, a two-group (fast group and thermal group) neutron space–time kinetics equation can generally be used. This set of equations has high computational accuracy, can compensate for the shortcomings of the point reactor neutron kinetics equation, and can reflect the space–time behavior of neutrons of different energies in the reactor. However, to meet computational accuracy, the spatial grid division of the reactor is fine, and the computation speed is relatively slow.

(b) Numerical Solution Method

The traditional space–time separation model separates the neutron flux into a shape function representing the spatial distribution of flux density and an amplitude function reflecting the average neutron flux density. The shape function is then normalized, and the finite difference method or nodal method is used for solution.

(c) Diffusion Equation of Space–Time Separation Variables

By applying space–time separation variables, the diffusion equation of neutron flux density space–time separation variables can be obtained.

(d) Solution of the Diffusion Equation of Space–Time Separation Variables

① The amplitude function is solved using the method of the point reactor neutron kinetics equation considering six groups of delayed neutrons;

② The shape function is discretized using the finite difference method to form a system of linear equations, which is then solved using an appropriate method.

(e) Quasi-Static Method Solution

The main approach is to perform a finite difference on the time derivative term and then solve it.

(3) Thermal–Hydraulic Analysis Model of Nuclear Power Devices

1) Basic Concepts

(a) Control Volume

The reactor and coolant pipes are divided into several independent volumes to describe the shape and physical state of different parts and spaces of the coolant system. These volumes are called control volumes. Currently, the best estimate transient thermal–hydraulic program or safety review program is used to perform a thermal–hydraulic transient analysis of a nuclear power device, which requires dividing the nuclear power device into several control volumes and solving the mass and energy conservation equations for each control volume. If the parameter change process within each control volume is clearly understood, the transient characteristics of the entire system can be accurately described.

Principles for dividing control volumes:

① The more control volumes are divided, the closer the system is to continuous description, and the more realistic the various state parameters of the system are expressed;

② The increase in the number of control volumes increases the computation time and affects the computation time step;

③ Therefore, the relationship between computation step, computation time, and computation accuracy needs to be considered comprehensively;

④ Parts with dramatic changes in thermal–hydraulic parameters should be divided finely, while those with small changes should be divided more coarsely.

The characteristic parameters of the control volume mainly include:

① Geometric parameters of the control volume: length, area, volume, surface roughness, equivalent diameter, elevation, etc.;

② State parameters: pressure, temperature (or internal energy), steam (or gas) content, etc.

(b) Streamlines

The pipelines connecting the control volumes are represented as streamlines. Streamlines do not have a concept of volume; they incorporate the volume into the connected control volumes. For each streamline, only the momentum conservation equation needs to be solved to determine the fluid velocity and flow rate.

The characteristic parameters of the streamline include:

① The flow area of the streamline, energy loss coefficient (local resistance coefficient), hydraulic diameter, etc.;

② State parameters: mass flow rate or velocity of the fluid.

2) Basic Equations of Thermal–Hydraulic

The basic content of the dynamic model includes the mass conservation equation, momentum conservation equation, and energy conservation equation, as well as the corresponding structural relationships. During the transient and accident processes in pressurized water reactors, the coolant may be in a two-phase state. The thermal–hydraulic process of two-phase flow is very complex, and the two-phase flow model used in computer programs greatly affects the calculation accuracy, the ability to describe transient processes, and the difficulty of solving. The two-phase flow mathematical models currently available mainly include two-fluid model, drift flux model, mixture model, and homogeneous flow model, each with its characteristics and applicable scenarios.

The two-fluid model lists the mass, momentum, and energy conservation equations for the vapor and liquid phases separately, and considers the momentum, energy, and mass exchange between the two phases, which can reflect the intrinsic mechanisms of various physical phenomena.

The drift flux model is established based on two structural parameters describing the bubble distribution and the relative slip between the vapor and liquid phases. Overall, it has the characteristics of a homogeneous flow model, is simple to solve, and can also represent the local characteristics of two-phase flow. In many cases, this model can achieve satisfactory results. Using different structural conditions of the drift model, various forms of two-phase flow models can be listed, and the conservation equations can have three, four, or five different forms.

The mixture model views two-phase flow as a whole mixture. The model is simple, with fewer equations and variables, so the calculation workload is not large. However, it is mainly used to determine the overall characteristics of two-phase flow and cannot accurately reflect the local parameters of two-phase flow other than heat, pressure, and pressure difference.

In addition to the basic mass conservation, momentum conservation, and energy conservation equations, corresponding structural relationships must be constructed. The structural characteristic relationships used in various models are to close the field equations. The main ones belonging to the structural relationship include boundary conditions, property (state) equations, momentum and energy exchange equations between the medium and the wall surface, various exchange equations between the two phases, and some relationships describing specific thermal–hydraulic phenomena. The choice of structural relationship varies with the different two-phase flow field equations, and whether it is chosen appropriately has a significant impact on the workload of problem-solving.

3) Two-Fluid Model

The two-fluid model is currently considered the most comprehensive and widely used two-phase flow mathematical model. It lists the mass, momentum, and energy conservation equations for the vapor and liquid phases separately, and considers the momentum, energy, and mass exchange between the two phases, which can reflect the intrinsic mechanisms of various physical phenomena. Therefore, in principle, this model can describe various complex conditions of two-phase flow, such as the flow in the lower chamber of the reactor, the reverse flow of steam and liquid, and the complex thermal–hydraulic process during re-flooding. However, due to the large number of field equations and the need to supplement many structural relationships, the solution is difficult and the calculation amount is large. In addition, due to insufficient understanding of certain laws, especially the laws at the interface between the two phases, some structural relationships are difficult to establish very accurately, which affects the calculation accuracy of this comprehensive mathematical model.

(a) Basic Field Equations

The basic field equations of the two-fluid non-equilibrium model consist of two-phase continuity equations, two-phase momentum equations, and two-phase energy equations. The basic field equations consist of one-dimensional mass conservation equations, momentum conservation equations, and energy conservation equations for the gas and liquid phases.

(b) Structural Relationships

For a set of flow field conservation equations, the main unknown parameters that need to be solved can be arbitrarily selected from the parameters appearing in the equations. We call the main unknown parameters "primary parameters". The number of primary parameters selected is generally the same as the number of conservation equations. For the six-equation one-dimensional two-fluid model, six main quantities can be selected as primary parameters. In addition to these six parameters, there are 23 other parameters appearing in the equations. To close the equation set, these 23 quantities must be expressed in terms of primary parameters, and the established expressions are called structural relationships. These expressed quantities are called secondary variables. For the solution of secondary variables, refer to relevant literature.

4) Wall Heat Transfer Model of Coolant

The heat transfer of coolant can mainly be divided into calculations for the heat transfer between the coolant and fuel elements, heat transfer tubes of steam generators, coolant pipes, and various types of heat exchangers. In the RELAP5/MOD3.3 program, the heat flux density from the surface of the thermal component to the coolant equals the heat exchange from the wall to the vapor phase plus the heat exchange to the liquid phase of the wall. The key here is to calculate the heat transfer coefficient. Different heat exchange areas have different heat transfer coefficients. Heat exchange areas are mainly divided into single-phase convective heat transfer areas, boiling heat transfer areas, and steam condensation heat transfer areas. The boiling area can be further divided into nucleate boiling areas, partial film boiling areas, boiling crisis points, transition areas, and film boiling areas, etc.

5) Interphase Heat Transfer Model of Coolant

The energy exchange between two phases in two-phase flow occurs through heat exchange and mass exchange processes, which are not independent of each other. For example, the mass exchange amount (flash evaporation or condensation) is calculated by the heat transfer across the phase interface. If there is a temperature difference at the phase interface, there is a heat transfer process, i.e., an energy transfer process. This energy is the energy required for phase change, and the mass exchange amount in the process can be calculated through latent heat. In a typical two-phase flow control volume, there are four possible thermodynamic states: superheated liquid, subcooled liquid, superheated vapor, and subcooled vapor. The values of the interfacial heat transfer coefficients under various flow patterns have different models and formulas, which should be calculated separately.

6) Main Hydraulic Model of Coolant

During the operation of the reactor, the working temperature of the fuel elements, the heat transfer rate and enthalpy rise of the coolant, the critical heat flux density, and the delivery power of the main circulation pump are closely related to the uniform flow, pressure drop, and flow rate of the coolant. That is to say, the thermal–hydraulic analysis of the reactor is closely related to the hydraulics of the reactor. The main hydraulic model of the coolant includes flow resistance (pressure drop), sonic speed and critical flow, and analysis models of vapor–liquid counterflow, etc.

The main components of the pressure drop calculation formula include: frictional pressure drop calculation model along the path; calculation formula for local resistance pressure drop. The analysis of sonic speed and critical flow mainly includes: single-phase subcooled critical flow model; two-phase critical flow model.

7) Heat Transfer Model of Thermal Components

In nuclear power devices, thermal components refer to components that exchange heat with the coolant. Thermal components in nuclear power devices can be classified according to their shape and function. According to the function of the thermal components, they can be divided into two categories: the first category is heating components, which are equivalent to heat sources, such as fuel rods and electric heaters; the second category is heat transfer components, such as the pressure boundary of the coolant and the heat transfer tubes of the steam generator. According to the shape of the thermal components, they can be divided into spherical, cylindrical, and square shapes. In thermal–hydraulic problems, in order to simplify the calculation, a one-dimensional heat conduction differential equation is generally used to describe the heat conduction of the thermal components, and the finite difference method is used to solve it.

6.1.3 Deterministic Safety Analysis Common Programs

With the development of nuclear power technology, people have reached a consensus that due to the large size, complex system, and strong radioactivity of nuclear power reactors, if actual devices are used to conduct full-scale experimental research on actual accidents or events, it would require a huge amount of manpower and financial resources, and could potentially bring about serious radioactive consequences. Therefore, deterministic safety analysis programs based on computer numerical simulation technology, integrating a large number of advanced theoretical research and experimental research results, have become the most important research tools in the field of nuclear energy technology, and are also the main technical means for deterministic safety evaluation of nuclear power. Currently, deterministic safety analysis programs are widely used in nuclear power plant design, safety review, procedure verification, operation event evaluation, operation management, maintenance decision-making, etc.

According to the purpose of the safety analysis program, it can be divided into two major categories. One is the safety review program, which is used to evaluate whether a reactor to be built meets the safety criteria stipulated by the standards. These programs all add several conservative assumptions to the dynamic model. The other is the best estimate program, which strives to simulate the complete characteristics of nuclear power devices and systems as accurately as possible, removing unnecessary conservatism.

According to the research object of the safety analysis program, it can be divided into reactor physics analysis programs, thermal–hydraulic analysis programs of nuclear power devices, severe accident analysis programs of reactors, radioactive material source term analysis programs, radioactive material migration and diffusion analysis programs, fuel element performance analysis programs, reactor containment integrity analysis programs, etc.

The reactor core physics analysis program is mainly used to simulate the steady and transient physical characteristic parameters of the reactor core, mainly including the fuel assembly homogenization physical parameter calculation program, the reactor criticality and burnup calculation program, the reactor neutron dynamics program (point reactor model, three-dimensional few-group model), etc. Closely related to this, it also includes nuclear databases, etc.

Nuclear power device thermal–hydraulic analysis programs can be divided into system analysis programs, single-channel calculation programs, or sub-channel calculation programs. Among them, system analysis programs such as RELAP5 and TRACE can be further divided into one-dimensional or three-dimensional thermal–hydraulic analysis programs (Chen et al. 2023a, b). Fuel element performance analysis programs are mainly used to analyze the heat transfer performance, mechanical performance, and chemical performance of fuel elements under normal operation and accident conditions. Reactor containment integrity analysis programs are mainly used to analyze the thermal–hydraulic behavior and mechanical performance of the reactor containment under large break loss-of-coolant accidents. Severe accident analysis programs, radioactive material source term analysis programs, and radioactive material migration and diffusion analysis programs are introduced in the severe accident analysis section.

The above programs can form an analysis platform composed of two or more analysis programs through multi-scale coupling, such as a point reactor model plus a system thermal–hydraulic safety analysis program, a space–time neutron kinetics program coupled with a one-dimensional or three-dimensional thermal–hydraulic safety analysis program, a system program and sub-channel coupling program, an integrated severe accident analysis program package, etc. With the integrated analysis program, deterministic safety analysis work can be carried out.

(1) Reactor Core Physics Analysis Program

The reactor core physics analysis program is mainly used to simulate the steady and transient physical characteristics parameters of the reactor core, mainly including the fuel assembly homogenization physical parameters calculation program, the reactor criticality and burnup calculation program, the reactor neutron kinetics program, etc., closely related to it also includes nuclear data calculation programs, etc. The basic components of the reactor physics program are shown in Fig. 6.2.

1) Nuclear Database

The selection and processing of nuclear data, especially cross-section data, is the starting point and important basis for reactor nuclear calculations. In order to improve the accuracy of reactor nuclear calculations, on the one hand, efforts should be made to improve the models and methods of calculation, and on the other hand, efforts should be made to improve the accuracy of original nuclear data and parameters. For nuclear engineers, it is very important to correctly understand and use these data, which is a key to obtaining correct calculation results.

Since nuclear calculations involve a large number of isotopes, and the relationship between neutron cross-sections and energy in various energy regions is complex, the amount of nuclear cross-section data needed for nuclear reactor calculations is very large. Since the 1950s, with the rapid development of reactors, accelerators, and measuring instruments, a large amount of neutron cross-section data has been gradually accumulated, and the collection and compilation of nuclear data has also

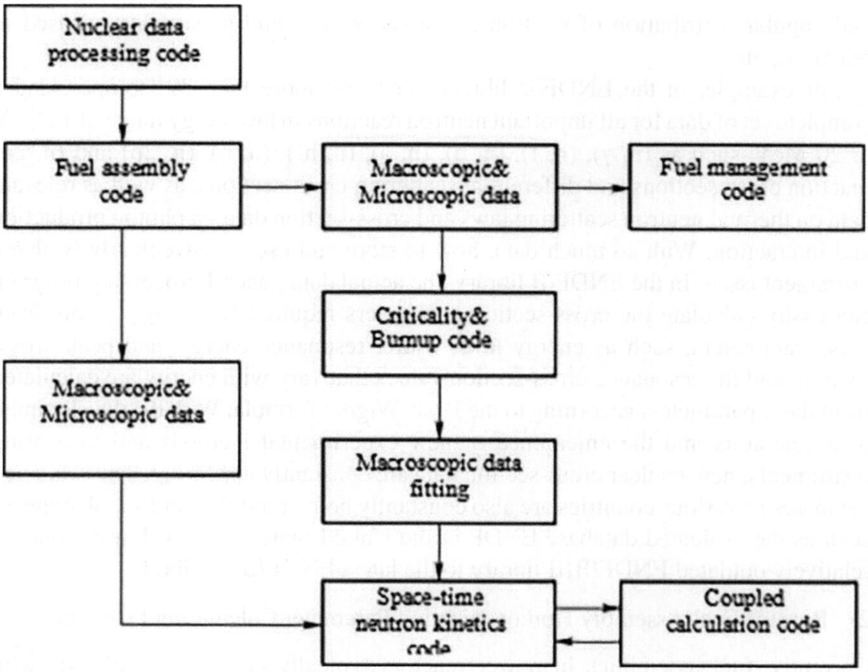

Fig. 6.2 Basic components of the reactor physics analysis program

been rapidly carried out. The "Neutron Cross-Section Compilation BNL-325" system has collected a wealth of experimental and theoretical calculation data. However, different experiments may give different values for the same cross-section, and there are still gaps in the cross-sections of certain energy domains and certain elements, so it is necessary to pay attention to the correct selection and evaluation of these cross-sections and other nuclear data when performing reactor nuclear calculations. With the development of nuclear technology, higher requirements have been put forward for the accuracy and energy range of data. In the past 20 years, many countries have been striving to establish a set of standard, evaluated nuclear cross-section databases. China also has this kind of data "Evaluated Neutron Data Compilation".

The evaluated database ENDF in the United States is a relatively complete and advanced database. The ENDF/B library is one of the standard nuclear database sources for nuclear reactor physics analysis, and it has a complete set of cross-section data required for nuclear physics calculations. In addition to ENDF, there are other national nuclear databases such as the European JEF library. Generally, these databases contain nuclear data in the possible domain used for nuclear reactor analysis for various isotopes, including: microscopic cross-sections of all possible neutron reactions; angular distribution of neutrons during elastic and inelastic scattering; energy distribution of fission neutrons; average number of secondary neutrons per fission; resonance parameters; yields and cross-sections of fission products; energy and angular distribution of neutrons, protons, γ-rays, and α particles released in reactions, etc.

For example, in the ENDF/B library, there are more than 80 isotopes, and a complete set of data for all important neutron reactions in the energy range of 10^{-5} eV to 20 MeV, such as (n, γ), (n, f), (n, p), (n, a), (n, n'), (n, n), (n, 2p) and (n, 2n) reaction cross-sections and differential scattering cross-sections, as well as relevant data on thermal neutron scattering laws and cross-section data on photon production and interaction. With so much data, how to store and use it conveniently is also a prominent issue. In the ENDF/B library, the actual data general processing program can easily calculate the cross-section parameters required for energy points from these parameters, such as energy level width, resonance energy, and peak cross-section, and the resonance cross-section values that vary with energy are calculated from these parameters according to the Breit-Wigner formula. With the development of accelerators and the emergence of new experimental methods and measuring instruments, new nuclear cross-section data are constantly appearing. The evaluated databases of various countries are also constantly being modified and supplemented, such as the evaluated database ENDF in the United States has developed from the relatively outdated ENDF/BIII library to the latest ENDF/BVII library.

2) Reactor Fuel Assembly Homogenization Parameter Calculation Program

Currently, fuel assemblies in power reactors typically consist of fuel rods with different enrichment levels, burnable poisons, control rods, water gaps, and cladding materials. Different materials have different reaction cross-sections. When calculating the physics of the reactor core, the assembly is usually homogenized first to obtain the homogenization parameters of the assembly. The calculation of the assembly's homogenization parameters is based on the aforementioned nuclear database, through neutron transport calculations, to obtain the fine distribution of neutron flux within the assembly, and then through the reaction rate conservation equation, to obtain the assembly's homogenization parameters. At present, according to different theoretical models, various countries have developed corresponding program systems, with TPFAP, WIMS-D4, and CASMO being widely used.

(a) Assembly Macroscopic Cross-Section Parameter Calculation Package TPFAP

TPFAP is a component calculation program developed by China. It uses the penetration probability method for component calculations and can be used for transport calculations of various lattice elements and fuel assemblies in pressurized water reactors (PWR). It first uses the collision probability method to complete the lattice element transport calculation of three or four-zone annular geometry under the library energy group structure (69 groups). The two-dimensional penetration probability method component calculation is carried out in the X–Y geometry, and the base mode calculation is used to consider the correction of neutron leakage, and the equivalent diffusion cross-section parameters are calculated according to the reaction rate conservation. TPFAP performs burnup calculations in each fuel rod and burnable poison rod, providing the average few-group cross-section, power distribution, etc., for each burnup step of the burnup assembly and lattice element.

TPFAP is also applicable to both pressurized water reactor (PWR) and boiling water reactor (BWR) component calculations. The program handles different geometric areas of two-dimensional fuel assemblies, including fuel rods with different compositions arranged in a square matrix, burnable poison absorption rods, flux measurement tubes, water gaps, boron compensation plates, and rod-shaped or cross-shaped control rods. The fuel rods can contain the absorber gadolinium. The program can also perform lattice element calculations.

(b) Assembly Macroscopic Cross-Section Parameter Calculation Program WIMS-D4

The WIMS program was developed by WINFRITH in the UK in the 1970s. It uses transport theory to calculate the lattice element program of flux as a function of energy and position in the lattice element. The program comes with a 69-group constant library and can calculate uniform, flat, single rod, and bundle rod lattice elements, providing various physical parameters of the reactor under steady-state and burnup conditions.

The WIMS-D4 program is suitable for calculating various physical parameters of the reactor under steady-state and burnup conditions, including: multiplication factor, curvature, energy spectrum, and temperature coefficient, corresponding to various nuclear densities at each time step. Uniform medium, flat and cylindrical rod lattice elements, and various physical quantities of the assembly, such as diffusion coefficients, various average reaction cross-sections, etc. The power peak factor of the assembly and the fine disadvantage factor inside the assembly. Resonance self-shielding cross-section, infinite dilution cross-section.

(c) Assembly Macroscopic Cross-Section Parameter Calculation Program CASMO

CASMO is a transport calculation program similar in function to TPFAP. It can be used for component burnup calculations of pressurized water reactors or boiling water reactors, and can also perform simple lattice element calculations. The program can handle different areas of two-dimensional geometric fuel assemblies, including fuel rods with different compositions arranged in different ways, burnable poison absorption rods, flux measurement tubes, water gaps, boron compensation plates, and rod-shaped or cross-shaped control rods.

3) Reactor Steady-State Criticality and Burnup Analysis Program

After performing various reactor component parameter calculations, in order to obtain different burnup distributions, critical rod positions, core power distributions, etc., at different burnup moments, critical burnup calculations need to be performed. At present, the critical burnup calculation programs widely used in China include CITATION, BM3DD, SIMULATE, etc.

(a) Fine Mesh Finite Difference Program CITATION

CITATION was developed by Oak Ridge National Laboratory in the United States in the 1970s. It uses diffusion theory for analysis and calculation, spatial discretization uses finite difference, it can handle three-dimensional problems under various spatial coordinates, adopts zero flux, symmetry, and extrapolation three types of boundary conditions, and can also be used to handle scattering of any energy group. The program can perform boron adjustment critical search burnup calculations or operation tracking calculations, providing the effective multiplication factor of the core at different burnup steps, power peak factor, absolute average flux of the assembly and block, relative normalized power distribution and burnup distribution in the radial and axial directions of the core.

The theoretical derivation process of the CITATION program is simple, and the calculation stability is good, so it has been widely used internationally. However, to achieve satisfactory accuracy, it must take a very small grid (1 ~ 2 cm), which is time-consuming to calculate. Currently, with the development of the nodal method, there is a trend to replace it.

(b) Fine Mesh Finite Difference Programs HEXFGD and BM3DD

HEXFGD and BM3DD are developed by the China Institute of Nuclear Power Engineering based on the CITATION model. They share the same physical calculation model and CITATION program. HEXFGD is primarily used for the design calculation of hexagonal fuel assembly reactors, while BM3DD is mainly used for the analysis calculation of square assembly reactors. BM3DD incorporates a thermal engineering model in its physical calculations, allowing it to consider thermal feedback effects.

(c) Nodal Analysis Program

To improve computational efficiency, research on modern coarse-mesh nodal methods began in the late 1970s. Representative nodal methods such as the Nodal Expansion Method (NEM), Analytical Nodal Method (ANM), Analytical Function Method (AFEN), and Green's Function Nodal Method (NGFM) were developed. These methods can achieve good accuracy under very coarse spatial grids (usually one grid per assembly). The advent of nodal methods greatly improved computational efficiency. The time required to solve steady-state problems using these methods is at least an order of magnitude shorter than the difference method, and the number of unknowns is greatly reduced, while still meeting the precision required for engineering. Therefore, modern coarse-mesh nodal method programs such as NEM, QUANDRY, NGFM, ANC, etc., have been widely used in steady-state diffusion calculations of reactors.

It should be noted that in the 1980s, Wagner, Smith, and others proposed an advanced nodal method using a nonlinear iterative strategy. This method increases the degree of freedom in low-order methods and uses the results of high-order methods to correct low-order methods to obtain high-order precision solutions. The entire process is mainly iterative solution of low-order equations, intermittently interspersed with high-order method calculations. This not only ensures the computational accuracy of high-order methods but also improves computational efficiency and reduces the need for computer storage. This method can be applied to any nodal method that uses nodal average flux and surface flow as unknowns. Currently, this method is widely used in the calculation of steady-state neutron diffusion equations and the numerical solution of space–time dynamics equations, such as the SIMULATE-3, SKECH-N, SPANDEX, and NESTLE calculation programs developed abroad in the 1990s, all of which use nonlinear nodal methods.

SIMULATE-3 is a three-dimensional steady-state power distribution nodal analysis calculation program for light water reactors. It can be used to study PWR reactor fuel cycle length, power distribution, control rod arrangement, and xenon transients. This program can calculate all meaningful feedback effects of light water reactors, including control rod insertion, moderator voids and density changes, fuel temperature, fuel changes, and transient changes in xenon and samarium.

It has the following features:

① It uses a two-group, three-dimensional advanced nodal method.

② It applies modern advanced homogenization theory and introduces discontinuity factors.

③ It can reconstruct the fine power within the core node.

④ The baffle or reflector layer is explicitly represented, with no need to adjust the reflectivity.

4) Core Neutron Dynamics Analysis Program

The core neutron dynamics analysis program is developed based on the above physical program calculations. It mainly calculates the reactivity, power, hot spot factor, and other parameter response characteristics under different transient or accident conditions.

From the perspective of reactor safety, in the design, accident analysis, and load tracking process of nuclear reactors, it is often desired to obtain the spatial distribution of core power or neutron flux density and its transient changes over time. This requires solving the three-dimensional space–time neutron dynamics equation group. In the early days, due to many restrictions on calculation methods and conditions, various approximation methods were adopted, the simplest of which is the "point reactor approximation model". The "point reactor model" assumes that the shape distribution of neutron flux density in the reactor core does not change over time during transients, and the reactor is like a point without spatial effect distribution, which greatly simplifies transient calculations and has been widely used. In recent years, with the development of calculation methods and computer levels, great attention has been paid to the research on the space–time dynamics problem of nuclear reactors internationally, and calculation models that directly solve the space–time neutron dynamics equation have been developed. It should be pointed out that when calculating reactivity accidents, the space–time neutron dynamics calculation model has functions that the point reactor model does not have. It can simulate the change of neutron flux density in the three-dimensional space of the reactor over time, give the change of the radial power unevenness factor, axial power unevenness factor, and local unevenness factor over time, and can simulate reactivity distribution abnormal accidents caused by the action of control rods. Therefore, internationally renowned system thermal–hydraulic analysis programs such as RELAP5, RETRAN, TRAC are gradually adopting the space–time neutron dynamics model to replace the original point reactor model.

5) Monte Carlo Analysis Program

The Monte Carlo method, also known as random sampling technique or statistical experiment method, differs significantly from general numerical calculation methods. As a branch of computational mathematics, it is a numerical calculation method developed based on probability and statistical theory. The Monte Carlo method can realistically describe the characteristics of things and the process of physical experiments, and can be used to solve some problems that are difficult to solve with numerical methods, hence its wide application. Its basic idea, in layman's terms, is to use the method of random experiments to calculate integrals, that is, to treat the integral to be calculated as the mathematical expectation of a random variable $g(r)$ that follows a certain distribution density function $f(r)$.

$$\langle g \rangle = \int g(r)f(r)dr$$

Through certain experiments, we can obtain N observation values $r_1, r_2, r_3 \ldots r_N$ (from a probability perspective, this is equivalent to drawing N sub-samples from the distribution density function $f(r)$), and take the arithmetic mean of the corresponding N random variable values $g(r_1), g(r_2), g(r_3) \ldots g(r_N)$ as the estimated value of the integral.

$$\hat{g} = \frac{1}{N}\sum_{i=1}^{N} g(r_i)$$

When using the Monte Carlo method, for parts of the problem that can be handled with analytical (numerical) methods, these methods should be used as much as possible. For parts that are difficult to handle with analytical (numerical) methods, the Monte Carlo method should be used. This approach can fully exploit its unique advantages in simulation calculations.

Common programs based on the Monte Carlo method include the MORSE program developed by Oak Ridge National Laboratory in the United States, the MCNP program developed by Los Alamos National Laboratory, the EGS program provided by the Stanford Linear Accelerator Center, the electron-photon SANDYL program and TIGER program series from Sandia National Laboratories, and the SUPERMC from the Institute of Nuclear Energy Safety of the Chinese Academy of Sciences.

MCNP (Monte Carlo N-Particle transport code System) is a large-scale three-dimensional transport general program, initially developed by the Monte Carlo group (X-5 group) of the Applied Theoretical Physics Department of Los Alamos National Laboratory. It can handle complex three-dimensional geometric structures under neutron, photon, electron or neutron-photon-electron coupled transport problems, and can also calculate the corresponding eigenvalue problems for critical systems.

The MCNP5 program was released in 2003, rewritten in standard Fortran90 language, and has strong versatility. It is relatively easy for users to use. It uses fine point cross-section data based on evaluated libraries such as ENDF/B-VI and all types of neutron reactions. For thermal neutrons, it can choose between the free gas and S(α, β) models. With the continuous release of new ENDF/B evaluation libraries, the temperature-related neutron cross-section library based on the ACE format is also updated accordingly. MCNP5 has carefully designed counting functions. In addition to the standard counts, users can also calculate according to the requirements of the calculated physical quantities. It also includes many techniques to reduce variance and a large amount of cross-section data.

The application range of the MCNP program is very wide, mainly with important applications in the field of nuclear physics calculations. It is commonly used in core design, critical safety calculations, reactor shielding calculations, radiation detection and protection, personal dose detection and physical health. It also has a wide range of applications in medical physics and radiotherapy, spent fuel and waste treatment.

(2) System Thermal Hydraulic Safety Analysis Program

The thermal–hydraulic safety analysis program is primarily used to analyze the response of thermal–hydraulic parameters in nuclear power devices under accident conditions. Currently, the widely recognized program systems outside China mainly include the core subchannel analysis program COBRA, the reactor system thermal–hydraulic analysis program TRAC, ATHLET, RETRAN-02, CATHARE, RELAP5, etc. The following is an introduction to each.

1) Core Subchannel Analysis Program COBRA (THAS)

The COBRA-IV program (known as THAS-1 in the Chinese version) uses subchannel analysis to determine the enthalpy and flow distribution within the fuel assembly under steady-state and transient conditions. It retains the steady-state and transient solutions used in COBRA-III as an option for implicit solutions. In addition, it introduces a new explicit solution that can calculate adverse transient conditions such as backflow, circulation, ejection, and re-injection, and adapts to pressure or flow boundary conditions. This program significantly reduces memory usage and computation time, allowing for the calculation of problems involving hundreds of subchannels.

The THAS-3 program is a reactor steady-state and transient thermal–hydraulic analysis program developed based on the THAS-1 program (the Chinese version of COBRA-IV). In addition to retaining all the functions of THAS-1, it enhances the analysis of plate-type core performance. Although the THAS-1 program has the function of analyzing the temperature field of plate-type fuel elements, it does not have the function of analyzing the critical heat flux density and flow stability of plate-type cores and other thermal–hydraulic characteristics. THAS-3 focuses on improving THAS-1 in this regard, enhancing the ability to analyze the thermal–hydraulic characteristics of plate-type cores.

2) Introduction to the TRAC Program

The main goal of developing the Transient Reactor Analysis Code (TRAC) is to obtain an advanced best-estimate system program for light water reactors, which should be able to reliably calculate the characteristics of various accidents in light water reactors. We aim to verify and evaluate the program's calculation results with extensive experimental data to enhance the program's credibility. To achieve these goals, the following principles were followed in the development of the TRAC program:

(a) The program should have sufficient calculation accuracy for various experimental conditions. When transitioning from one experiment to another, users do not need to change and adjust simulation schemes and parameter variables.
(b) Important physical phenomena should be simulated as much as possible from basic principles to facilitate extrapolation beyond the parameter range covered by experiments. This modeling method also helps to provide more detailed results of system thermal–hydraulic characteristics.
(c) The program should have sufficient flexibility to simulate various designs of light water reactors and related experimental configurations. The TRAC version developed according to these principles is called the "detailed calculation" version. Developed versions include TRAC-P1, TRAC-PIA, TRAC-PD2, etc. Another goal of the TRAC program research is to provide a "fast calculation" version. It can be used for parameter studies, display calculations, safety reviews, and long transient processes (such as small break loss of coolant accidents). In this version, empirical models are usually used to shorten the calculation time. Currently developed fast calculation versions include TRAC-PFl, TRAC-PFl/MOD1, and TRAC-PFl/MOD2, etc.

The following introduces some important features of TRAC-PFl and its new versions. TRAC-PFl has the following features:

(a) Multidimensional fluid dynamics model

The program uses a one-dimensional model for components outside the reactor pressure vessel, and a complete three-dimensional equation inside the pressure vessel: this can accurately calculate many complex multidimensional flow conditions inside the reactor pressure vessel, such as bypass phenomena of emergency cooling water, precooling during core rewetting, and oscillations of liquid level in the lower plenum, dehydration of the upper plenum during reflooding and water fallback, etc., can be modeled by the three-dimensional fluid dynamics equation itself. These processes play an important role in determining the behavior of the accident process.

When modeling accident conditions, certain specific flow path boundaries within the vessel can be blocked to simulate internal structures such as downcomers; internal resistance can be specified to simulate structures such as core support plates; and one-dimensional components can be connected to any side of the pressure vessel's internal grid cells (including internal grid cells) to simulate corresponding loop pipes, etc. Under appropriate conditions, the three-dimensional treatment inside the pressure vessel can be simplified to two-dimensional treatment.

(b) Non-uniform unbalanced two-phase flow model

The program uses a complete two-fluid field equation to describe two-phase flow problems. An incompressible gas mass conservation equation is added to the vapor phase, and it is assumed that the mixture of steam and incompressible gas is a balanced uniform flow. Thus, the model has a total of seven basic field equations. This two-phase flow model can describe important physical phenomena such as reverse flow of vapor and liquid, interaction of underheated water and superheated steam, etc.

(c) Structural characteristic equation package related to flow conditions

The structural equations describe the mass, momentum, and energy transfer between the vapor and liquid phases and the interaction between the two phases and system components. Since these interaction characteristics are greatly related to flow conditions, a structural characteristic equation package related to flow conditions is included in the program.

(d) Comprehensive Heat Transfer Analysis Capability

The heat transfer model in TRAC includes a thermal conduction model for calculating the temperature field within structural materials and fuel rods, as well as a model describing heat transfer between structural components and coolant. Two-phase heat transfer is calculated using a complete boiling curve, which is composed of many heat transfer relationships based on local surface conditions and flow conditions.

The fuel rod thermal conduction model takes into account changes in gap thermal conductivity, metal-water reactions, and quenching phenomena. The axial direction of the fuel rod can be divided into more refined grid nodes to accurately model re-submergence heat transfer and track the quenching front within the fuel rod channel. In the early versions of TRAC, the advancement speed of the quenching front was described using empirical relationships. However, it was found that these relationships had significant errors when simulating low submergence rates. Therefore, a new re-submergence model was incorporated into TRAC-PFl, which directly describes the axial heat conduction problem near the quenching front.

The total power level of the reactor core is either obtained from lookup tables or derived from point reactor neutron kinetics equations that include decay heat. The spatial power distribution is determined by the radial and axial power distribution shape factors within the reactor core, plus the radial distribution within the fuel rod. In each grid cell of the reactor core, only one "average fuel rod" and one "hot rod" are used for heat transfer calculations. The so-called average fuel rod is obtained by averaging all the rods in that grid.

(e) Modular Structure

The structure of the TRAC program is entirely modular. The program contains many component modules. These modules are assembled through input data and can effectively simulate any pressurized water reactor structure or experimental configuration. This structure allows the TRAC program to adapt to various uses. The program can also improve, modify, or add some component modules without affecting other modules. Existing modules in the program can simulate safety injection tanks, pipes, regulators, pumps, steam generators, T-shaped components, valves, pressure vessels, and their internal components, etc.

The functionality of TRAC is also modular. That is, the main computational functions are implemented in some independent modules. For example, basic one-dimensional fluid dynamics solutions, wall temperature field solutions, heat transfer coefficient selection, and other functions are performed in some subroutines. All component modules access these subroutine groups. When better relationships or experimental data are available, this modular structure makes the program easy to improve.

(f) Strong Stability Numerical Method

The handling of the program field equations includes two parts: one part is the three-dimensional semi-implicit difference treatment inside the pressure vessel, and the other part is the one-dimensional finite difference treatment. TRAC-P uses a one-dimensional strong stability two-step method (SETS) numerical method. This method consists of basic steps and stabilization steps, and correspondingly generates basic equation groups and stable difference equation groups during the difference treatment. This new time integration approximation method breaks the traditional Courant stability condition restriction, allowing the use of larger time steps, improving the stability of the solution, and greatly speeding up the calculation speed.

(g) Coordinated Analysis of the Entire Accident Process

The program can perform a coordinated and consistent continuous analysis of the entire accident process, including initial conditions. Its steady-state initialization function provides coordinated and consistent initial conditions for subsequent transient calculations. There is also no need to artificially divide the process into several stages for calculation during transient calculations. If necessary, a single run can complete both steady-state initialization and transient calculations.

(h) Features of TRAC-PFl/MOD1 and TRAC-PFl/MOD2

TRAC-PFl/MOD1 adds an eighth field equation to track the flow in the liquid phase, such as the boron concentration problem in water. In addition, it has improved the condensation model and the wall heat transfer model in the film boiling zone, and made many improvements in functionality. The steam generator model has been modified to better simulate steam pipe rupture accidents; a secondary side turbine component has been added to better simulate secondary loop systems; it allows users to input re-submergence curves obtained from experiments; a new steam-water separation model has been added. In summary, it is believed that the improved MOD1 version has significantly improved modeling capabilities.

TRAC-PPI/MOD2 adopts a new void fraction distribution model and improves heat transfer calculations. Using a newly developed universal thermal structure, it can simulate the heat conduction path between any two fluid lattice elements. Using a two-dimensional heat conduction equation, it can simulate the heat conduction problem of any shaped object in the reactor. It also adopts a three-dimensional strong stability two-step method (3DSETS) numerical method, which increases the time step used in calculations, doubling the calculation speed. The newly adopted vectorization method also greatly improves the speed of multidimensional calculations.

3) ATHLET

The ATHLET program is a thermal–hydraulic system analysis program developed by the German Nuclear Safety Center (GRS). The program has undergone comprehensive and detailed verification and continuous improvement, and is internationally recognized as one of the reliable analysis tools for thermal–hydraulic simulation and safety evaluation of nuclear energy systems. The ATHLET program has been approved by many nuclear safety authorities in Germany and Europe and can be used as a safety analysis program for nuclear facility licensing applications. To date, many national research departments and institutions are still carrying out verification and evaluation work applying this program to advanced reactors (such as improved VVER, EPR, ESBWR, and heating reactors).

The ATHLET program uses a five-equation drift flux model, and the program is mainly composed of thermal-fluid mechanics modules, heat conduction modules, water accumulation fallback modules, etc. The program can calculate complex conditions such as the up and down oscillation of the liquid level in the lower chamber of the core and the re-flooding of the core. The structure of the ATHLET program adopts a highly modular design, which is convenient for modification.

4) RETRAN-02

The RETRAN program is funded by the Electric Power Research Institute (EPRI) and compiled by Energy Inc. (EI). It was developed based on RELAP4, mainly describing the thermal–hydraulic transient characteristics of light water reactor systems. It can calculate and analyze many accidents and operating transients of nuclear power plants and is an optimal estimation program. It uses 11 calculation models including reactor fission energy and decay energy calculation, zirconium-water reaction, field equations, heat conduction, material properties, heat transfer, critical flow, steam-water separator, centrifugal pump characteristics, jet pump characteristics, and valve characteristics. The program also includes steady-state initialization, one-dimensional (two-dimensional) space—time neutron kinetics equation solution, turbine characteristics calculation, pressure stabilizer characteristics calculation, control system simulation, etc.

This program can serve as a dual-use program for nuclear power plant owners and safety approval. The program is very useful for nuclear power plant design, safety analysis, operation transient calculation, and safety approval. It is a relatively ideal and complete nuclear power plant safety analysis program.

5) CATHARE

CATHARE is a large-scale thermal–hydraulic calculation program jointly developed by the French Atomic Energy Commission, Framatome, and Electricité de France, and the latest version was released in April 1999.

The application of CATHARE mainly involves the design of nuclear power plant systems and components, the definition and evaluation of emergency operation procedures, research on new core management, design of new reactors and systems, and pre-budget analysis and post-experiment refinement evaluation of experimental projects. The program is also a mathematical method used to calculate the uncertainty of program prediction results when developing safety analysis programs.

The CATHARE program can simulate any experimental device and pressurized water reactor (including Western-type pressurized water reactors and WWER), and can also be used for other types of reactors (fusion reactors, RBMK reactors, boiling water reactors, and research reactors). Currently, the application scope of the CATHARE program is expanding towards gas reactors (high-temperature gas-cooled reactors, gas turbine modular helium-cooled reactors, etc.), and a dedicated calculation module for gas turbines and gas compressors will soon be developed. The CATHARE program can also be used in conjunction with CRONOS (three-dimensional neutron physics program), FLICA (core thermal–hydraulic analysis program), ICARE2 (severe accident program for analyzing fission product release and core meltdown), etc.

The CATHARE program is based on a two-fluid six-equation model. By adding 1 to 4 transport equations, the transport characteristics of incompressible gases (such as nitrogen, hydrogen, air) that appear in the system can be simulated. The program basically simulates all flow types of two-phase flow, and only two transition flow types are explicitly written into several closed relationship formulas in the program. These two transition flow types are used to describe the transition from separated flow to dispersed flow; annular flow and counter-current flow are described using the CCFL model (counter-current flow limitation). The program can calculate various heat transfer processes between the coolant and the wall and fuel rods; the interface heat and mass transfer model can not only describe the evaporation caused by superheated steam and the condensation phenomenon caused by subcooled liquid, but also describe the quasi-steady-state superheated steam or superheated liquid caused by vapor phase condensation and liquid flash evaporation phenomenon.

The numerical methods of the CATHARE program have the following characteristics:

(a) It adopts a first-order finite volume, finite difference method.
(b) For time discretization, zero-dimensional and one-dimensional modules use a fully implicit format, while three-dimensional modules use a semi-implicit format.
(c) The mass equation and energy equation are discretized in a conservative form.
(d) The problem of phase generation and disappearance is correctly solved by using the residual volume fraction and the approximate conditions of interface mass and heat transfer.
(e) The wall heat conduction problem is solved in conjunction with the hydraulic calculation.

The numerical method adopted by the CATHARE program has been validated through a series of numerical standard experiments. The validation calculations demonstrate that this numerical method can handle all two-phase flow problems described by the two-fluid six-equation model with or without non-condensable gases over a wide range of pressures and temperatures. It can solve incompressible and compressible flow problems under conditions of extensive interfacial mass transfer and high flow rates, ensuring accurate conservation of mass and energy, as well as precise solutions for the propagation of pressure waves and density waves. The CATHARE program ensures that CPU runtime meets engineering needs.

The CATHARE program uses the Newton–Raphson iterative method to solve nonlinear equation systems, with each iteration including the following three steps:

(a) The increment of internal variables in each cell is eliminated as a function of the increment of interface variables.
(b) Calculate all interface variable increments to obtain new variable values.
(c) Update all interface variable increments and calculate convergence.

A distinctive feature of the CATHARE program is its three-dimensional module model, initially designed to simulate the three-dimensional state of pressure vessels under reactor accident conditions. It is mainly used to calculate stages such as blowdown, refill, and reflooding during large break loss-of-coolant accidents where three-dimensional phenomena are prominent. It can also be used to calculate other transient conditions of pressurized water reactors (including low-pressure transients), boiling water reactor transients, and containment calculation problems. Additionally, the three-dimensional module can be used in experimental analyses in other fields, such as external cooling of pressure vessels in severe accidents, pool-type heat exchangers, etc. The three-dimensional module can also be coupled with neutron physics programs, core thermal–hydraulic programs, and severe accident programs.

6) FRAPCON-2

FRAPCON-2 is a steady-state overall analysis program for fuel rod elements in light water reactor nuclear power plants. It is mainly used to analyze the behavior of fuel elements under long-term irradiation conditions, including fuel and cladding temperature distribution, cladding stress and strain, cladding oxidation, fuel irradiation swelling and densification effects, fuel rod internal pressure, generation and release of fission gases in fuel, cladding failure rate, etc., and can perform uncertainty analysis of output parameters as per user requirements. This program is also a supporting program for the transient element analysis program FRAP-T6, providing initial conditions for FRAP-T6 and system analysis programs like RELAP4/MOD7. Therefore, this program is not only an effective tool for the design and testing of nuclear power plant fuel elements but also an indispensable means for safety review of nuclear power plants.

7) Introduction to the RELAP5 Program

The RELAP5 computer program, developed by the Idaho National Engineering Laboratory (INEL) for the U.S. Nuclear Regulatory Commission (NRC), is a light water reactor transient analysis program that can be used in various aspects such as procedure formulation, review calculations, evaluation of accident mitigation measures, operator procedure evaluation, and analysis of experimental plans. RELAP5 has also become the foundation of nuclear power plant analyzers. RELAP5 can simulate events in light water reactor systems such as large break loss-of-coolant accidents, small break loss-of-coolant accidents, anticipated transient without scram

(ATWS), as well as loss of feedwater, loss of offsite power, station blackout, turbine trip, etc., covering almost all thermal–hydraulic transients and accident spectra in nuclear power plants.

Although RELAP5 still uses the traditional RELAP name, it should be noted that RELAP5 is significantly different from RELAP4; it is a completely new program. RELAP5 is based on the non-homogeneous and non-equilibrium fluid dynamics model of two-phase systems. Its basic field equations are the six equations of two fluids and the mass equation of non-condensable gases, solved using a fast semi-implicit numerical method. In addition, RELAP5 includes models of various component processes, such as pipes, pumps, safety injection tank, valves, electric heaters, turbines, separators, accumulators, control system components; it also establishes analytical models for some special processes, such as sudden changes in flow area, branches, obstructed flow, boron tracking, effects caused by non-condensable gas migration, etc.

From the start of its development in 1967 to the implementation of RELAP5/MOD3.3, RELAP5 has been in use for over 40 years. RELAP5 has been validated and compared with a large number of instances and experimental data from many experimental plans in various countries. The development of RELAP5 is still ongoing to expand its functionality, enabling its application in advanced pressurized water reactors and Eastern European WWER reactor types. In the latest version, RELAP5/MOD3.4, the research results obtained in two-phase flow theory research, numerical solution methods, computer programming techniques, and various scale experiments are concentrated.

RELAP5/SCDAP/MOD3.2 combines the models of RELAP5/MOD3.2 and SCDAP. RELAP5 includes models of reactor coolant system thermal-hydraulics, control system actions, neutron kinetics, and non-condensable gases, while SCDAP is used for simulating core behavior under severe accident conditions. This program can simulate and calculate the thermal–hydraulic transient behavior of pressurized water reactors, boiling water reactors, research reactors, and general thermal–hydraulic devices.

RELAP5-3D is developed based on RELAP5/MOD3.2 and includes a three-dimensional thermal–hydraulic model of the flow field. It can analyze the three-dimensional flow field of important components such as reactors and steam generators under accident conditions. In addition to the point reactor model, it can provide an interface with three-dimensional space–time neutron kinetics in the core.

The program structure of RELAP5/MOD3.3 includes:

(a) Program Input

The input cards of the program are used to describe the basic composition of the computational research object and the computational task. The input cards mainly include descriptions of the computational object and task, identification of a new transient task or program restart task, selection of computational units, selection of non-condensable gases, selection of computational time steps, input of logical triggers and controls, descriptions of various control volumes and flow parameters, descriptions of thermal component parameters, selection of neutron kinetics models, etc.

The input cards of the program need to be completed according to the division of the computational object's control volumes, flow junctions, and thermal components. The division of control volumes, flowjunctions, and thermal components is crucial for the completion of the computational task, which requires experienced computational personnel to obtain through certain computational sensitivity analysis.

(b) Steady-State Initialization and Transient Calculation

The steady-state initialization and transient calculation functions of the RELAP5 series are mainly carried out by selecting appropriate computational time steps for the thermal–hydraulic, control logic, and neutron kinetics of nuclear power devices.

Steady-state initialization is a re-evaluation and calculation of the state parameters of each control volume and flow junctions in the input card. In the initial input card, the initial parameters of each control volume and flow junction may not necessarily satisfy the convergence of each field equation solution, and thus cannot proceed to the next step of transient calculation. Therefore, before performing transient calculation, it is necessary to perform steady-state initialization calculation of the nuclear power device.

Transient calculation is used to calculate the expected transients or accidents provided by the user. During transient calculation, the user needs to arrange various logical triggers according to the event sequence of the calculated transient or accident, and complete the calculation analysis of the entire event through the restart file generated by the transient calculation.

(c) Computational Result Processing

Computational result processing mainly involves graphically plotting various transient parameters in the computational results or transferring data to other program modules.

(3) Multi-Physics Field Multi-Scale Coupled Simulation

Research on the physical field of light water reactor nuclear reactors involves multiple professional fields, mainly including nuclear fuel, nuclear reactor physics, reactor thermal-hydraulics, radiation protection, engineering thermophysics, materials science, mechanics, chemistry, automatic control, electrical, etc.

Research on the physical field of nuclear reactors can be divided into local microscopic scale, component and equipment scale, system scale, and device scale in terms of geometric scale; physical phenomena and characteristics can be divided into second-level, minute-level, hour-level, day-level, year-level, and even hundreds of years or more in terms of time scale.

The multi-physics field multi-scale research of nuclear reactors will inevitably have great differences in content, focus, and methods. To completely and accurately describe the real response behavior of nuclear power plants or nuclear power devices under various operating conditions, it is necessary to conduct multi-physics field multi-scale coupling research, achieve the goal of digital reactors and digital nuclear power plants through the construction of integrated simulation programs, which is also the current hot spot and focus of academic research.

1) Coupling of Reactor Point Model and One-Dimensional Thermal–Hydraulic System Program

In the early days and even now, many system programs such as RELAP, the nuclear physics calculation model of the reactor still uses the point model, numerical calculation uses the fourth-order Runge–Kutta method or fully implicit difference method, fuel temperature feedback, coolant temperature feedback, coolant density feedback, reactor average channel axial power distribution, reactor thermal channel axial and radial power distribution, reactor shutdown curve, control rod worth curve and other physical parameters are calculated by the nuclear physics program as input, the feedback coefficient, control rod worth, shutdown reactivity and other data in the thermal–hydraulic calculation process are generally obtained in the form of fitting formulas or data interpolation, and the power distribution factor remains unchanged. This method is relatively simple and can meet the analysis accuracy needs in most conditions. However, it has serious deficiencies in simulating reactivity accidents.

2) Coupling of Reactor Three-Dimensional Space–Time Neutron Kinetics Model and One-Dimensional Thermal–Hydraulic System Program

By using the three-dimensional space–time neutron kinetics model of the reactor, the nuclear physics calculation can provide the main parameters at a fixed burnup moment in each calculated time step, including normalized nuclear power, normalized power distribution in the three-dimensional space of the reactor, normalized peak power factor in the average tube radial direction, normalized peak power factor in the average tube axial direction, normalized peak power factor in the thermal tube radial direction and local non-uniformity factor, normalized peak power factor in the thermal tube axial direction and local non-uniformity factor, etc. The reactor thermal–hydraulic program uses these parameters for thermal–hydraulic calculation and safety analysis. Some three-dimensional space–time neutron kinetics programs of the reactor contain dynamic burnup calculation models that can perform multi-burnup point dynamic calculations. These space–time calculation programs are usually based on a component scale; neutron kinetics equations usually use diffusion calculation models or transport calculation models.

3) Coupling of Reactor Three-Dimensional Space–time Neutron Kinetics Model and Three-Dimensional Thermal–Hydraulic System Program

The three-dimensional thermal–hydraulic system program can simulate the three-dimensional thermal–hydraulic characteristics of a reactor, based on the coupling calculation of neutron kinetics and thermal-hydraulics at the component scale. It can simulate both the uneven spatial power distribution and the uneven flow distribution and component mixing effects, providing more accurate simulation of local hot spots.

4) Coupling of the three-dimensional space–time neutron kinetics model and the core sub-channel program

Compared to the system program, the core sub-channel program has certain advantages in simulation accuracy and range. The sub-channel program can be precise down to 1/4 of the component or even finer. At this time, the three-dimensional space–time neutron kinetics model requires a smaller grid. The transport calculation program or Monte Carlo calculation program can simulate a single fuel rod, coupled with the sub-channel program, to achieve PIN BY PIN full core neutron physics and thermal–hydraulic coupling calculation.

5) Coupling of the one-dimensional thermal–hydraulic system program and the core single-channel program

The system program calculation provides the boundary conditions needed for the single-channel calculation program. The core single-channel calculation program calculates parameters such as the temperature field, pressure field, void fraction, gas content, DNBR, etc., for the average component, hot component, average tube channel, and hot tube channel. This coupling is unidirectional.

6) Coupling of the one-dimensional thermal–hydraulic system program and the core sub-channel program

The system program calculation provides the boundary conditions needed for the sub-channel calculation program. Core sub-channel calculation programs such as the COBRA series, VIPRE series, FLICA series, etc., can calculate parameters such as the temperature field, pressure field, void fraction, gas content, DNBR, etc., for the entire core or 1/4 of the core. This coupling can be unidirectional or bidirectional.

7) Coupling of the thermal–hydraulic system program, sub-channel program and CFD

The reactor thermal–hydraulic system program can study the entire reactor loop system and even nuclear devices from a system scale, such as RELAP5, TRACE, CATHARE, etc. These programs usually assume that fluid flow is one-dimensional and use a lumped parameter method to quickly calculate the entire system. The sub-channel program simulates components with porous medium characteristics, such as the reactor core, steam generator, or heat exchanger, from a component scale. The hydraulic model in these programs adds a lateral flow equation, which can better simulate the mixing phenomenon between channels. CFD can study large spatial areas with strong three-dimensional flow from a local scale, such as FLUENT, CFX, STAR-CD, etc. These programs directly solve the basic equations of fluid mechanics and contain various mature turbulence models, which can perform detailed three-dimensional calculations on space and obtain three-dimensional distributions of parameters. In traditional analysis methods, the three scales of simulation are almost decoupled, usually with the system program providing the boundary conditions required for sub-channel and CFD analysis, and the results of sub-channel and CFD calculations are not fed back into the system program. This often requires the introduction of many assumptions, thereby affecting the accuracy and authenticity of the analysis. Therefore, they can be coupled through their mutual interaction relationships, making comprehensive use of the advantages of each program, i.e., multi-scale coupled simulation. From the perspective of numerical solution, multi-scale coupling can be divided into strong coupling and weak coupling. Strong coupling is the simultaneous solution of the basic equations of each scale, also known as solution-based coupling; weak coupling refers to the use of separate programs to simulate each scale, with coupling achieved through data exchange at the interface, so it is also called data-based coupling or inter-program coupling. Strong coupling will result in very complex nonlinear equations, which are generally difficult to solve numerically. Weak coupling can make full use of existing programs, and the implementation of coupling is relatively easy. However, this exchange based solely on boundary conditions may bring stability and convergence problems. At present, almost all existing research adopts the weak coupling method.

8) System program coupling

This mainly includes the coupling of the reactor system program with the secondary loop fluid network system program, the primary loop auxiliary system program; the system analysis program with the control system, electrical system program; the system program with the containment program; the design basis accident analysis program with the severe accident program, such as RELAP/SCADP, RELAP/MAPP, RELAP/MECOLE; the thermal–hydraulic program with the mechanical analysis program; the thermal–hydraulic program with the water chemistry analysis program, etc.

6.2 Probabilistic Safety Assessment Method

In 1972, Probabilistic Safety Assessment (PSA) was first applied to nuclear power plants, with a milestone report published in 1975 titled "Reactor Safety Study: An Assessment of Accident Risks in U.S. Nuclear Power Plants" (WASH-1400). Since the publication of WASH-1400, the PSA method has greatly evolved and has become a standardized tool for safety assessment in nuclear power plants, leading to the emergence of commonly used PSA software. The Three Mile Island nuclear accident occurred in the United States in 1979, and it was found that the WASH-1400 report had accurately predicted the occurrence and development of this nuclear accident. Since then, PSA has been widely recognized and applied in the field of nuclear power safety. In the process of analyzing and designing for weak links in severe accidents at nuclear power plants, people can use PSA to identify these weak links, and then improve their design, operation management, safety supervision, fault diagnosis, and maintenance strategy, thereby enhancing the ability of nuclear power plants to handle severe accidents. Currently, PSA is gradually developing into an important tool for safety decision-making and safety assessment in various engineering fields.

6.2.1 Overview

Probabilistic Safety Assessment, also known as Probabilistic Risk Assessment (PRA), is a systems engineering method that developed after the 1970s. Based on probability theory and reliability engineering, PSA can comprehensively analyze the possible occurrence and development of accidents in various complex engineering systems, and provide a comprehensive evaluation based on the probability of potential accidents and the magnitude of their consequences.

(1) Comparison of Probabilistic Safety Assessment and Deterministic Methods

Currently, there are two methods for evaluating the safety of nuclear power plants: one is the deterministic method, and the other is the PSA method. In the early stages of nuclear power development, the deterministic method was used for a long time. PSA is a new safety assessment method developed in recent years, based on probability risk theory and using systems engineering methods, it provides a comprehensive and integrated analysis and evaluation of the safety of nuclear power plants from different perspectives.

The deterministic method adopts some conservative assumptions and combines many years of application experience to artificially divide accidents into "credible" and "incredible" categories. For example, in the most widely used pressurized water reactor nuclear power plants, the double-ended guillotine break of the main coolant pipe cold leg is usually set as the maximum credible accident. This is carefully considered during design, and emergency core cooling systems and other safety facilities are used as stringent defensive measures. Even if such a severe initiating event occurs, due to the stringent defensive measures, the reactor may not necessarily result in severe consequences. For accidents with lighter consequences, such as small break loss-of-coolant accidents, and operational transients that occur during operation, the deterministic method has not conducted in-depth research. In addition, the deterministic method does not give due attention to the operation management and personnel training of nuclear power plants. In fact, the main cause of the Three Mile Island nuclear accident was that people did not fully understand the transitional conditions and small break loss-of-coolant accidents, which led to operator misjudgment, causing the originally minor accident to escalate repeatedly, eventually developing into a serious core damage accident in nuclear power history. The occurrence of the Three Mile Island nuclear accident exposed the shortcomings of the deterministic method.

The PSA method believes that there is no absolute "credible" or "incredible" accident, it is just that the probability of an accident occurring varies, thus introducing the concept of risk. Research on the risk of nuclear power plants shows that the main factor for the release of radioactive materials to the environment is core meltdown, and the main causes of core meltdown are small break loss-of-coolant accidents and operational transients. The PSA method can not only determine the accident sequences caused by various different initiating events, but also reflect the frequency of accidents and the consequences they cause as realistically and systematically as possible. The lessons from the Three Mile Island nuclear accident indicate that it is necessary to use the PSA method to conduct safety assessments of nuclear power plants.

The deterministic method uses whether the radiation dose received by residents living around the nuclear power plant during a design-basis accident exceeds the allowable limit as the evaluation standard for the safety of the nuclear power plant. However, for most of the public, when the irradiation dose exceeds the allowable limit, they are not clear about the extent of the harm it will cause, so the public cannot directly compare the harm of nuclear power plants with the harm of thermal power plants and hydropower plants. The PSA method introduces the concept of risk with quantitative significance, so that PSA can be applied to various fields of social activities. In this way, a common platform is provided, which can compare the social risks caused by human factors or nuclear power plant accidents on the same basis, and can also compare the social harm caused by thermal power plants or hydropower plants.

In the deterministic method, people study the physical processes of nuclear power plants under accident conditions through mechanistic procedures. When analyzing accidents, it is usually based on the basic assumption of a single failure, without considering how likely the accident is to occur, or considering operator intervention after the accident occurs. The PSA method, on the other hand, is a systematic evaluation technique, based on probability theory and rigorous mathematical logic reasoning, providing a comprehensive, structured approach to finding credible accident sequences, and evaluating and describing the probability of accidents and the consequences they cause.

It should be noted that the process of analyzing a nuclear power plant using the PSA method is essentially a comprehensive review and understanding of the plant. The PSA method does not limit its analysis to design-basis accidents, but rather considers a broader spectrum of events as much as possible. It conducts a comprehensive analysis of the occurrence and development of these events, and further quantifies the risks based on this. Therefore, the PSA method can realistically reflect the actual state of the nuclear power plant. At the same time, the PSA method is not limited to the basic assumption of a single failure, but considers the possibility of various systems and equipment failing during the event process. It also takes into account the possibility of human intervention failure during the event and the correlation between systems, equipment, and personnel. The main differences between the deterministic method and the PSA method are shown in Table 6.1.

As a system evaluation tool, although PSA provides a lot of useful information, its numerical results' limitations and uncertainties should also be recognized. For example, it is difficult to quantitatively describe human behavior and human-induced destruction in the PSA evaluation process.

Therefore, for a specific nuclear power plant, the principles of in-depth defense and multiple barriers should be adhered to during design to prevent accidents and mitigate their consequences. For some aspects that the deterministic method cannot handle well, the PSA method can be used for in-depth analysis, so that the two complement each other. The "Nuclear Power Plant Design Safety Regulations"

Table 6.1 Comparison of deterministic method and PSA method

Evaluation method	Deterministic method	PSA method
Basic assumption	Single failure No consideration of human error	Consideration of multiple failures Consideration of human error
Model & parameters	Conservative assumptions	As realistic as possible
Analysis procedure	Mechanistic	Logical
Correlation	Difficult to consider systematically	Able to consider systematically
Final result	Meets quantitative acceptance criteria	Checks final risk

(HAF102) clearly states that deterministic methods and probabilistic methods are "complementary technologies". The "Final Policy Statement on the Use of Probabilistic Safety Assessment Methods in Nuclear Management Activities" issued by the U.S. NRC in 1995 also mentioned that "by considering risks in a more orderly and complete manner, probabilistic methods are an extension and improvement of traditional management methods" and "Probabilistic Safety Assessment methods play a supplementary role in relation to deterministic methods".

(2) Steps of Probabilistic Safety Assessment Analysis

According to the "PSA Implementation Guidelines" (NUREG/CR-2300), a typical PSA analysis includes the following steps after systematic summarization and induction.

1) Collection of Initial Information

PSA is a comprehensive task that requires a large amount of information, which can be divided into three categories: ① Information on plant design, site, and operation; ② General data and specific plant data; ③ Documents and reports on the PSA method.

2) System Analysis

This task includes determining accident sequences, analyzing each system and its operating characteristics, forming a database about initiating events, component failures, and human errors, and evaluating the frequency of accident sequences. System analysis includes several subtasks: ① Constructing event trees; ② System modeling; ③ Analysis of human reliability and procedures; ④ Formation of the database; ⑤ Quantitative analysis of accident sequences.

3) Containment Analysis

Containment analysis is part of the second and third level PSA analysis content. It includes two subtasks: physical process analysis and analysis of the release and migration of radioactive nuclides.

4) Migration and Consequence Evaluation of Radioactive Nuclides in the Environment

Based on the source term provided by the containment analysis results, using the local topographic information and specific meteorological data of the site, analyze the migration of radioactive nuclides in the environment, calculate the radiation dose received by residents around the nuclear power plant and the resulting health effects, and give various consequences of the release of radioactive nuclides.

5) External Event Analysis

Many nuclear power plants internationally have conducted analyses of external events such as fires, earthquakes, or flooding, and have developed relatively mature analysis methods. The results of the external event analysis should be included in the accident sequence analysis.

6) Uncertainty Analysis

Regardless of the scope of the analysis, uncertainty analysis is a necessary component of Probabilistic Safety Assessment (PSA). Uncertainty issues exist in every step of the PSA analysis, whether it's qualitative or quantitative analysis. The uncertainty of data, model assumptions, and the completeness of the analysis should all be considered. To the extent possible, these uncertainties will be propagated through the analysis.

7) Sensitivity and Importance Analysis

Sensitivity and importance are often considered additional parts of the classic PSA analysis, but the vast majority of PSA reports include these two analyses. Sensitivity analysis is a supplement to uncertainty analysis, which can obtain the possible range of calculation results; importance analysis is an effective method to find the most likely failed equipment and systems. The results can help power plant designers and managers focus their improvement, maintenance, and inspection efforts on key equipment and systems.

8) Obtaining and Interpreting Results

Listing the accident sequences and frequencies that contribute significantly to risk, establishing the cumulative distribution function of the power plant, and determining the distribution function that can reflect the uncertainty of the accident sequence are important contents of obtaining PSA results.

9) Forming Result Documents

Forming a complete result document is an important step in the PSA analysis.

6.2.2 Common Methods of Probabilistic Safety Assessment

Traditional PSA analysis usually uses event tree analysis and fault tree analysis to establish the path from the initiating event to the accident consequences, calculate the probability of different accident sequences, and clarify the causes and probabilities of system failure through fault tree analysis. With the continuous expansion of system safety analysis requirements and the continuous development of PSA technology, some new PSA analysis methods have emerged accordingly.

(1) Event Tree Analysis

Event Tree Analysis (ETA) is a common deductive analysis method in system engineering, originating from decision tree analysis. This method starts from a given initiating event, uses a tracking method according to the time process, analyzes the status of each mitigation system item by item until the final consequence, thus building a horizontal tree diagram to express various accident sequences and conduct related analysis.

A nuclear power plant is a very complex and large system, in which several safety measures are set to limit the adverse consequences of the initiating event l. After a specific accident type l occurs, the accident process and its consequences will inevitably vary due to the different degrees of failure of the safety measures. Therefore, to calculate the probability value P_l of the occurrence of accident type l, it is necessary to examine the accident sequences I that may occur after the occurrence of accident type l. Each sequence has a probability of occurrence P_{li}, so the total probability of accident type l can be calculated using the following formula:

$$P_I = \sum\nolimits_i P_{li} \tag{6.1}$$

Generally speaking, for a certain initiating event, if there are N kinds of safety measures to limit adverse consequences, and each safety measure can only take the normal or failure state, then the total number of accident sequences caused by this initiating event is 2^N. However, considering the dependency between these sequences, many sequences are impossible or meaningless, so in fact, the accident sequences that should be analyzed are much less than 2^N.

From the known initiating event, the method of deductive reasoning is often used to find all possible consequences, commonly using event tree analysis. In the risk analysis of nuclear power plants, event tree analysis is often used to find all possible accident sequences for a specific initiating event.

An event tree is a representation of all possible consequences that may arise from a specific initiating event, these possible consequences depend on all possible paths of development after the initiating event. In principle, the event tree is similar to the "decision tree" widely used in enterprise decision-making. Since the event tree starts from an initiating event, it represents a deductive logical reasoning process.

Figure 6.3 outlines an event tree, where the initiating event is a reactor primary circuit pipe rupture, and P_A is the probability of this event. The probability P_A can be obtained by the fault tree analysis method to be introduced later.

The second step is to consider whether the specially designed safety facilities can operate with power supply. The probability of power failure is P_B. This can also be estimated from the corresponding fault tree. If the power supply (including auxiliary power) fails, the active safety facilities cannot operate, resulting in damage to the

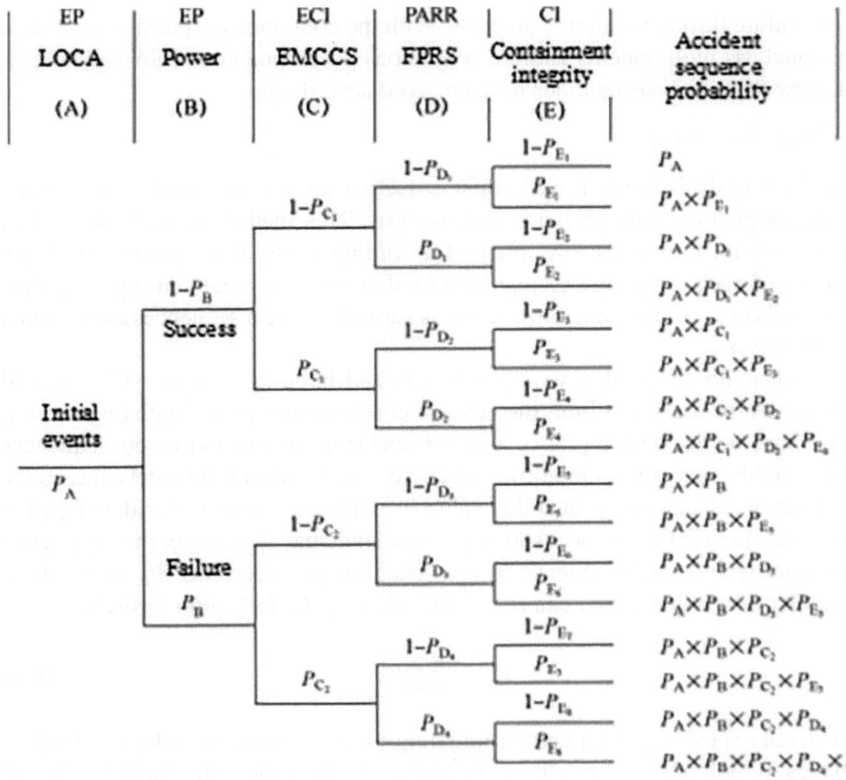

Fig. 6.3 Loss of coolant event tree for pressurized water reactor nuclear power plant

reactor core and a large amount of radioactive release. Therefore, as shown in the figure, the probability of this sequence of events is $P_A \times P_B$.

If the power supply does not fail, the next event to consider is whether the Emergency Core Cooling System (ECCS) can operate, with the probability of ECCS failure being P_C. The possible developments after this event are whether the fission product removal system and the integrity of the containment fail; assuming their failure probabilities are P_D and P_E respectively. Note that the probability of not failing at each link is 1-P, where $P = P_B$, P_C, etc.

The total probability of the event chain given at the right end of Fig. 6.3 is equal to the product of the conditional probabilities of each link in the chain. Therefore, the total probability in the penultimate row should include the factor 1-P_{E8}, and the third row from the bottom should include (1-P_{D2}) (1-P_{E4}). However, since the probability P_i is very small, the value of 1-P_i is taken as 1 in each row.

Obviously, if the failure probabilities of each subsystem in the event tree, i.e., P_A, P_B, and P_C, etc., are known, then the total probability of each failure sequence can be calculated from this.

If the subscript i is used to represent the sequence on the event tree, then the amount of radioactive release represented by the right side of Fig. 6.3 can be represented as C_{li}. Thus, the possible amount released in the l-type accident can be represented by the following formula:

$$P_l C_l = \sum_i P_{li} C_{li} \tag{6.2}$$

According to Eq. (6.2), the total risk of the reactor generated by all types of accidents can be represented as follows:

$$R = \left(\sum_l \sum_i P_{li} C_{li}\right) D \tag{6.3}$$

It must be pointed out that the simplified event tree described here is only to clarify the universal principles of risk assessment. In fact, several systems in Fig. 6.3 can be further subdivided into several subsystems, each with its own event tree.

(2) Fault Tree Analysis

From Eq. (6.2) and Fig. 6.3, the problem of calculating the total risk is reduced to calculating the conditional probabilities of each link on the event tree, i.e., calculating the probability of the initial event P_A and the probability of failure of the reactor safety system or facility P_i. The problem of calculating the system failure probability must know the system's failure rate. If the system's failure rate can be directly selected based on experience, then the subsequent work is simple. Unfortunately, it is extremely difficult to directly determine these failure rates $\theta(t)$ based on reactor operating experience, even in the near future. This is because the nuclear industry is still young, and there is little accumulated data on the failure rates of safety systems; secondly, because reactors require special reliability, the failure rates of more important initial events and safety systems are very small, and it is difficult to directly estimate their failure rates.

However, the equipment used in the reactor safety facilities or systems, such as valves, pumps, transmission devices, switches, and circuit breakers, have many years of application experience in some industries, and a large amount of selectable failure rate data has been accumulated; with these data, the system's failure rate and failure probability can be calculated. There are many methods available to calculate the failure probability of the entire system from the failure rates of the equipment used in the system, but most of these methods apply the fault tree analysis method. Therefore, the fault tree analysis method is introduced next.

A fault tree system is a logical diagram that describes a series of component failure modes that lead to system failure. First, it is assumed that the system experiences a certain accidental event or failure state, which is the so-called "top event"; then, from this, we trace back step by step to find the sequence of combinations of various other events (or failures) that lead to this top event. After many steps of such backward tracing, a set of primary failure events can be identified, which cannot (or need not) be further traced back to their causes. In many cases, the failure of a complex system is often due to the failure of several subsystems. At this time, a fault tree analysis can be performed on each subsystem separately, and then these results are input as the "primary" part of the overall system fault tree.

Figure 6.4 shows the first few levels of a fault tree. The top event shown in the figure is insufficient flow of cooling water spray in the containment of a pressurized water reactor. The reactor has two redundant systems A and B, each of which can provide all the water for spraying independently. Therefore, the premise for the occurrence of the above top event is the simultaneous failure of the two systems. This situation is represented by the logical symbol "AND gate" (a box with a round top), which connects the secondary events with the top event. The direct cause of the failure of the spray system is insufficient flow in the injection tank connected to the spray head, so this should be the second-level event in the fault tree.

In the third level, four types of events are found, each of which is sufficient to cause the above second-level event. Therefore, the events of the third level are connected to the second level with an "OR gate" (a box with a pointed top). Events represented by circles or diamond-shaped boxes do not need to be further traced back to their causes. Events in circles are events for which failure probabilities can be obtained or calculated, while events in diamond-shaped boxes are general failures, and the tracing of causes is stopped due to their less importance or lack of data. Events in rectangular boxes are events that must be traced down. Figure 6.4 does not show the fault tree after the third level; in fact, in some cases, it can be traced back to more than twenty levels.

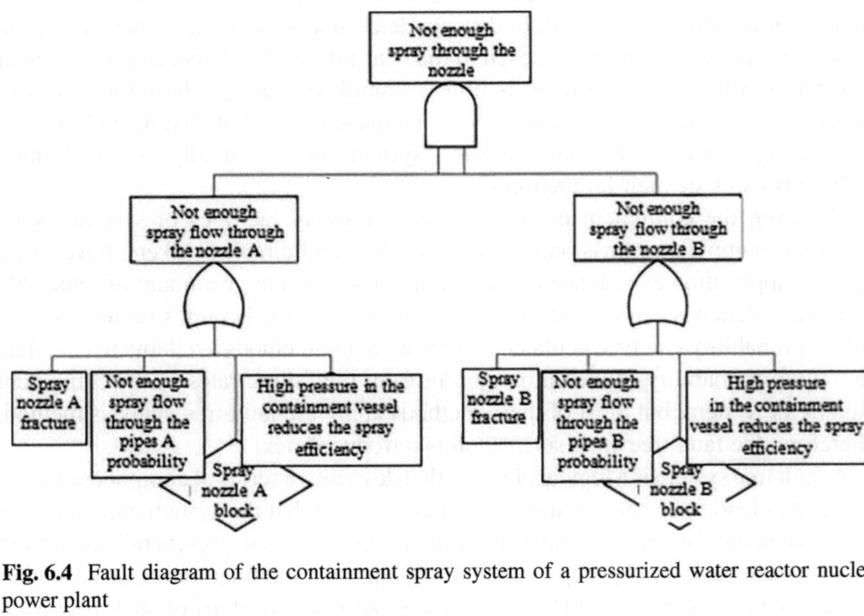

Fig. 6.4 Fault diagram of the containment spray system of a pressurized water reactor nuclear power plant

6.2.3 *Commonly Used Programs in Probabilistic Safety Assessment*

Since the twentieth century, the Probabilistic Safety Assessment (PSA) method has made significant progress, prompting the vigorous development of PSA software. The development and marketization level of PSA software internationally is relatively high, and the developed PSA software has its own characteristics. With the large-scale development of nuclear power plants in China, the pace of localization of China's PSA software has also accelerated. From the perspectives of advanced core algorithms, outstanding software performance, and strong practicality, the strength of China's PSA software has reached an international advanced level.

(1) Software Development History

With the deepening application of PSA technology, PSA software has gradually begun to popularize. Looking back at the development history of international PSA software, it can be roughly divided into the following three stages (as shown in Fig. 6.5).

1) Initial Stage (1970s)

Starting from the 1970s, PSA gradually gained widespread application in the fields of nuclear energy and aerospace. Due to the reliability analysis needs of large complex systems, a small amount of PSA software began to appear, such as Elraft, FAUNET, Supkit, Bit-Frantic. However, most of the PSA software at this stage only had fault tree analysis functions, and the user interface was relatively simple. Some software

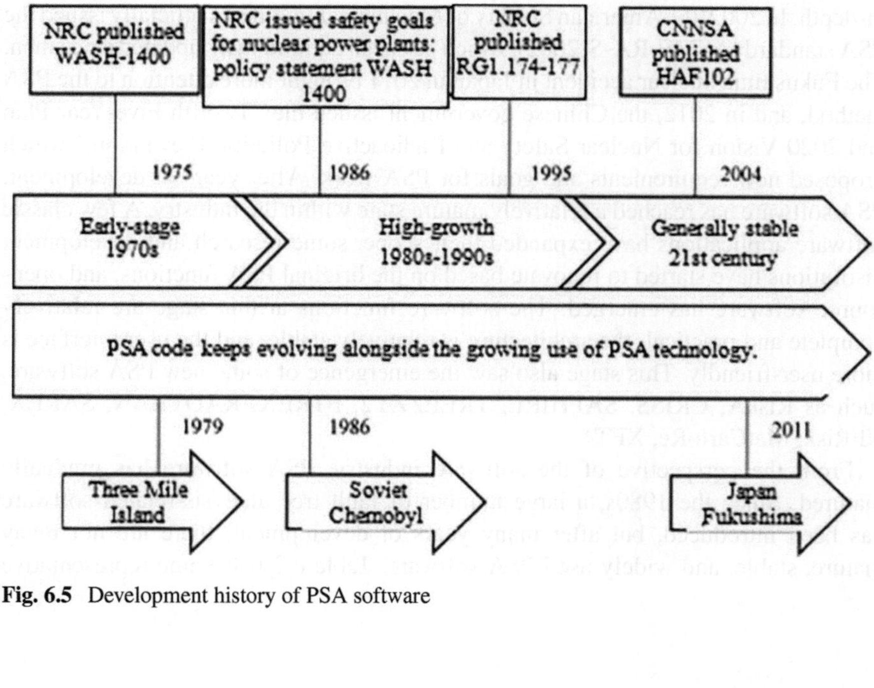

Fig. 6.5 Development history of PSA software

even had no user interface and relied solely on command parameters to run the program.

2) Rapid Growth (1980s–1990s)

The Three Mile Island nuclear accident in 1979 and the Chernobyl nuclear accident in 1986 had far-reaching impacts on the entire nuclear industry. The U.S. Nuclear Regulatory Commission (NRC) published the "PSA Implementation Guidelines" (NUREG/CR-2300) and its supporting reports, "Nuclear Power Plant Operational Safety Objectives: Policy Statement", "Severe Accident Risk: Evaluation of Five U.S. Nuclear Power Plants" (NUREG-1150) and its supporting reports, etc., after the two nuclear accidents, which played an important guiding role in the development of PSA. PSA technology began to be used worldwide to conduct safety evaluations of nuclear power plants. In 1986, the China National Nuclear Safety Administration organized the first PSA study on nuclear power plants under design and construction, and completed the first PSA report for operating nuclear power plants between 1995 and 1999. As PSA gradually matured, PSA software developed rapidly, and many classic PSA software entered the market, forming a situation where hundreds of schools contended, such as Ftap, Aftp, Paft-F77, Ffta, Cafta + , RiskMan, ITEMtoolkit, Fault-Tree + , PTCWindchillFTA, Reliasoft, LESSEPS, RiskSpectrum, Kirap, LIPSAS, MetaPrime, Ftaes, FAMOCUTN&CUTQN, DIFtree, FORTE, PROFAT, Galileo, etc. The software functions of this stage were expanded a lot, basically having all the functions of PSA analysis (including user interface), and some countries' nuclear energy management agencies also began to develop their own review PSA software.

3) Overall Stability (21st Century)

At this stage, the application of PSA technology is becoming more widespread and in-depth. In 2002, the American Society of Mechanical Engineers officially issued the PSA standard (ASME-RA-S-2002), which has been continuously updated since then. The Fukushima nuclear accident in Japan in 2011 brought more attention to the PSA method, and in 2012, the Chinese government issued the "Twelfth Five-Year Plan and 2020 Vision for Nuclear Safety and Radioactive Pollution Prevention," which proposed new requirements and goals for PSA work. After years of development, PSA software has reached a relatively mature state within the industry. A few classic software applications have expanded their scope, some research and development institutions have started to innovate based on the original PSA functions, and open-source software has emerged. The software functions at this stage are relatively complete and practical, the architecture is relatively stable, and the user interface is more user-friendly. This stage also saw the emergence of some new PSA software, such as RiskA, CRISS, SAPHIRE, TREEZZY2, FTREX, RADYBAN, SAREX, NFRisk, MatCarloRe, XFTA.

From the perspective of the software industry, PSA software has gradually matured. Since the 1980s, a large number of fault tree analysis-related software has been introduced, but after many years of development, there are not many mature, stable, and widely used PSA software. Table 6.2 lists some representative

PSA software in the history of international PSA software development and their characteristics.

(2) Introduction to Typical PSA Software

RiskA is the first large-scale commercial PSA software with independent intellectual property rights developed by the FDS team of the Institute of Nuclear Energy Safety Technology, Chinese Academy of Sciences. Compared to other units in China, it has the advantages of high reliability and mature technology. RiskA has core functions such as fault tree analysis, event tree analysis, uncertainty analysis, sensitivity analysis, and importance calculation, and also provides various auxiliary functions such as intelligent conversion of mainstream models, interactive graphic modeling, report output, multi-threaded collaborative computing, and network deployment. The software is highly versatile and has been applied in more than 600 units. It has been successfully applied in many fields including nuclear energy, aerospace, electronics, defense, etc., including risk monitoring of China's Qinshan Third Nuclear Power Plant, safety assessment and construction permit certification of the International Thermonuclear Experimental Reactor ITER, reliability analysis of the National

Table 6.2 Representative PSA analysis software

Name	Year	Country	Institution	Features
RiskMan	1980s	USA	ABS Consulting	Integrated expert system, training module, drawing module, etc
ITEM toolkit	1985	UK	ITEM	Basic PSA functions, integrated standard analysis packages from different fields
Risk Spectrum	1985	Sweden	Relcon AB	Comprehensive PSA functions integrated in the form of a software series
SAPHIRE	1993	USA	NRC & INL	Composed of IRRAS and SARA, covering reliability and PSA
R&R Work Station	1995	USA	DS&S EPRI	Friendly and intuitive fault tree modeling analysis interface (CAFTA module), can be used as a PSA workstation
RiskA	1999	China	Institute of Nuclear Energy Safety Technology, Chinese Academy of Sciences·FDS Team	Integrates all PSA functions, friendly fault tree drawing interface, independently released high-speed fault tree calculation engine

Large Science Project—EAST Superconducting Tokamak Nuclear Fusion Experimental Device, safety analysis of the ADS Transmutation System of the Strategic Priority Research Program of the Chinese Academy of Sciences, safety analysis of the LIDAR system, independent verification and calculation of nuclear safety review by the National Ministry of Ecology and Environment, and other major safety analysis and monitoring projects.

The RiskSpectrumPSA software developed by Sweden's Scandpower company is currently widely used in nuclear power plants. RiskSpectrumPSA is one of the safety analysis software in the RiskSpectrum series, capable of performing fault tree analysis and event tree analysis. Its integrated analysis tool, RSAT, can handle large-scale fault tree models, offering functions such as MCS, BDD, sensitivity, importance, and time-variant analysis.

Discussion Questions.

(1) Explain the basic logic and specific steps of deterministic safety analysis.
(2) Discuss the specific methods of deterministic safety analysis and their interrelationships.
(3) What are the characteristics and applicable scopes of the distributed parameter method and the lumped parameter method?
(4) Analyze the main challenges encountered by nuclear power numerical simulation technology.
(5) Discuss the prospects of combining nuclear power plant operational experience feedback with modern information technology.
(6) Analyze the composition and operation process of simplified reactor kinetics models.
(7) Analyze the solution process of the space–time neutron kinetics equation in reactors.
(8) What are the concepts and division principles of control volumes and streamlines in the thermal–hydraulic analysis model of nuclear power devices?
(9) Analyze the characteristics of the two-fluid model in thermal–hydraulic analysis.
(10) Analyze the composition and functions of the core physics analysis program.
(11) What are the characteristics of using the nodal method and the fine mesh finite difference method for reactor physics calculations?
(12) Analyze the functions and features of the nuclear physics calculation program MCNP series.
(13) What are the main components of the nuclear power thermal–hydraulic system analysis program? What are the functional features of advanced thermal–hydraulic analysis programs?
(14) Compare probabilistic safety assessment and deterministic analysis methods.
(15) Explain the basic analysis steps of the probabilistic safety assessment method.
(16) Explain the functions and basic steps of using event tree and fault tree analysis methods.

References

Chen WZ, Zhang LM, Xiao HG. (2019) Principles of Nuclear Reactor Engineering (Volume I). Beijing, China Atomic Energy Press.

Chen WZ, Hao JL, Li HF, Li MR. (2023). Neutron Dynamics in Nuclear Reactors. Beijing, China Atomic Energy Press.

Chen WZ, Yu L, Hao JL. (2023). Thermal-Hydraulics of Nuclear Power Systems. Beijing. Science Press.

Wu HC. (2023). Advanced Reactor Physics. Beijing. Science Press.

Zhu JZ, Shan JQ. (2019), Nuclear Reactor Safety Analysis. Xi'an, Xi'an Jiaotong University Press.

Chapter 7
Nuclear Safety Legal System and Regulation

Nuclear safety is the prerequisite and foundation for the sustainable development of nuclear energy. To ensure nuclear safety, a legally authorized agency must represent the government to issue and implement nuclear safety regulations, regulate various nuclear activities through legal norms, and establish independent nuclear safety review and supervision mechanisms. Nuclear safety regulation is a state supervision and management activity implemented to ensure the safety of nuclear facilities. The government provides a legal basis for regulation by establishing nuclear safety laws and regulations; the nuclear safety regulatory agency exercises independent supervisory and management powers to regulate nuclear facilities. As summarized by the International Atomic Energy Agency (IAEA) from nuclear power practice experience, as pointed out in its nuclear safety implementation regulations "Government Agencies Managing Nuclear Power Plants": for a member country implementing a nuclear energy plan, it must establish a nuclear safety management agency and formulate relevant regulations before starting to build the first nuclear facility. Internationally, all countries with existing or planned nuclear energy have established nuclear safety supervision and management agencies and have formulated or plan to formulate nuclear safety regulations. This chapter introduces regulations related to nuclear safety, international organizations and institutions, and nuclear safety supervision and management.

7.1 Nuclear Safety Regulations

The development of nuclear energy is premised on nuclear safety (International Atomic Energy Agency 1991). To reduce risks to public and environmental safety, nuclear activities must be regulated by law, and safety supervision and oversight must be conducted through the issuance and implementation of safety regulations (International Atomic Energy Agency 2020). The nuclear safety legal system refers

L. Yu et al., *Nuclear Technology and Safety*, Nuclear Science and Technology,
https://doi.org/10.1007/978-981-95-2856-1_7

to the interconnected, complementary, and mutually restrictive unity composed of various legal norms that adjust the social relations arising from activities related to the field of nuclear safety.

Internationally, all of the more than 30 countries with existing or planned nuclear power have established nuclear safety supervision and management agencies and have formulated or plan to formulate nuclear safety regulations.

The nuclear safety legal systems of different countries have similarities and their own characteristics. For example, in terms of levels, the U.S. nuclear safety legal system is divided into five levels; Japan and South Korea are divided into four levels; Russia and Canada are divided into three levels; French nuclear safety-related laws are distributed in different legal documents, without a systematic level. From the perspective of basic laws in the nuclear field, the United States, Japan, Russia, and South Korea all have Atomic Energy Laws, which serve as the starting point for further detailed development and play a general role in the formulation of subsequent related laws and regulations; Canada also has an Atomic Energy Law, but it only targets the use and development of nuclear energy and is not the basic law in the nuclear field, while the Nuclear Safety and Control Act is its most important nuclear safety law; France currently does not have a comprehensive Atomic Energy or Nuclear Safety Law, etc. Despite the different legislative systems of various countries, there are some commonalities in nuclear safety legislation: nuclear safety legislation is part of national law and matches the country's legal system; the focus of legislation is how the state effectively reviews, supervises, inspects, penalizes violations, criminally handles violations, and takes emergency measures in the event of accidents for nuclear facilities and units using radioactive materials; the content of the law is referenced by countries around the world, covering all areas where radioactive harm may occur, with detailed provisions and good operability. The following is a brief introduction to this issue.

7.1.1 Overview of International Nuclear Safety Legislation

Currently, 36 countries and regions have enacted the "Atomic Energy Act" or "Radiation Protection Act". These include both developed nuclear industry countries such as the United States and Russia, as well as developing countries with lower levels of nuclear industry development such as Myanmar and Indonesia. In addition, at least nine countries have enacted a "Nuclear Safety Act", and 12 countries have issued decrees establishing Atomic Energy Commissions or Radiation Commissions. The nuclear safety legislation in the United States, France, Russia, and Japan will be briefly introduced below.

(1) United States

The nuclear safety laws and regulations in the United States are centered around the "Atomic Energy Act" signed by the President in 1954. This act established the regulatory framework for all subsequent nuclear facilities and is the only basic atomic energy law in the world that regulates nuclear energy. The law aims to ensure public safety and national defense security, providing overall principles and concepts, leaving detailed issues to the U.S. Nuclear Regulatory Commission (NRC). The NRC, based on the powers granted by the Atomic Energy Act and the Energy Reorganization Act, has formulated a series of regulations including license applications, supervision and management, nuclear facility site selection criteria, design criteria, radiation protection criteria, etc. There are 200 sections in the tenth chapter of the U.S. Federal Regulations applicable to the license approval of all nuclear facilities under the jurisdiction of the NRC. The operation and implementation procedures of the NRC are closely related to 23 laws such as the Administrative Procedure Act, the Clean Air Act, the Resource Conservation and Recovery Act, and the Toxic Substances Control Act. U.S. nuclear law also includes the "Nuclear Safety Research, Development, and Demonstration Act" of 1980 and the "Nuclear Waste Policy Act" of 1982. These laws encourage the development and research of nuclear materials (including fuel) and facilities, while also focusing on regulating activities in all areas of the atomic energy industry, making detailed provisions for organizational management, licensing systems, information management, nuclear damage liability, intellectual property, judicial procedures, import and export systems, and mining and supply, among others.

(2) Russia

The Russian nuclear safety legal and regulatory system mainly consists of the Russian Federation Constitution, federal laws, and government laws and regulations and technical regulations implemented by presidential decree. The "Russian Federation Atomic Energy Utilization Act" is the basic law for the use of atomic energy in Russia, with the legislative purpose of promoting the development of atomic energy science and technology and protecting public and environmental safety. Presidential decrees or federal government orders mainly provide detailed regulations for the use of nuclear energy. Presidential orders include the "Export Control Act for Nuclear Materials, Nuclear Devices, and Nuclear Technology", "Regulations for Nuclear Power Utilities", and "Management of Nuclear Enterprises in Privatization and Market Economy", among others. Government orders mainly include "Document Regulations for Exporting Nuclear Energy Devices, Materials, and Technology", "Measures to Protect the Safety of Residents Near Nuclear Installations", and the "Russian Federation Nuclear Energy Enterprise Restructuring Order", among others. There are also special technical regulations in the field of atomic energy utilization issued and implemented by the Russian nuclear safety regulatory authority, such as federal standards and regulations, administrative management regulations, guidance documents, and safety guidelines.

(3) Japan

Japanese laws related to nuclear energy and the application of nuclear technology consist of four parts: basic laws and regulations, regulations on nuclear raw materials, nuclear fuel materials and reactors, and prevention of radioactive hazards caused by radioactive isotopes, etc. For example, the basic law "Atomic Energy Basic Act" regulates all areas of nuclear facilities and activities; the "Regulation of Nuclear Raw Materials, Nuclear Fuel Materials, and Nuclear Facilities" provides a comprehensive licensing mechanism, licensing the enrichment of nuclear raw materials, the manufacture and use of nuclear fuel, the construction, operation, and retirement of nuclear facilities, the disposal of radioactive waste, and the international control of other nuclear materials. It also implements licensing management for the operation of nuclear-powered ships, as well as supervision and inspection of the construction, operation, and retirement of nuclear facilities; the "Radiation Protection Act" mainly regulates the use, sale, rental, and disposal of radioactive isotopes and devices that produce ionizing radiation, and has the power to suspend or revoke licenses for applications that do not comply with relevant legal provisions; the "Atomic Energy Damage Compensation Act" establishes the sole and unlimited liability of operators; the "High-Level Radioactive Waste Disposal Act" stipulates Japan's policy and final disposal plan for high-level waste treatment, the conditions and procedures for site selection, the entities responsible for implementation, and the funding mechanism for disposal activities, and so on.

(4) France

France has a well-developed nuclear industry, with nuclear power accounting for 80% of total electricity, and a comprehensive legal and regulatory system. Nuclear safety regulations are divided into mandatory and non-mandatory categories: mandatory regulations include bills passed by the parliament, decrees and ministerial orders, as well as regulatory decisions by the French Nuclear Safety Authority (ASN); non-mandatory regulations mainly include ASN's basic safety requirements and guidelines. Key nuclear-related laws include: in terms of radiation protection, the 1966 "Decree on the Revised European Atomic Energy Basic Standards for Radiation Protection", revised in 1988, now replaced by the 2002 "Decree on Comprehensive Protection of the Human Body from Ionizing Radiation"; in terms of nuclear facility control, the 1963 "Important Nuclear Facilities Decree", revised in 1973, 1990, and 1993, with implementing rules formulated in 1999; in terms of radioactive materials and waste control, the 1991 "Bataille Law" and "Nuclear Waste Management Research Law"; in terms of nuclear power and environmental protection, as well as nuclear damage liability, the 1960 "Paris Convention", the 1963 "Brussels Convention" and its two protocols in 1982. There are also the "Nuclear Civil Liability Law", "Nuclear Materials Protection and Control Law", "Information Transparency and Nuclear Safety Law", and "Sustainable Management of Radioactive Materials and Waste Law". The French nuclear safety system covers a wide range of areas, including radiation protection, nuclear facility supervision, radioactive material management, radioactive medicine, nuclear irradiation processing, nuclear trade exports, third-party nuclear liability, nuclear waste management, nuclear accident emergency response, etc. At the same time, France's environmental protection law, water pollution prevention law, air pollution prevention law, and public health law all contain provisions regulating the use of nuclear power.

(5) Canada

Canadian nuclear safety-related laws are mainly enacted and issued by the parliament, as well as regulations approved and issued by the Governor General, licenses approved by government departments, license documents, and regulatory documents issued by the Canadian Nuclear Safety Commission (CNSC). The former includes the "Nuclear Safety and Control Act", "Nuclear Damage Compensation Act", "Atomic Energy Act", "Radioactive Waste Act", "Environmental Impact Assessment Act", etc. The latter includes legal documents supporting the explanation of the "Nuclear Safety and Control Act", such as the "Comprehensive Regulations on Nuclear Safety and Control", etc. There are also guidance documents provided for license holders on how to meet the regulations set by the CNSC.

(6) South Korea

South Korean nuclear safety-related laws include the "Atomic Energy Act", "Protection and Radiation Emergency Act", "Nuclear Liability Act", "Nuclear Damage Compensation Act", "Amendment to the Nuclear Damage Compensation Act", and "Environmental Policy Framework Agreement", etc. The "Atomic Energy Act", as the basic law in the field of nuclear safety in South Korea, plays a general role in other atomic energy-related laws and provides a basic institutional framework for the management and supervision of nuclear energy utilization activities. The other five laws are authorized and issued by the president, detailing the "Atomic Energy Act", including the implementation steps, methods, and other necessary provisions of the "Atomic Energy Act".

7.1.2 Chinese Nuclear Safety Regulations

China has gradually promulgated nuclear safety regulations since 1986, based on extensive collection and careful study of the nuclear safety laws and regulations of advanced nuclear power countries, and referring to the basic rules, safety requirements, and safety guidelines of the International Atomic Energy Agency, thereby gradually establishing China's nuclear safety regulatory system (National Nuclear Safety Administration 2020). This includes more than 10 laws, several administrative regulations, more than 20 departmental rules, more than 70 related guidelines, more than 180 technical documents, comprehensive reports and nuclear safety translations, and 5 international conventions, etc. China's nuclear safety regulatory system can be divided into five levels: laws, administrative regulations, departmental rules, guidance documents, and standard norms, as shown in Fig. 7.1.

(1) Laws

The first level consists of laws approved by the Standing Committee of the National People's Congress and issued by the President of the State. Laws are used to adjust the development of nuclear energy and its safety as well as environmental protection, and are legally binding documents established by the state legislative body to establish a nuclear safety regulatory body and establish a basic regulatory system.

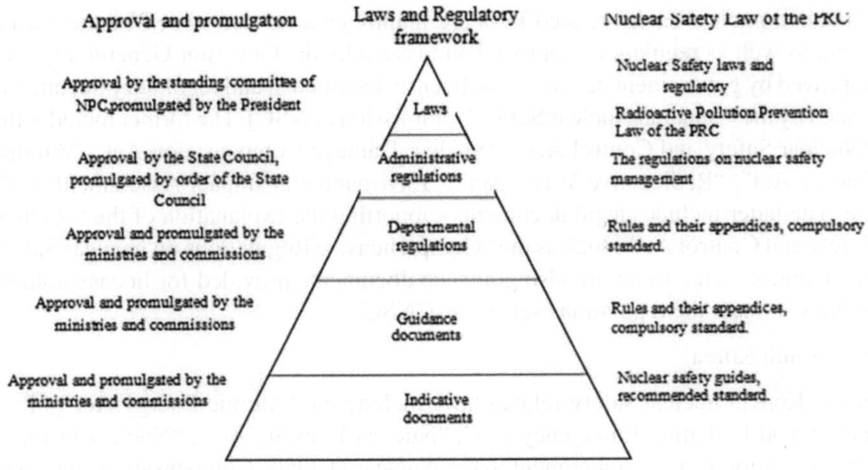

Fig. 7.1 Diagram of the levels of China's nuclear safety regulatory system

They are passed and issued by the Standing Committee of the National People's Congress. They lay the legal foundation for the regulatory body to formulate regulations, guidelines, and implement independent supervision. In some countries, they are also referred to as decrees, orders, regulations, etc. The existing national laws applicable to the field of nuclear safety mainly include the following:

1) "Nuclear Safety Law of the People's Republic of China".

The "Nuclear Safety Law of the People's Republic of China" was passed by the 29th session of the Standing Committee of the 12th National People's Congress on September 1, 2017, and was promulgated by Order No. 73 of the President of the People's Republic of China. The "Nuclear Safety Law of the People's Republic of China" consists of 8 chapters and 94 articles, including general provisions, nuclear facility safety, nuclear material and radioactive waste safety, nuclear accident emergency, information disclosure and public participation, supervision and inspection, legal responsibility, and supplementary provisions (National People's Congress of China 2017). This law is necessary to improve the nuclear safety legal system, is beneficial to enhance the public's understanding and confidence in nuclear safety, strengthens nuclear safety supervision work, and promotes China's international cooperation in nuclear safety (Lu et al. 2018).

The "Nuclear Safety Law of the People's Republic of China" incorporates the concept of " rational, coordinated and balanced " into law, establishing it as a guiding principle. "Rational" aims to highlight the dialectical relationship between the safety and development of the nuclear industry. Currently, China's energy structure is in a transition period, with the proportion of nuclear energy utilization gradually increasing. The demand for nuclear energy is urgent for economic development. China's nuclear power construction accounts for about 40% of the world's nuclear power construction. Ensuring nuclear safety is China's unshirkable development responsibility and historical mission. However, the development of the nuclear industry must be based on ensuring safety. Any nuclear development that sacrifices safety is unsustainable and not true development. "Coordinated" aims to clarify the relationship between national autonomy and international cooperation in the field of nuclear safety. The primary responsibility for nuclear safety should be borne by governments, which need to strengthen their awareness of nuclear safety, enhance mechanism construction, improve technical levels, adopt the most suitable nuclear safety policies and measures according to their national conditions, and fulfill relevant international obligations. Nuclear safety is also a global issue, and more countries should be encouraged to join the international nuclear safety process, strengthen exchanges, mutual learning, and sharing. Multilateral mechanisms and initiatives should be coordinated and jointly pursued. The international nuclear safety process should be promoted based on respect for the rights and interests of all countries, and universal nuclear safety should be sought through mutual benefit and win–win cooperation. "Balanced" aims to clarify the equal emphasis on four aspects: development and safety, rights and obligations, autonomy and cooperation, and addressing both symptoms and root causes. The efforts made by the international community in the field of nuclear safety, including capacity building, international cooperation, and the formulation and implementation of international norms, are more of a responsive effort. However, attention should be paid to the fundamental factors causing these problems and their solutions, improve nuclear safety policy measures, adhere to the balance of nuclear material supply and demand, deepen international cooperation in combating nuclear terrorism, create a peaceful and stable international environment, eliminate or minimize threats and challenges from the field of nuclear safety, and achieve lasting safety and development of nuclear energy.

2) Other Laws

The "National Security Law of the People's Republic of China". With the development of China's nuclear energy and nuclear technology, nuclear safety has received more widespread attention and has become an important part of the national security strategy. Article 31 of the "National Security Law of the People's Republic of China" promulgated in July 2015 clearly states: The state insists on the peaceful use of nuclear energy and nuclear technology, strengthens international cooperation, prevents nuclear proliferation and improves non-proliferation mechanisms; strengthens the safety management, supervision and protection of nuclear facilities, nuclear materials, nuclear activities and nuclear waste disposal; strengthens the nuclear accident emergency system and emergency capacity building, and prevents, controls and eliminates the harm of nuclear accidents to citizens' lives and health and the ecological environment.

The "Radioactive Pollution Prevention Law of the People's Republic of China", passed by the third session of the Standing Committee of the Tenth National People's Congress on June 28, 2003, and promulgated by Order No. 6 of the President of the

People's Republic of China, is a special law formulated to prevent radioactive pollution, protect the environment, safeguard human health, and promote the development and peaceful use of nuclear energy and nuclear technology.

The "Environmental Protection Law of the People's Republic of China", passed by the eleventh session of the Standing Committee of the Seventh National People's Congress on December 26, 1989, and revised by the eighth session of the Standing Committee of the Twelfth National People's Congress on April 24, 2014, is formulated to protect and improve the environment, prevent pollution and other public hazards, safeguard public health, promote the construction of ecological civilization, and promote sustainable economic and social development.

(2) Regulations

The second level consists of government regulations or ordinances issued by the State Council. Regulations generally refer to documents with administrative legal binding force formulated by the government or authorized by the government, and nuclear safety regulatory agencies to establish a licensing approval system and establish basic requirements for nuclear safety. Some countries also list licenses and license conditions as regulatory documents. The regulations on nuclear safety management stipulate major issues such as the scope of management, management agencies and their powers, supervision and management principles and procedures, are issued by the State Council, belong to administrative regulations, and are documents with legal binding force. For example, the "Implementation Rules for Nuclear Safety Management Regulations" are rules that specify specific implementation methods for a certain aspect according to the Nuclear Safety Management Regulations, are issued by the National Nuclear Safety Administration, belong to departmental rules, and are documents with legal binding force. The "Nuclear Safety Rules" are rules that stipulate nuclear safety goals and basic safety requirements, are approved and issued by the State Council or the National Nuclear Safety Administration, also belong to departmental rules, and are documents with legal binding force.

In summary, the administrative regulations issued by the State Council are rules that stipulate major issues such as the scope of management, management agencies and their powers, supervision and management principles and procedures, have legal binding force, and are issued in the form of management regulations.

The current nuclear safety management regulations in China mainly include the following:

The "Regulations on the Safety Supervision and Management of Civil Nuclear Facilities of the People's Republic of China" (issued by the State Council on October 29, 1986) are regulations formulated to ensure safety in the construction and operation of civil nuclear facilities, protect the health of workers and the public, protect the environment, and promote the smooth development of the nuclear industry.

The "Regulations on the Control of Nuclear Materials of the People's Republic of China" (June 15, 1987, issued by the State Council) were established to ensure the safe and legal use of nuclear materials, prevent theft, destruction, loss, illegal transfer, and illegal use, protect the safety of the nation and the public, and promote the development of nuclear energy.

"The Regulations on Emergency Management of Nuclear Accidents in Nuclear Power Plants" (August 4, 1993, issued by the State Council, revised on January 8, 2011) were established to strengthen the emergency management of nuclear accidents in nuclear power plants and control and reduce the harm of nuclear accidents.

"The Regulations on the Safety and Protection of Radioactive Isotopes and Radiation Devices" (August 31, 2005, issued by the State Council, revised in March 2019) were established to strengthen the supervision and management of the safety and protection of radioactive isotopes and radiation devices, ensure the health and safety of radiation workers and the public, protect the environment, and promote the application and development of radioactive isotopes and radiation technology. Related regulations include "Regulations on the Safe Transportation of Radioactive Items" and "Regulations on the Safe Management of Radioactive Waste", etc.

(3) Departmental Regulations

The third level consists of regulations formulated by departments under the State Council. Departmental regulations mainly include implementation details and safety regulations. The implementation details are regulations that specify specific implementation methods based on nuclear safety management regulations, issued by relevant national departments, and have legal binding force. Nuclear safety regulations are divided into 8 series according to their technical fields, and their standard format is "HAF xxx/yy/zz", where: HAF is the abbreviation of "nuclear safety regulations" in Chinese Pinyin; the first digit of "xxx" is the code for each series, and the second and third digits are the sequence number; "yy/zz" is the code for the corresponding implementation details and their attachments of nuclear safety regulations or provisions.

The arrangement of each series of nuclear safety regulations is as follows:

HAF 0xx/yy/zz—General series;

HAF 1xx/yy/zz—Nuclear power plant series;

HAF 2xx/yy/zz—Research reactor series;

HAF 3xx/yy/zz—Nuclear fuel cycle facility series;

HAF 4xx/yy/zz—Radioactive waste management series;

HAF 5xx/yy/zz—Nuclear material control series;

HAF 6xx/yy/zz—Civil nuclear pressure equipment supervision and management series;

HAF 7xx/yy/zz—Radioactive material transportation management series.

Existing implementation details include "Regulations on the Safety Supervision and Management of Civil Nuclear Facilities of the People's Republic of China" (HAF001), "Detailed Implementation Rules of the Regulations on the Control of Nuclear Materials of the People's Republic of China", "Regulations on Emergency Management of Nuclear Accidents in Nuclear Power Plants", etc.

(4) Guiding Documents

The fourth level consists of guiding documents in the field of nuclear safety. Guiding documents mainly explain or supplement nuclear safety regulations and recommend methods and procedures for implementing safety regulations. Nuclear safety guidelines are recommended, and the methods and procedures that should be adopted in implementing nuclear safety technical requirements administrative management regulations can be adopted, or equivalent alternative methods and procedures can be adopted. Safety guidelines are formulated by nuclear safety regulatory agencies, but they do not have legal binding force. They are guiding documents that suggest how to meet basic requirements and principles based on practical experience. Nuclear safety guidelines are guiding documents that explain or supplement nuclear safety regulations or recommended methods and procedures. When adopting other methods and procedures without following the guidelines, it is necessary to demonstrate their safety to the National Nuclear Safety Administration.

(5) Reference Documents

The fifth level consists of reference documents in the field of nuclear safety. Reference documents are mainly technical documents of nuclear safety regulations, which show the views of nuclear safety authorities on specific technical or administrative management issues, and are referred to in application. They include technical documents such as national standards, industry standards, and international standards related to nuclear safety.

7.2 International Organizations Related to Nuclear Safety

The international organizations or institutions related to nuclear safety supervision in the world mainly include IAEA and OECD/NEA. Although these international organizations do not have mandatory constraints on the nuclear safety supervision of various countries, they have played a certain role in promoting the peaceful use of nuclear energy in the world. At the same time, countries with nuclear energy have concluded a series of international conventions through international cooperation and other means to ensure the safe use of nuclear energy worldwide.

7.2.1 International Atomic Energy Agency

In 1953, U.S. President Eisenhower delivered a speech titled "Peaceful Use of Atomic Energy" at the United Nations General Assembly, advocating the establishment of the International Atomic Energy Agency (IAEA), a proposal that was unanimously approved by the 81 participating countries.

After four years of effort by countries around the world, the International Atomic Energy Agency (IAEA) was officially established on July 29, 1957, with its headquarters in Vienna, Austria. The IAEA is not a specialized agency of the United Nations (it is the specialized agency for nuclear science and technology within the UN system), but an independent intergovernmental organization within the UN system.

This agency has signed agreements with the United Nations and has direct links with the General Assembly, the Economic and Social Council, and the Security Council. This is mainly manifested in respecting the United Nations Charter, implementing General Assembly resolutions, submitting annual reports to the General Assembly, and reporting to the Security Council when necessary. The agency is also the only intergovernmental international organization promoting the peaceful use of nuclear energy.

The IAEA Statute stipulates that the agency has two main objectives: "to seek to accelerate and enlarge the contribution of atomic energy to peace, health, and prosperity throughout the world" and "to ensure, so far as it is able, that assistance provided by it or at its request or under its supervision or control is not used in such a way as to further any military purpose." Accordingly, the IAEA's activities are divided into two parts: promoting the peaceful use of nuclear energy and implementing safeguards.

Any country, whether a member of the United Nations or a specialized agency of the United Nations, can become a member of the agency by depositing an instrument of acceptance of the IAEA Statute after being recommended by the Board of Governors and approved by the General Conference.

The organizational structure of the IAEA includes: the General Conference (consisting of all member states), the Board of Governors, and the Secretariat, as shown in Fig. 7.2.

The General Conference and the Board of Governors are the decision-making bodies of the IAEA. They jointly formulate the IAEA's work plan, execute the budget, and appoint and dismiss the Director General of the IAEA. The Secretariat, led by the Director General, is the executive body of the IAEA. The Secretariat is responsible

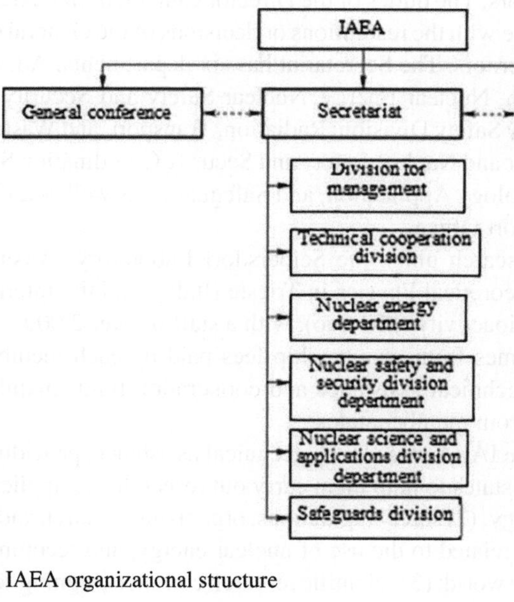

Fig. 7.2 IAEA organizational structure

for organizing the completion of reports on important issues formed by the IAEA, and the Director General reports to the Board of Governors and the General Conference, and forms resolutions on important issues. These resolutions are binding on the Secretariat and all member states.

The General Conference is held once a year, usually in September, for a week. Special meetings are held when necessary. The General Conference also has a Plenary Committee and a General Affairs Committee, whose main functions include: electing members of the Board of Governors, approving applications for membership, reviewing annual reports, budgets and final accounts proposed by the Board of Governors, submitting reports to the United Nations, approving the appointment of the Director General, and making decisions on matters submitted to the General Conference by the Board of Governors.

The Board of Governors consists of 35 member states, 13 of which are designated members and 22 are elected members. The designated members are the ten most advanced countries in the world in nuclear technology (including raw material production) and the three most advanced countries in the relevant regions, with a term of one year. However, in fact, except for the designated members from Western Europe (excluding the UK, Germany, and France) and Latin America, who take turns, the other designated members are permanent, because these countries are designated as members every year. China, the UK, France, Russia, and the US are all designated members. Elected members are elected by the General Conference based on the principle of regional balance, with half of them being re-elected each year for a term of two years.

The Secretariat is the daily executive office, with the Director General as the administrative head, under the jurisdiction of the Board of Governors. The Director General is appointed by the Board of Governors and approved by the General Conference, with a term of four years. The duties of the Director General of the IAEA are to perform duties in accordance with the resolutions or decisions of the General Conference and the Board of Governors. The Secretariat has six departments: Administration, Technical Cooperation, Nuclear Energy, Nuclear Safety and Security (which includes the Nuclear Facility Safety Division; Radiation, Transport, and Waste Safety Division; Emergency Center and Nuclear Safety and Security Coordination Section), Nuclear Science and Technology Application, and Safeguards, as well as an Internal Audit and Evaluation Support Office.

The IAEA has three research units: the Seibersdorf Laboratory (Austria), the International Center for Theoretical Physics in Trieste (Italy), and the International Laboratory for Marine Radioactivity (Monaco), with a staff of over 2,000.

The IAEA's funding comes from membership fees paid by each member state according to their share. Technical assistance and cooperation funds mainly come from voluntary donations from member states.

The main activities of the IAEA include: (1) technical assistance, providing technical assistance to member states to help them carry out research and application of peaceful use of nuclear energy; (2) safety regulations, organizing research and formulation of safety regulations related to the use of nuclear energy, and recommending them to countries around the world; (3) scientific research contracts, signing scientific research contracts with relevant member states or specialized international organizations; (4) scientific and technological conferences and information exchange, convening various scientific and technological conferences; (5) organizing information exchange on the peaceful use of atomic energy through the establishment of information networks, libraries, and publication of books and periodicals; (6) safeguards, signing "safeguards agreements" with relevant countries and international organizations, implementing safeguards on the following projects to ensure that these projects are not used for any military purpose: technical assistance projects provided by the agency itself or introduced by it, and projects supervised by member states or other international organizations and entrusted with non-proliferation obligations.

Furthermore, the IAEA has established the International Nuclear Safety Advisory Group (INSAG), founded in 1985. The advisory group consists of authoritative experts in the field of nuclear energy, recommended by countries worldwide. Its primary function is to exchange information on significant nuclear safety issues at the international level and provide nuclear safety advice to the IAEA and the Director-General. INSAG periodically writes reports on important international nuclear safety topics, which are published by the IAEA. For example, after the 1986 Chernobyl nuclear accident, the group organized post-accident evaluation expert meetings and accident cause analysis reports; in 1988, it organized the drafting of the Basic Safety Principles for Nuclear Power Plants; in 1991, it organized the writing of the Safety Culture report; in 1999, it organized the drafting of the Principles for the Safety Management of All Radiation Sources. As of 2016, 26 INSAG reports have been published.

The IAEA's nuclear safety work can be divided into the formulation of safety standards, international peer reviews, knowledge management and capacity building, and the organization of technical conferences.

Formulating and promoting safety standards is an important task of the IAEA. The IAEA's safety standards are divided into three levels vertically, from high to low: Fundamental Safety Principles, Safety Requirements, and Safety Guides. The Fundamental Safety Principles are programmatic documents, Safety Requirements are standards that must be met to ensure safety, and Safety Guides are recommended. The IAEA's safety standards are not mandatory for member states, but many countries' safety regulatory authorities have adopted the IAEA's safety standards, making them legally binding in China.

In terms of peer review, the IAEA organizes various international expert peer reviews, conducts expert assessments in a certain field of a member state, and provides constructive opinions. The review opinions are not mandatory but have a significant influence. Currently, all IAEA peer reviews are based on the voluntary principle of member states. After the Fukushima nuclear accident, some countries suggested that member states must regularly accept comprehensive regulatory review services.

In terms of knowledge management and capacity building, the IAEA, as the only intergovernmental organization in the international nuclear field, has accumulated a large number of databases, special reports, training materials, and conference records over the years, forming a huge knowledge base.

In terms of organizing technical conferences, the IAEA is a platform for discussing various technical issues and maintains close contact with high-level experts in various fields worldwide. The important technical conferences organized have a significant influence globally. For example, after the Fukushima nuclear accident, the series of expert meetings discussing accident lessons learned organized by the IAEA under the framework of the Nuclear Safety Action Plan has become the most influential platform for discussing the Fukushima nuclear accident.

In September 1983, the Chinese government officially applied to join the IAEA and officially became a member of the IAEA from January 1, 1984. Joining the IAEA has given China more opportunities to understand the development trends of the world's nuclear industry and carry out cooperation and exchanges with advanced international nuclear technology countries. At the same time, it also provides convenience for the IAEA and countries worldwide to understand the development of China's nuclear industry. As a member state, China also has more opportunities to participate in the drafting and revision of international atomic energy laws, representing the country's interests with other developing countries, and promoting the balanced development of nuclear energy peaceful use and security mechanisms, coordinating various positions, and playing a role in world peace and development.

Since the establishment of the IAEA, under the framework of international atomic energy law, it has done a lot of work in the establishment of security mechanisms and the field of peaceful use of nuclear energy. It has presided over and formulated 11 international treaties related to radiation safety, radioactive waste management, and nuclear damage liability systems; based on this, the organization has formulated a large number of nuclear safety guidelines and standards and actively promoted them internationally; at the same time, the IAEA has also played an excellent coordinating role in promoting international nuclear safety supervision and scientific and technological cooperation in the peaceful use of nuclear energy. For example, the "Convention on Early Notification of a Nuclear Accident," "Convention on Assistance in the Case of a Nuclear Accident or Radiological Emergency," and "Convention on the Physical Protection of Nuclear Material," which came into effect in February 1987, were approved by China in September 1987 and January 1989, respectively; and the "Convention on Nuclear Safety," which came into effect on October 24, 1996, was officially submitted to the Director-General of the International Atomic Energy Agency, Blix, by the Chinese Permanent Representative to the International Atomic Energy Agency in April 1996, making China the 18th country to submit the ratification.

7.2.2 *Organization for Economic Cooperation and Development and Nuclear Energy Agency (OECD/NEA)*

(1) Organization for Economic Cooperation and Development (OECD)

The Organization for Economic Cooperation and Development, abbreviated as OECD, is an intergovernmental international economic organization engaged in economic and social development policy research. There are currently more than 70 non-member countries and regions. According to statistics, OECD countries account for 80% of the world's nuclear power generation.

The predecessor of the Organization for Economic Cooperation and Development (OECD) was the European Organization for Economic Cooperation, established in April 1948. The OECD was officially established in Paris, France, in 1961. Its primary focus has always been on strengthening policy coordination among its member countries. In the 1990s, the organization changed its traditional working methods and began actively developing relationships with non-member countries, especially some "dynamic non-member economies," with the aim of playing a larger role in global economic development through dialogue and cooperation. Among non-member countries and regions, China, Russia, Brazil, India, and Indonesia, known as the "Big Five," are the main targets for OECD dialogue and cooperation. The OECD has over 200 specialized committees and working groups; its work involves economic and social issues, including macroeconomics, trade, education, development, science, and innovation. It is primarily known for its publications and statistics. China is not a member of the OECD. China formally established a policy dialogue and cooperation relationship with the OECD in July 1995.

(2) Nuclear Energy Agency (NEA)

The Nuclear Energy Agency (NEA), formerly the European Nuclear Energy Agency (ENEA), was established in 1958. Initially, only European countries participated, but later five of the six OECD member countries outside Europe were allowed to participate, and South Korea was allowed to join in May 1993. This marked the NEA's transformation into a global organization promoting nuclear cooperation and development, with its headquarters in Paris. The main functions of the NEA are to develop international research and development plans, exchange scientific and technical experiences and information, and promote the peaceful use of nuclear energy; in cooperation with the IAEA, it continuously inspects the economic and technical issues of world uranium resources, production, demand, and nuclear fuel cycle.

The NEA has seven specialized committees (See Fig. 7.3): the Committee on the Safety of Nuclear Installations, the Committee on Nuclear Regulatory Activities, the Radioactive Waste Management Committee, the Committee on Radiation Protection and Public Health, the Nuclear Science Committee, the Committee on the Economics of the Development and the Fuel Cycle of Nuclear Energy, and the Nuclear Law Committee.

The Committee on the Safety of Nuclear Installations assists member countries in better assessing the safety of nuclear reactors and nuclear fuel cycle facilities. The committee is composed of representatives from each country's nuclear safety regulatory department or technical research and development department, or senior scientists and engineers responsible for safety technology and research projects.

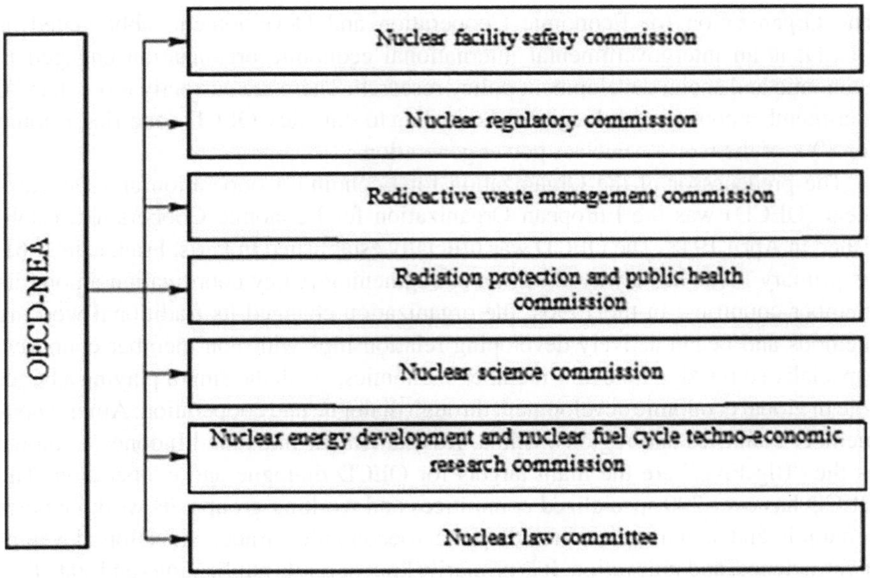

Fig. 7.3 Organization chart of OECD-NEA

The Committee on Nuclear Regulatory Activities is responsible for guiding the NEA's activities in the field of nuclear facility safety regulations, licensing, and supervision. It is composed of senior representatives from member countries' nuclear safety regulatory agencies, with specific tasks including exchanging information and experiences of regulatory agencies, assessing the latest developments that may affect nuclear safety regulation, and evaluating current practices and operational experiences.

China is currently not a member of the NEA. Current cooperation mainly involves participation in the activities of the Committee on the Safety of Nuclear Installations, the Committee on Nuclear Regulatory Activities, and the Multinational Design Evaluation Program.

7.2.3 Other National Nuclear Organizations

In addition to the important international nuclear organizations mentioned above, there are some other international nuclear organizations, including the World Nuclear Association (WNA), the International Radiation Protection Association (IRPA), the International Radiation Research Association (IRRA), the International Commission on Radiological Protection (ICRP), the United Nations Scientific Committee on the Effects of Atomic Radiation (UNSCEAR), the International Nuclear Transport Organization, the World Health Organization (WHO), the World Nuclear University, the Pacific Nuclear Council, etc. In addition, the world's major nuclear countries have also established relevant nuclear management agencies.

(1) United States

The main ones are the U.S. Department of Energy (DOE), the American Nuclear Society (ANS), the U.S. Nuclear Regulatory Commission (NRC), and the Nuclear Energy Institute (NEI). The U.S. Nuclear Regulatory Commission (NRC), also known as the Nuclear Management Commission, is the U.S. nuclear safety management authority. Its predecessor was the U.S. Atomic Energy Commission (USAEC), established in 1946 and renamed the Nuclear Management Commission in 1975. Its function is to independently supervise and manage civilian nuclear facilities and nuclear materials across the U.S. The commission has one chairman and four commissioners, each serving a five-year term, appointed by the president. It currently has more than 2,900 employees. The main functional departments include: the Office of Nuclear Reactor Regulation (NRR), the Office of Nuclear Material Safety and Safeguards (NMSS), the Office of Nuclear Regulatory Research (RES), the Advisory Committee on Reactor Safeguards (10 members), the Advisory Committee on Nuclear Waste, and the Advisory Committee on Nuclear Safety and Licensing Applications.

The Nuclear Energy Institute (NEI) is a research institution on nuclear energy policy, headquartered in Washington, D.C., and often provides various information on U.S. nuclear energy policy.

(2) Japan

Under the Cabinet Office, there is the Atomic Energy Commission, with one chairman and four commissioners; and the Nuclear Safety Commission, with five commissioners. The main responsibilities are to plan, review, and decide on nuclear safety regulatory policies, develop safety review guidelines, license reviews, safety regulations for nuclear fuel and reactors, hazard prevention, investigations into hazards caused by radioactive materials, etc., and directly report to and advise the Prime Minister and administrative chiefs.

The Ministry of Economy, Trade and Industry oversees the Nuclear and Industrial Safety Agency (NISA), which employs over 600 people. Its primary responsibilities include the regulation of energy facilities and their industrial activities for nuclear and industrial safety, preventing accidents in various fields including nuclear energy, electricity, civil gas, heating, explosives, liquefied petroleum gas, and mining.

Additionally, the Ministry of Education, Culture, Sports, Science and Technology (MEXT) is responsible for nuclear safety regulation pertaining to other types of R&D reactors, experimental research reactors, and radiation hazards caused by radionuclides.

(3) France

In France, the main agency is the French Atomic Energy and Alternative Energies Commission (CEA), which employs over 16,000 people. Formerly known as the French Atomic Energy Commission, it was renamed in 2009. It represents the

French government in international nuclear affairs with organizations such as the International Atomic Energy Agency, leads the international development of French nuclear power companies, and actively conducts international scientific research cooperation. It is also responsible for managing various nuclear research and activities in France, publishing various news and information about its affiliated research centers, laboratories, and nuclear industrial activities.

(4) Canada

In Canada, the main agency is the Canadian Nuclear Safety Commission (CNSC), which is responsible for nuclear safety supervision and management. It is a committee composed of seven members. Under it are the advisory organization—the Radiation Protection and Reactor Safety Advisory Committee, and the executive agencies—the Reactor Regulation Bureau (DRR) and the Nuclear Fuel Cycle-Nuclear Materials Regulation Bureau (DFCMR). The main function of the CNSC is to be accountable to Parliament, manage various nuclear-related activities through regulations, focus on safety issues related to nuclear energy applications, impacts on public health, environmental protection, etc., and issue licenses for various nuclear activities.

7.3 Nuclear Safety Supervision and Management System

Regarding the provisions of the national nuclear safety regulatory system and management system, Article 6 of the "Nuclear Safety Law of the People's Republic of China" states: The State Council's nuclear safety supervision and management department is responsible for nuclear safety supervision and management. The State Council's nuclear industry department, energy department, and other relevant departments are responsible for relevant nuclear safety management work within their respective responsibilities. The state establishes a nuclear safety work coordination mechanism to coordinate relevant departments to advance related work.

The nuclear safety supervision and management system mainly refers to the distribution of responsibilities for nuclear safety supervision and management. International laws in the field of nuclear safety emphasize independent regulation. The "Nuclear Safety Convention", "Joint Convention on the Safety of Spent Fuel Management and on the Safety of Radioactive Waste Management", and the International Atomic Energy Agency's "Basic Safety Principles" that China has joined all emphasize that the national nuclear safety regulatory system is the cornerstone of the institutional framework for the development of national nuclear energy and nuclear technology utilization. For example, Article 8 of the "Nuclear Safety Convention" stipulates that (1) each contracting party should establish or designate a regulatory body, entrust it to implement the legislative and supervisory management framework described in Article 7, and give it the appropriate powers, functions, and financial and human resources needed to fulfill its responsibilities. (2) Each contracting party should take appropriate steps to ensure that the functions of the regulatory body are effectively separated from the functions of any other institution or organization involved in promoting or utilizing nuclear energy. Article 20 of the "Joint Convention on the Safety of Spent Fuel Management and on the Safety of Radioactive Waste Management" also clearly stipulates that each contracting party should establish or designate a regulatory body, entrust it to implement the legislative and supervisory framework mentioned in Article 19, and grant it sufficient powers, functions, and financial and human resources to fulfill its responsibilities.

China's nuclear safety system has undergone several reforms to form the current pattern, and China's nuclear safety management system has its own characteristics, which are the result of years of experience accumulation. According to the long-standing and effective division of responsibilities of the State Council departments, the law confirms that the State Council's nuclear safety supervision and management department is responsible for nuclear safety supervision and management. The Nuclear Safety Law mainly regulates the management of civilian nuclear safety. Other relevant departments of the State Council are responsible for relevant nuclear safety management work within their respective responsibilities.

Nuclear safety-related activities not only involve the State Council's nuclear safety supervision and management department, but also involve public security, transportation, safety production and other relevant departments. These departments exercise relevant management functions in safety protection, transportation, safety production, etc. for nuclear energy development and utilization activities. In addition, local people's governments also have an obligation to take certain measures to ensure nuclear safety within their jurisdiction.

7.3.1 Main Nuclear Regulatory Agencies Abroad

Overseas nuclear safety regulatory agencies have many years of regulatory experience, most of which are directly led by the government or national leaders, and have strong independence. Independence is precisely an important direction in the development of nuclear safety regulatory agencies. Why establish a relatively independent nuclear safety supervision and management system? For a normal person, regardless of their sense of responsibility, their initial intention of work will not want to put safety in an uncontrolled state. However, they will be affected by their living environment, and there will inevitably be some factors that affect their investment in safety, such as companies may excessively reduce costs to expand market share, or relax strict requirements for product quality due to excessive demands for construction period and capital turnover period. Technicians may reduce the experimental inspection of products due to heavy workloads. The occurrence of nuclear accidents is often a series of unfortunate combinations, and any negligence can put safety in an uncontrolled state. The first duty of safety supervision is to exclude unqualified units and individuals from nuclear-related activities, and constantly remind practitioners that neglecting safety will lead to the loss of everything in the end, which is more important than their awakening after an accident; another duty is to remind practitioners to pay attention to attempted events, establish normative order, pay attention to early investment in safety, rather than "make up for the loss". Therefore, in order to strengthen the independence of supervision, countries such as the United States and Japan have experienced the continuous division and reorganization of regulatory agencies, from agencies with dual functions of supervision and promotion of development, to independent agencies for supervision and promotion; France has strengthened the independence of regulatory agencies and introduced nuclear safety and transparency bills to establish independent regulatory agencies; Russian regulatory agencies are directly under the jurisdiction of the federal government and are not subject to the leadership of other departments; in order to ensure its independence, the Canadian nuclear regulatory agency has also been granted judicial power beyond its regulatory duties; South Korea has reorganized the regulatory agency under the Ministry of Science and Technology to be directly led by the president in order to strengthen the independence of the regulatory agency.

(1) United States

The U.S. nuclear safety regulatory agency originated from the U.S. Atomic Energy Commission (AEC) established in 1946. Its main responsibility is to promote the peaceful use of nuclear energy in science and technology, without regulatory responsibility. The Atomic Energy Act Amendment passed by the U.S. Congress in 1954 gave the AEC the power of nuclear safety regulation. It can be seen that the AEC has to implement strict control, protect public and environmental safety, and provide incentive conditions to promote the development of the nuclear energy industry. This responsibility of both managing production and being responsible for supervision makes the AEC both a "referee" and an "athlete".

In the process of nuclear safety regulation, the AEC, in order not to hinder the development of the nuclear industry, did not interfere too much with the management of nuclear facilities, which caused a lot of controversy in the United States. In the 1960s, the AEC was increasingly accused of lax regulation in terms of radiation standards, reactor safety, power plant site selection, and environmental protection. Therefore, in 1974, Congress passed the Energy Reorganization Act, which reorganized the AEC's business. The U.S. Nuclear Management Commission (NRC) is responsible for nuclear safety regulation; the Energy Research and Development Administration (DOE) is responsible for promoting nuclear energy development and utilization.

The NRC, as an independent agency of the U.S. government, mainly exercises nuclear safety regulatory activities on civilian nuclear facilities and nuclear materials (including fuel) across the United States. It has the characteristics of separation of politics and supervision. Its purpose is to better protect the health of workers and the public and the environment, and ensure national security. The NRC's main regulatory scope includes commercial nuclear power plants, research reactors, experimental reactors, nuclear fuel cycle facilities, medical and industrial radioactive isotopes, and the treatment, disposal, and transportation of radioactive waste.

The NRC has two permanent institutions, the "Nuclear Safety Advisory Committee" and the "Nuclear Waste Advisory Committee". The former is responsible for reviewing and supervising nuclear power plants and related nuclear facility activities, and providing advisory opinions on important nuclear safety issues and related research projects. The latter is responsible for providing advisory opinions on nuclear waste storage, disposal facilities, and related safety issues. In addition, the NRC also has a permanent institution, the "Nuclear Safety and Licensing Review Group", which is mainly responsible for various hearing activities in the NRC's nuclear safety and nuclear waste safety supervision and management activities.

(2) Russia

Russia's nuclear safety regulation is carried out by the Federal Environmental, Industrial and Nuclear Safety Supervision Bureau (referred to as the Russian Supervision Bureau), which was established in 2004 and is under the Russian Federal Government. The Russian Supervision Bureau is entrusted by the federal government to formulate and implement national policies and development plans, formulate laws and regulations in the field of nuclear and radiation safety supervision (in the aspect of civilian nuclear energy utilization), and implement comprehensive nuclear safety supervision on nuclear energy utilization.

The Russian Supervision Bureau consists of a headquarters, regional nuclear and radiation safety supervision and management departments, technical support units, and training centers. The headquarters has 7 functional departments, 2 nuclear safety supervision departments, and 6 industrial and energy supervision departments. The regional nuclear and radiation safety supervision and management department includes 6 local supervision and management bureaus. At the same time, 31 local industrial safety supervision bureaus also participate in regional nuclear and radiation safety supervision work. Russia has 2 technical support units and 1 analysis laboratory in the field of nuclear safety supervision.

(3) Japan

Before the Fukushima nuclear accident, Japan's nuclear regulatory activities were overseen by three departments: the Atomic Energy Safety Commission, the Nuclear Safety Commission, and the Nuclear and Industrial Safety Agency. The Atomic Energy Safety Commission and the Nuclear Safety Commission were under the Cabinet Office, with the former responsible for research, development, and utilization of nuclear energy, and the latter for matters related to nuclear safety.

The Nuclear and Industrial Safety Agency, under the Ministry of Economy, Trade, and Industry, was responsible for nuclear power licensing, nuclear power generation review, and nuclear accident prevention. However, as the Ministry's responsibilities also included energy supply and promoting the development of nuclear energy, the Agency, while ostensibly exercising independent regulation of nuclear safety, struggled to avoid the suspicion of having a dual role as both "referee" and "player".

After the Fukushima nuclear accident, in order to streamline the system and clarify responsibilities, the Japanese government restructured the nuclear regulatory agencies in 2012, merging the three departments into a single independent nuclear regulatory agency (NRA). The NRA, consisting of a Secretariat, Nuclear Regulation Department, Radiation Protection Department, local organizations, and cooperative agencies, is under the Ministry of the Environment and is specifically responsible for nuclear safety regulation.

(4) France

To strengthen nuclear safety regulation, in 2002 the French government merged the Nuclear Safety Bureau of the former Ministry of Industry and the Radiation Protection Department of the former Ministry of Health to establish the General Directorate for Nuclear Safety and Radiation Protection (DGSNR). The DGSNR's main responsibilities include formulating and implementing safety-related policies and measures in the nuclear field outside the military domain, and it has the power to represent the state in exercising nuclear and radiation protection safety supervision. The DGSNR is composed of direct units, regional supervision stations, the Institute for Radiation Protection and Nuclear Safety, and other technical support units.

In 2006, to further strengthen the authority and regulatory power of the nuclear safety regulatory agency, the French Nuclear Safety Authority (ASN) was established. The ASN has 11 regional supervision agencies and 8 management departments, responsible for the supervision of civilian nuclear facilities and activities in France. The 11 regional supervision agencies are located in cities such as Paris, Lyon, and Marseille, and the 8 management departments are the Nuclear Power Plant Department, Nuclear Pressure Equipment Department, Transport and Source Department, Waste and Fuel Cycle Facilities Department, Ionizing Radiation and Health Department, Environment and Emergency Department, International Relations Department, and Public Communication and Communication Department.

(5) Canada

Canada's nuclear safety regulation is overseen by the Canadian Nuclear Safety Commission (CNSC), established in 2000, with 6 departments, including Legal Services, Operational Regulation, Technical Support, Regulatory Services, and Collaborative Services. The CNSC has supervisory authority over nuclear safety affairs, with its main responsibilities including the development, production, and use of nuclear energy in Canada; the possession and use of radioactive materials; the management and control of nuclear devices and nuclear information; avoiding unreasonable risks to national defense security; conveying objective, scientific, technical, and regulatory information to the public, as well as information about the Commission's activities and effects, and fulfilling international commitments. In addition, the CNSC has judicial power, can summon and question witnesses, inspect products, review records, and enforce compliance when necessary. The Nuclear Safety Commission has quasi-legislative power, and any decision or order of the Commission can be regarded as a rule, order, or judgment of the Federal Court or the Provincial Superior Court, and has the same effect, and is binding on the Federal Court and the Provincial Superior Court.

(6) South Korea

In 1959, South Korea established the Atomic Energy Bureau to oversee nuclear safety regulation, under the Ministry of Education, Science, and Technology, with the

Radiation Safety Department, Nuclear Safety Emergency Department, and Nuclear Safety Department under its jurisdiction. To strengthen nuclear safety regulation, in 1997, South Korea established the National Nuclear Safety Commission, with the Nuclear Safety Special Committee and Special Investigation Committee under it, primarily responsible for national decision-making on major nuclear safety policies and affairs. In 2011, to strengthen the execution of nuclear safety regulation, the South Korean government integrated the Nuclear Safety Commission and the Atomic Energy Bureau to establish the Nuclear Safety and Security Commission (NSSC). The Commission reports directly to the President and covers nuclear safety regulation, nuclear security, and non-proliferation. The NSSC has a Nuclear Regulation Bureau and a Radiation Emergency Bureau, with the former having a Nuclear Safety Policy Division, Nuclear Safety Division, and License Inspection Division; the latter has a Radiation Safety Division, Environmental Radiation Safety Division, Nuclear Emergency Division, Nuclear Assurance and Export Control Division.

7.3.2 China's Nuclear Safety Regulatory Agency

The "Nuclear Safety Law of the People's Republic of China" stipulates that the State Council's nuclear safety regulatory department is responsible for the supervision and management of nuclear safety, and legally implements full life cycle, full process supervision of nuclear facilities and nuclear activities. According to the "Regulations on the Safety Supervision and Management of Civil Nuclear Facilities in the People's Republic of China" (HAF001), the nuclear safety management of mainland China's nuclear facilities is divided into three levels: the nuclear safety regulatory agency, the nuclear industry authority, and the nuclear facility operating unit. The following mainly introduces the first two levels.

The State Council's nuclear safety supervision and management department is responsible for the supervision and management of nuclear safety. The State Council's nuclear industry authority, energy authority, and other relevant departments are responsible for relevant nuclear safety management work within their respective responsibilities. The state establishes a nuclear safety work coordination mechanism to coordinate and promote related work among relevant departments.

(1) Nuclear Safety Regulatory Agency

1) National Nuclear Safety Administration

The National Nuclear Safety Administration (NNSA) was established in October 1984, originally under the jurisdiction of the National Science Commission, and is now under the Ministry of Ecology and Environment. As the State Council's nuclear safety regulatory department, the NNSA reviews and issues corresponding licenses for different stages of nuclear facilities, including site selection, design, construction, operation, and decommissioning. It ensures comprehensive control over the entire process of nuclear and radiation safety supervision. The NNSA's six regional supervision stations conduct regular and irregular nuclear and radiation safety supervision on the construction quality, equipment manufacturing quality, personnel qualifications, and operational safety of nuclear facilities within their jurisdiction, ensuring effective supervision of nuclear facilities.

The NNSA supervises the impact of nuclear facility operations on the environment through monitoring radioactive emissions from nuclear facilities and the surrounding environment. In the event of an accident, the NNSA also urges the responsible units (operators and carriers) of nuclear facilities and nuclear activities to effectively activate emergency mechanisms to control and mitigate the consequences, with the aim of protecting the environment and public health.

The NNSA carries out comprehensive safety supervision of nuclear facilities and nuclear activities, and does not exclude other relevant departments of the State Council from carrying out supervisory law enforcement work according to their respective responsibilities. Other relevant departments of the State Council, such as the public security department, health department, environmental protection department, transportation department, and safety production department, supervise and manage activities related to nuclear facilities, nuclear materials, and radioactive waste within their respective responsibilities, including environmental protection, industrial safety, occupational health, road transport safety, fire management, special equipment safety, etc.

The basic responsibilities of the NNSA can be summarized as follows:

a) Responsible for the supervision and management of nuclear safety and radiation safety. Formulate policies, plans, laws, administrative regulations, departmental rules, systems, standards, and norms related to nuclear safety, radiation safety, electromagnetic radiation, radiation environmental protection, and nuclear and radiation accident emergencies, and organize their implementation.
b) Responsible for the unified supervision and management of nuclear safety, radiation safety, and radiation environmental protection of nuclear facilities.
c) Responsible for the supervision and management of the licensing, design, manufacturing, installation, and non-destructive testing activities of nuclear safety equipment, and the safety inspection of imported nuclear safety equipment.
d) Responsible for the supervision and management of nuclear material control and physical protection.
e) Responsible for the supervision and management of radiation safety and radiation environmental protection of nuclear technology utilization projects, uranium (thorium) mines, and associated radioactive mines. Responsible for radiation protection work.
f) Responsible for the supervision and management of the safety and radiation environmental protection of radioactive waste treatment and disposal. Responsible for the supervision and inspection of radioactive pollution prevention and control.
g) Responsible for the supervision and management of the safety of radioactive material transportation.

h) Responsible for the supervision and management of electromagnetic radiation devices and electromagnetic radiation environment of power transmission and transformation facilities and lines, signal stations, etc.
i) Responsible for nuclear and radiation emergency response and investigation and handling. Participate in the prevention and handling of nuclear and radiation terror incidents.
j) Responsible for the qualification management of reactor operators, special process personnel of nuclear equipment, etc.
k) Organize radiation environmental monitoring and supervisory monitoring of nuclear facilities and key radiation sources.
l) Responsible for the fulfillment of international conventions related to nuclear and radiation safety.
m) Guide the related business work of nuclear and radiation safety supervision stations.

The NNSA has 3 regulatory departments, 6 nuclear power supervision stations, and 5 technical support units.

2) Ministry of Ecology and Environment

The Ministry of Ecology and Environment is under the State Council and implements unified supervision and management of national ecological and environmental protection work. Its main responsibilities are:

a) Responsible for the formulation and supervision of environmental management regulations and standards for nuclear power plants.
b) Responsible for the approval of environmental impact reports for nuclear power plants.
c) Responsible for radioactive environmental monitoring.
d) Responsible for the supervision and management of radioactive waste.
e) Participate in emergency response.

3) National Health Commission

The National Health Commission is under the State Council and implements unified supervision and management of national health work. Its main responsibilities are:

a) Responsible for the formulation of health regulations and standards related to the health of nuclear facility workers and the public.
b) Responsible for the supervision of radiation doses received by radiation workers and the public.
c) Responsible for the approval of evaluations of the impact of nuclear pollution on human health.
d) Responsible for the prevention and treatment of radiation injuries.

(2) Nuclear Industry Regulatory Department

From the perspective of the main departments responsible for nuclear energy development and utilization, there are primarily the State Council's departments related to the nuclear industry and energy, as well as other relevant departments. Nuclear power development planning, admission conditions, technical standards, nuclear power layout, major project approval, scientific research, and management of the nuclear industry may all have certain impacts on nuclear safety. For example, a too rapid development pace may put pressure on the capacity for nuclear safety supervision, which is not conducive to the safety of nuclear power plants. The safety of the introduced technology for nuclear power plants also directly affects the safety level of the nuclear power plants. The layout of nuclear power plants is related to the environment, population distribution, energy demand, etc., especially the site selection in the layout of nuclear power plants, which involves many safety issues such as nuclear accident emergency response. The scientific research related to the development and utilization of nuclear energy will also have an impact on China's nuclear safety level.

1) State Administration of ScienceTechnology and Industry for National Defense

The State Administration of ScienceTechnology and Industry for National Defense is under the Ministry of Industry and Information Technology. According to the current division of departmental responsibilities, the main responsibilities of the Administration in the supervision of nuclear energy development and utilization are to formulate policies and development plans for nuclear industry production and technology, implement management of the nuclear industry, review and issue nuclear material licenses, etc. Its specific related responsibilities include:

a) Management of affiliated nuclear facilities: Responsible for the safety management of affiliated nuclear facilities; ensuring necessary support is provided to the operating units of affiliated nuclear facilities; conducting supervision and inspection.
b) Formulation of nuclear safety regulations and standards: Participating in the drafting and formulation of relevant nuclear safety regulations; organizing the formulation of technical standards related to nuclear safety, and filing with the National Nuclear Safety Administration.
c) Emergency management: Organizing the formulation and implementation of on-site emergency plans for affiliated nuclear facilities; participating in the formulation and implementation of off-site emergency plans.
d) Training and assessment: Responsible for the technical training and assessment of various personnel in affiliated nuclear facilities.
e) Scientific research organization: Organizing scientific research work on nuclear safety in the development of nuclear energy.

2) China Atomic Energy Authority

The basic function of the China Atomic Energy Authority (CAEA) is to be responsible for the development of China's peaceful use of atomic energy, the formulation of relevant regulations, and the control of nuclear materials; representing the Chinese government in participating in the International Atomic Energy Agency and its activities. Its responsibilities include:

a) Responsible for intergovernmental exchanges and cooperation in the nuclear field and with international organizations, and leading the national nuclear accident emergency management work.
b) Responsible for researching and formulating policies and regulations for China's peaceful use of nuclear energy.
c) Responsible for researching and formulating development plans, plans, and industry standards for China's peaceful use of nuclear energy.
d) Responsible for the demonstration, approval, supervision, and coordination of the implementation of related projects for China's peaceful use of nuclear energy (excluding nuclear power).
e) Responsible for nuclear security and nuclear material control.
f) Responsible for nuclear import and export review and management.
g) Responsible for intergovernmental exchanges and cooperation in the nuclear field and with international organizations, representing the Chinese government in participating in the affairs of the International Atomic Energy Agency.
h) Undertaking the daily work of the National Nuclear Accident Emergency Management Office, responsible for researching and formulating national nuclear accident emergency plans and organizing their implementation.
i) Responsible for the management of nuclear facility decommissioning and radioactive waste.

3) National Energy Administration.

The main responsibilities of the National Energy Administration include: responsible for nuclear power management, formulating nuclear power development plans, admission conditions, technical standards and organizing their implementation, proposing nuclear power layout and major project review opinions, organizing, coordinating and guiding nuclear power scientific research work, organizing nuclear power plant nuclear accident emergency management work.

4) Other relevant departments

In addition to the above-mentioned main departments, there are also relevant departments such as the public security department and the transportation department. The former has relevant management responsibilities in the safety protection of nuclear facilities, nuclear materials, and the use of nuclear technology, nuclear accident emergency response, and the safety of radioactive material transportation. The latter has relevant management responsibilities in the safety of radioactive material transportation. In addition, there is the safety production department, which has supervisory and management responsibilities for general industrial safety in the development and utilization of nuclear energy in accordance with the provisions of the Safety Production Law, such as supervising the safety of special equipment in nuclear power plants.

According to the principle of "managing the industry must manage safety", the above-mentioned government departments must make nuclear safety the primary goal in the activities of formulating industrial policies, industrial development plans, industrial production plans, project initiation, progress control, and acceptance of results, and bear management responsibilities for the nuclear safety of their jurisdictional units.

7.3.3 Main Methods of Nuclear Safety Management

The main methods of nuclear safety management in China are the implementation of a licensing system (license application and approval procedures), and conducting necessary nuclear safety reviews and nuclear safety supervision.

(1) Licensing System

The "Nuclear Safety Law of the People's Republic of China" stipulates in Article 22 that the state establishes a nuclear facility safety licensing system. Nuclear facility operators must apply for a permit from the State Council's nuclear safety supervision and management department for activities such as site selection, construction, operation, and decommissioning of nuclear facilities. If a nuclear facility operator wishes to change the conditions stipulated in the permit document, they must apply for approval from the State Council's nuclear safety supervision and management department.

Nuclear safety is directly related to national security, environmental safety, and public health. A strict licensing system must be adopted throughout the process to strengthen control. Licensing management is one of the pillars of China's nuclear safety supervision. China uses a safety licensing system for nuclear safety supervision of nuclear facilities. For activities that may pose potential hazards to personnel and environmental safety, nuclear facility operators must apply for a permit from the State Council's nuclear safety supervision and management department in advance.

Countries around the world generally regard the nuclear facility safety licensing system as the primary legislative goal and legal principle in legislation related to the use of nuclear energy, that is, to fully establish a licensing system based on general prohibition, and clearly stipulate that any unlicensed nuclear energy utilization activities are illegal. This regulatory system is a common practice in nuclear safety supervision in nuclear countries around the world. The "Convention on Nuclear Safety" also requires that the legislative and regulatory framework of the contracting parties should include "the implementation of a licensing system for nuclear facilities and the prohibition of unlicensed operation of nuclear facilities".

1) Nuclear Facility Construction Permit

The owner of a nuclear facility must provide the National Nuclear Safety Administration with the relevant permit application documents twelve months before the construction of the nuclear facility, to apply for a nuclear facility construction permit. After receiving the nuclear facility construction permit issued by the National Nuclear Safety Administration, the owner of the nuclear facility can pour the first batch of concrete and start construction.

2) Nuclear Facility Operation Permit

The owner must provide the National Nuclear Safety Administration with the relevant permit application documents twelve months after the full-power operation of the nuclear facility, to apply for a nuclear facility operation permit. After receiving the nuclear facility operation permit issued by the National Nuclear Safety Administration, they can carry out operations according to the conditions stipulated in the permit.

3) Nuclear Facility Operator License

The main operators of nuclear facilities must obtain a nuclear facility operator license issued by the National Nuclear Safety Administration. The operator license includes an operator license and a senior operator license. The operator license examination must be conducted under the strict supervision of the National Nuclear Safety Administration.

4) Nuclear Material Permit

The owner of a nuclear facility must submit a nuclear material permit application report to the "National Nuclear Material Control Office" six months before receiving and holding nuclear materials. After approval by the National Nuclear Safety Administration, a nuclear material permit is issued.

5) Nuclear Pressure Equipment Activity Qualification Permit.

The Nuclear Pressure Equipment Activity Qualification Permit includes: Nuclear Pressure Equipment Design Qualification Permit, Nuclear Pressure Equipment Manufacturing Qualification Permit, and Nuclear Pressure Equipment Installation Qualification Permit. The applicant submits the permit application to the National Nuclear Safety Administration, and after review by the National Nuclear Safety Administration, the permit is issued. Design, manufacturing, and installation units can only carry out corresponding activities with a permit.

6) Other Documents Requiring Approval.

Certain important operations or activities of nuclear facilities must be carried out after obtaining permission from the National Nuclear Safety Administration. The approval documents issued by the National Nuclear Safety Administration for corresponding activities mainly include: Nuclear Power Plant Site Selection Review Opinion, Nuclear Power Plant First Loading Approval, Nuclear Power Plant First Criticality Approval, Nuclear Power Plant Decommissioning Approval, etc.

7) Environmental Impact Report.

The National Nuclear Safety Administration is responsible for approving the environmental impact reports at various stages of the nuclear power plant. The approval document of the environmental impact report is one of the necessary conditions for issuing a permit.

(2) Permit Application and Approval Process

1) Site Selection

Regarding the regulations on the safety permit application for the site selection stage of nuclear facilities, Article 23 of the "Nuclear Safety Law of the People's Republic of China" stipulates that nuclear facility operators should scientifically assess factors such as geology, earthquakes, meteorology, hydrology, environment, and population distribution. On the premise of meeting the requirements of nuclear safety technical evaluation, they should submit a nuclear facility site selection safety analysis report to the State Council's nuclear safety supervision and management department. After review and compliance with safety requirements, they can obtain the Nuclear Facility Site Selection Review Opinion.

Some basic design parameters of nuclear facilities, including certain safety design criteria, depend on site characteristics. In the event of an accident, site characteristics will also directly affect the migration of radioactive materials and the implementation of emergency plans. Therefore, it is necessary to establish safety permits for the site selection of nuclear facilities. The safety permit for nuclear facility site selection is the first step in the phased safety permit for nuclear facilities. Its purpose is to confirm the suitability between the selected site and the proposed nuclear facility, and to avoid unfavorable site conditions that are difficult to resolve through engineering and management measures. This includes the impact of site conditions on the safety of nuclear facilities and the impact of nuclear facilities on the surrounding population and environment, thereby ensuring the safety of nuclear facilities and protecting the public and the environment from excessive radiation exposure.

2) Construction

Currently, China adopts a phased licensing system for nuclear safety. After obtaining a permit for the site of a nuclear facility, the operating unit must also obtain a construction permit before officially starting the construction of the nuclear facility. The main purpose of reviewing the application for a construction permit for a nuclear facility is to confirm that the design of the nuclear facility complies with relevant legal requirements, and that the design of safety–critical structures, systems, and equipment, as well as safety–critical procedures, can ensure nuclear safety.

According to the regulations on the submission of construction applications by nuclear facility operators, Article 25 of the "Nuclear Safety Law of the People's Republic of China" states that before the construction of a nuclear facility, the nuclear facility operator should submit a construction application to the State Council's nuclear safety supervision and management department, and submit the following materials: a nuclear facility construction application; a preliminary safety analysis report; an environmental impact assessment document; quality assurance documents; and other materials stipulated by laws and administrative regulations.

According to this provision, the operator of a nuclear facility must apply for a construction permit before commencing construction; without such a permit, no construction activities may proceed. Among these requirements, the Preliminary Safety Analysis Report is a comprehensive evaluation, based on the operator's completed preliminary design of the nuclear facility, of whether the proposed facility can comply with nuclear safety regulations and standards. This report serves as a critical basis for the National Nuclear Safety Administration (NNSA) to issue the construction permit. The Environmental Impact Assessment (EIA) document evaluates the environmental effects of the construction and operation of the nuclear facility. The primary objective of the EIA during the construction phase is to demonstrate whether the engineering design meets environmental protection requirements, ensuring that environmental protection measures are incorporated into the design. The Quality Assurance document refers to the "Nuclear Facility Quality Assurance Program (Design and Construction Phase)". The NNSA has issued relevant guidelines to provide specific instructions for drafting the Quality Assurance document and to standardize its content and requirements.

3) Operation

Regarding the regulations for the application for a nuclear facility operation permit, Article 27 of the "Nuclear Safety Law of the People's Republic of China" states that before the first fuel loading of a nuclear facility, the nuclear facility operator should submit an operation application to the State Council's nuclear safety supervision and management department, and submit the following materials: a nuclear facility operation application; a final safety analysis report; quality assurance documents; an emergency plan; and other materials stipulated by laws and administrative regulations. After obtaining the nuclear facility operation permit, the nuclear facility operator should operate in accordance with the provisions of the permit.

The purpose of the nuclear facility operating license is to confirm that the facility has been constructed in accordance with the approved design and quality assurance requirements, that commissioning results meet design specifications, that operational procedures comply with safety management requirements, and to establish the licensing conditions to be followed after initial fuel loading (operation). Among these, the emergency preparedness plan refers to the on-site emergency plan developed by the operator. After review and approval by the nuclear safety regulatory authority under the State Council, this plan must be filed with the competent nuclear industry department, the energy administration department under the State Council, and the relevant departments designated by the provincial, autonomous region, or municipal government, as stipulated in Paragraph 2, Article 55 of the Nuclear Safety Law.

Additionally, regarding the extension of an operating license's validity period, Article 27 of the "Nuclear Safety Law of the People's Republic of China" specifies that if a nuclear facility operating license is due to expire but continued operation is required, the operator must submit an extension application to the nuclear safety regulatory authority under the State Council at least five years before expiration. The application must include a demonstration and verification of compliance with nuclear safety standards. Continued operation is permitted only after review and approval.

4) Decommissioning

Decommissioning of a nuclear facility is the action taken after a nuclear facility has reached or exceeded its design life or has ceased operation for other reasons.

Through decommissioning, the harmful effects of residual radioactive materials and other harmful substances can be reduced to an acceptable level, achieving limited use or unrestricted opening and use of the site. The purpose of the nuclear facility decommissioning permit is to confirm that the decommissioning arrangements and the actual status during the decommissioning process meet the relevant nuclear safety and environmental requirements. For example, nuclear power plant decommissioning can take three forms: immediate dismantling and removal of radioactive materials from the reactor after ceasing operation; sealing the reactor for several decades and dismantling it after its radioactivity has naturally decayed; or building a concrete shell around the reactor for long-term containment.

Decommissioning of a nuclear facility is a relatively long process, and the cost required for decommissioning is usually high. To do a good job in decommissioning nuclear facilities, nuclear facility operators should study decommissioning plans as early as the site selection and construction stages, and continuously improve them according to operation and actual conditions until the decommissioning activities are completed.

The regulations for the decommissioning of nuclear facilities, as stated in Article 25 of the "Nuclear Safety Law of the People's Republic of China", indicate that before decommissioning a nuclear facility, the operating unit of the nuclear facility should submit a decommissioning application to the State Council's nuclear safety supervision and management department, along with the following documents: a decommissioning application form; a safety analysis report; an environmental impact assessment document; quality assurance documents; and other materials required by laws and administrative regulations.

After obtaining approval for the decommissioning of the nuclear facility, the operating unit can then commence the decommissioning activities.

Regarding the decision to grant permission, Article 32 of the "Nuclear Safety Law of the People's Republic of China" states that the State Council's nuclear safety supervision and management department should, in accordance with statutory conditions and procedures, organize a safety technical review of the nuclear facility safety permit application. If the nuclear safety requirements are met, a decision to grant permission should be made within twenty days from the completion of the technical review. When approving the construction and operation permit applications for nuclear facilities, the department should seek opinions from relevant departments of the State Council and the provincial, autonomous region, or municipal people's government where the nuclear facility is located. The units consulted should respond within three months.

The process of permit application and approval is shown in Fig. 7.4. The applicant submits the civil nuclear facility permit application documents and license application documents to the National Nuclear Safety Administration. The National Nuclear Safety Administration first conducts a formal review. If a technical review is needed after the formal review, the National Nuclear Safety Administration will organize technical support units to conduct a technical review and raise review questions. The applicant should respond to the review questions in a timely manner, provide explanations, or make necessary additions or modifications to the materials. If the

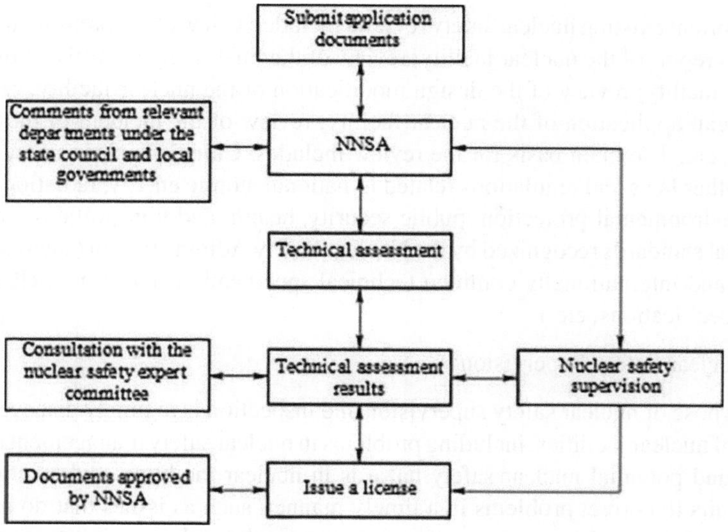

Fig. 7.4 Permit application and approval process

technical issues raised in the review are addressed and essentially resolved, the technical support unit will compile a technical evaluation report and submit it to the National Nuclear Safety Administration. If necessary, the National Nuclear Safety Administration will seek opinions from relevant departments of the State Council and the provincial people's government where the nuclear facility is located, organize a review by the Nuclear Safety and Environmental Expert Committee, and form advisory opinions. For construction projects that meet the conditions, the National Nuclear Safety Administration will make an approval decision, notify the applicant in writing, and copy the relevant departments. For construction projects that do not meet the conditions, the National Nuclear Safety Administration will make a decision not to approve and notify the applicant in writing.

(3) Nuclear Safety Review

During the permit application and approval process, a nuclear safety review must be conducted. The National Nuclear Safety Administration is responsible for organizing the nuclear safety review, and the Nuclear Safety Center conducts the technical review. If necessary, domestic and foreign experts and organizations can be asked to provide advisory opinions.

The purpose of the nuclear safety review is to determine whether the safety status and management of the nuclear facility meet the requirements of nuclear safety regulations through a technical review of the design, construction, debugging, operating conditions, management, and incidents of the nuclear facility. The review conclusion is the main basis for issuing various permits and approving related activities of the nuclear facility.

Important existing nuclear safety reviews include: review of the preliminary safety analysis report of the nuclear facility; review of the final safety analysis report of the nuclear facility; review of the design modification of the nuclear facility; review of the special application of the nuclear facility; review of the incident of the nuclear facility, etc. The main basis for the review includes: China's nuclear safety regulations; other laws and regulations related to national atomic energy, radiation protection, environmental protection, public security, health, and transportation; national technical standards recognized by the Nuclear Safety Administration (such as the GB series) and internationally common technical specifications (such as SRP, ASME, RCC specifications, etc.).

(4) Nuclear Safety Supervision

The purpose of nuclear safety supervision and inspection is to timely discover safety issues of nuclear facilities, including problems in nuclear safety management requirements and potential nuclear safety hazards in nuclear facilities, and to urge operating units to correct problems in a timely manner, such as issues that do not meet nuclear safety regulations, issues that do not meet the technical document requirements approved by the National Nuclear Safety Administration, etc. If necessary, compulsory measures can be taken, such as stopping operation.

The scope of supervision and inspection includes: reviewing whether the submitted safety documents or materials are consistent with reality; supervising whether construction is carried out according to the approved design; supervising whether management is carried out according to the approved quality assurance outline; supervising whether the construction and operation of nuclear facilities comply with relevant nuclear safety regulations, technical specification standards, and the requirements of various documents approved by the National Nuclear Safety Administration; examining whether the operating unit has the ability to operate safely and implement emergency plans.

Supervision is divided into routine supervision and routine/non-routine inspections. Routine supervision is the main form of nuclear safety supervision and inspection, and its purpose is to timely grasp the safety status of nuclear facilities and discover safety issues. Routine supervision and inspection is when regional supervision stations dispatch supervision (staff) groups to reside at the nuclear power plant site and conduct on-site inspections of nuclear facilities. The work carried out by the supervisor on each working day at the nuclear power plant site includes: tracking supervision, inspecting important activities, items, and records that affect safety; discovering safety issues in nuclear facilities; handling according to nuclear safety regulations, etc.

Routine/non-routine inspections are planned nuclear safety inspections carried out by the Nuclear Safety Administration or regional supervision stations organizing inspection teams, according to a pre-established inspection plan, and following certain procedures for the safety-important activities at various stages (including site selection, design, construction, debugging, operation, decommissioning, etc.) of the operating unit.

Discussion Questions

1) What are the similarities and characteristics of the nuclear safety regulatory systems in major countries? Please briefly describe.
2) What are the levels of the nuclear safety regulatory system in China? What are the main national laws related to nuclear safety that have been enacted so far?
3) In which year was the “Nuclear Safety Law of the People’s Republic of China” passed and implemented? What content does it include?
4) What is the concept of nuclear safety defined by the “Nuclear Safety Law of the People’s Republic of China”? Please briefly explain its connotation.
5) What are the important international organizations related to nuclear safety?
6) How is the Chinese national nuclear safety regulatory and management system stipulated?
7) What are the basic duties of the China National Nuclear Safety Administration?
8) What are the main responsibilities of the State Administration of Science Technology and industry for National Defense in the supervision of nuclear energy development and utilization?
9) What are the main methods of nuclear safety management in China?
10) What is the purpose of nuclear safety supervision and inspection? What does the scope of supervision and inspection include?

References

International Atomic Energy Agency. (1991). Safety Culture, INSAG-4.

International Atomic Energy Agency. (2020). Handbook on Nuclear Law. https://www-pub.iaea.org/MTCD/Publications/PDF/Pub1160_web.pdf.2020-04-15.

Lu H, Liu H, Wang YR, et al. (2018). Interpretation of the Nuclear Safety Law of the People’s Republic of China. Beijing, China Legal Publishing House.

National Nuclear Safety Administration. (2020). Policies and Regulations. http://nnsa.mee.gov.cn/zcfg_8964/fg/fgbmgz/index.html.2020-01-06.

National People’s Congress of China. (2017). Nuclear Safety Law of the People’s Republic of China. Beijing, Law Press China.

Chapter 8
Radiation Safety and Protection

Since the discovery of radioactivity at the end of the nineteenth century, the application of nuclear radiation and radioactive isotopes has a history of over a hundred years. The related research, development, and application have been widely and rapidly developed, bringing great convenience to people's production and life. However, it is a dual-impact phenomenon that can also harm human health when used improperly (Chen et al. 2019). Shortly after Roentgen discovered X-rays in 1895, researchers found that people exposed to X-rays for a long time would experience symptoms such as skin redness, swelling or ulcers, eye pain or conjunctivitis, hair loss, and leukopenia. Later, it was found that natural radioactive substances such as radium could also cause malignant diseases such as skin cancer and lung cancer, which led to people's attention and research on radiation hazards. In order to both ensure people's health and safety and the safety of the environment, and to carry out radiation application work smoothly, it is necessary to understand the basic effects of radiation on the human body and the effects it causes, so that people have a correct understanding of the harm of nuclear radiation, to eliminate unnecessary fear and to pay great attention; it is necessary to establish the basic principles of radiation protection, formulate necessary radiation protection standards, actively carry out radiation monitoring work, and take effective measures to reduce or avoid unnecessary harm (Editorial Board of Registered Nuclear Safety Engineer Training Series 2004).

8.1 Radiation Sources

All organisms are constantly exposed to ionizing radiation that always exists in nature. There are two types of radiation sources that irradiate the human body, namely natural radiation sources and artificial radiation sources. Throughout the entire historical stage of the formation of life on earth and the birth and life of humans, they are always exposed to the radiation emitted by cosmic rays and radioactive

L. Yu et al., *Nuclear Technology and Safety*, Nuclear Science and Technology,
https://doi.org/10.1007/978-981-95-2856-1_8

substances existing in the earth's environment. This natural radioactivity is objectively present and is usually referred to as natural background radiation. Natural background radiation is the main source of radiation exposure to humans so far.

For more than half a century, due to medical radiation and the development and application of nuclear energy and nuclear technology, as well as nuclear power production, nuclear tests, etc., many new radioactive substances and radiation exposures have been produced. This type of exposure is called artificial radiation source exposure.

8.1.1 Natural Radiation Sources

Natural radiation sources can be divided into three categories according to their sources: cosmic rays, terrestrial radiation sources, and human body radiation sources (Chen et al. 2021). These radiation sources constantly act on the human body in various ways, forming natural background radiation.

(1) Cosmic Rays

Cosmic rays can be divided into primary cosmic rays and secondary cosmic rays. The rays entering the earth's atmosphere from outer space are called primary cosmic rays, the main components of which are protons and alpha particles, as well as a small amount of heavy particles. The secondary particles or radiation produced by the interaction of primary cosmic rays with the atmosphere are called secondary cosmic rays, including muons, electrons, photons, and neutrons, etc.

Most of the primary cosmic rays are absorbed by the atmosphere, and almost all of the cosmic rays reaching the earth's surface are secondary cosmic rays.

Due to the blocking and absorption of the atmosphere, the intensity of cosmic rays increases with the increase of altitude. For example, in cities near the equator at sea level, the dose contribution is about 0.35 mSv per year, while in plateau cities like Denver, it is about 0.9 mSv per year. The highest living area on earth receives up to 30 mSv per year, and the top of Mount Everest receives about 80 mSv per year. Cosmic rays are deflected from the equator to the poles due to the influence of the earth's magnetic field, so the intensity of cosmic rays in the equatorial region is slightly smaller than that in the polar regions.

(2) Terrestrial Radiation Sources

Earth's radiation sources refer to the natural radionuclides contained in the Earth's atmosphere, rocks, soil, and other substances, which are divided into cosmogenic radionuclides and primordial radionuclides.

Cosmogenic nuclides are radionuclides produced by the interaction of cosmic rays and target atomic nuclei in the atmosphere, mainly including ^{3}H, ^{7}He, ^{14}C, ^{22}Na, ^{24}Na, etc. However, their contribution to the ground dose is minimal.

Primordial nuclides are natural radionuclides that existed at the beginning of the Earth's formation. Given that the Earth's lifespan has reached over 4 billion years, only those with extremely long half-lives and their decay products remain, mainly ^{238}U (half-life of about 4.5 billion years), ^{232}Th (half-life of about 12 billion years), ^{235}U (half-life of about 700 million years), ^{40}K (half-life of about 1.28 billion years), ^{87}Rb (half-life of about 47.5 billion years), etc. These are widely distributed throughout the Earth's crust. All naturally occurring materials contain a certain amount of radioactive impurities, mainly natural radionuclides such as uranium, thorium, and potassium in the soil. Due to their extremely long half-lives and generally very weak radioactivity in various substances, their impact on human survival and health is almost unobservable. However, in some high-demand experiments, it is necessary to reduce their impact on measurements.

(3) Radiation Sources in the Human Body

Due to the need for food and respiration, there is material exchange and circulation between the human body and nature, inevitably containing some radionuclides. However, the content is very small, the dose rate contribution is extremely low, and the impact on health can be completely ignored. The human body contains a higher proportion of elements such as carbon, hydrogen, and potassium. The most abundant radionuclides are mainly ^{40}K, ^{14}C, ^{3}H, etc. Uranium, thorium and their decay products are also present in certain amounts. The decay product radium has chemical properties similar to calcium and usually accumulates in the bone marrow.

(4) Natural Background Radiation

The combined effect of the above types of radiation sources constitutes the unavoidable natural background radiation on Earth. It has always existed in nature and is relatively stable, with its intensity basically not changing over time. However, the natural background radiation varies in different regions. The natural background radiation level in the plains is lower, even lower on the ocean surface, and higher in mountainous areas with various rocks.

The dose of natural radiation received by residents is related to their place of residence, type of housing, altitude, etc. The dose variation range for residents in most areas of the world is narrow, with only a few areas having high natural backgrounds. The main reasons are either living in areas with high radioactivity in soil and rocks, living in buildings with high radon concentrations, living in high-altitude areas, or consuming food with high radioactivity.

8.1.2 Artificial Radiation Sources

Artificial radiation sources refer to radioactive substances or devices manufactured by humans, as well as natural radiation sources that have been artificially altered. These mainly include reactors, accelerators, X-ray machines, medical radiation, X and gamma ray flaw detection nuclide radiation sources, irradiators, atomic tracer nuclide radiation sources, alpha, beta, gamma, X and neutron sources used for industrial product analysis, measurement and control, consumer product radiation, and radioactive waste, etc.

Human activities, practices, and events involving radiation sources have led to the release of radioactive substances into the environment and radiation exposure to humans and animals. Today, the main artificial radiation sources causing human exposure are nuclear energy production, medical radiation, and nuclear explosions.

(1) Nuclear Energy Production

According to data released by the International Atomic Energy Agency in August 2020, as of the end of 2019, there were a total of 442 nuclear power units in operation worldwide, with a total installed capacity of 402.3 GW, nuclear power generation of 2657 TWh, and nuclear power accounting for about 17% of the world's electricity supply. As of August 21, 2020, there were 441 operable nuclear reactors worldwide, with a total installed capacity of 391.7 GW; 54 nuclear reactors under construction, and 3 nuclear power units permanently closed. Nuclear power has been developed for several decades, which has accordingly generated a large number of artificial radiation sources.

The production of electric energy with nuclear reactors is predicated on the nuclear fuel cycle. The nuclear fuel cycle includes the mining and hydrometallurgy of uranium ore, transforming it into different chemical forms; ^{235}U. Enrichment of U isotopes; manufacturing of fuel elements; energy production through nuclear reactions; storage and post-processing of irradiated fuel; recycling and recovery of fissionable and useful materials in spent fuel; transportation of radioactive materials between different stages and facilities of the nuclear fuel cycle; disposal of radioactive waste. Although almost all man-made radioactive isotopes produced in nuclear power generation remain in the irradiated nuclear fuel, each of the above processes releases a small amount of radioactive material into the environment. Since most of these radioactive isotopes have short half-lives and migrate slowly in the environment, the radioactive materials released into the environment mostly only have local or regional impacts. However, some radioactive isotopes with long half-lives or fast dispersion in the environment can spread globally, causing radiation and pollution to humans and the environment worldwide. The annual collective effective dose and the per capita effective dose caused by nuclear power production are not large. In 1980, the per capita equivalent dose caused by nuclear power production was only five ten-thousandths of the natural radiation exposure level, and even by 2500, it will only be one percent of the natural radiation exposure level.

(2) Medical Radiation

Today, medical radiation is the leading source of artificial radiation exposure to the world's population, as the use of ionizing radiation in the medical field has been increasing for over a hundred years and has become an important means of diagnosis and treatment. Medical radiation comes from X-ray diagnostic examinations, nuclear medicine diagnoses involving the introduction of radioactive isotopes into the body, and radiation therapy processes.

When patients go to the hospital for medical examinations, such as undergoing X-rays, CT scans, PET scans, MRI, isotope therapy, etc., they are exposed to certain levels of radiation. These radiations are inherently harmful, but it doesn't matter if they are not used frequently. For example, the radiation dose for a lung fluoroscopy is 0.02 mSv, the exposure dose for a chest X-ray is 0.12 mSv, and the exposure dose for a CT scan is 10 mSv.

(3) Nuclear Testing

Atmospheric nuclear testing is the main reason why artificial radiation sources in the environment expose the global public to radiation. The artificial radioactive materials formed by nuclear tests initially enter the upper part of the atmosphere, then slowly move down from the upper atmosphere, and finally fall to the ground, known as fallout. It can cause external and internal exposure through inhalation. Nuclear testing began in 1945, with a large number of tests conducted in the atmosphere between 1954–1958 and 1961–1962. The last one was in October 1980, with a total of 543 atmospheric nuclear tests, equivalent to 440Mt. There were 1876 underground nuclear tests, equivalent to 90Mt, which caused relatively less environmental pollution. Although nuclear tests can produce hundreds of radioactive isotopes, most of them either have a small yield or a short half-life. Only seven contribute more than 1% to the effective dose of the world's residents. In decreasing order of exposure levels to the human body, they are ^{14}C, ^{137}Cs, ^{95}Zr, ^{90}Sr, ^{106}Ru, ^{144}Ce, and ^{3}H. However, since 1965, ^{137}Cs and ^{90}Sr have been the main isotopes in the remaining cumulative deposition. The level of exposure to fallout varies depending on the latitude of the residence, with residents in the southern hemisphere generally receiving less exposure than those in the northern hemisphere. The main route of exposure to residents from nuclear tests is ingestion, followed by external exposure. By the end of 1980, the collective effective dose caused by atmospheric nuclear tests was equivalent to the world's population receiving about four years of additional natural background radiation exposure. In terms of the annual per capita dose caused by nuclear tests, 1963 was the highest, equivalent to seven percent of the average annual dose caused by natural radiation sources. By 1966, it had dropped to two percent, and it is currently less than one percent.

8.2 Properties and Hazards of Common Radiation

This section introduces the properties and hazards of common ionizing radiation, including γ rays, β rays, α rays, and neutron rays.

8.2.1 Properties and Hazards of γ-Rays

Fundamentally, γ-rays are a form of extremely high-energy electromagnetic radiation, typically emitted when an atomic nucleus transitions from a higher energy state to a lower one (γ-decay). γ-photons are electrically neutral, have zero rest mass, and the energy of γ-photons released from γ-decay or nuclear fission generally does not exceed 20 MeV. Their interactions with matter primarily involve three mechanisms: the photoelectric effect, Compton scattering, and electron–positron pair production, while other interaction probabilities are negligible (Guo et al. 2020).

Compared to charged particles such as α and β-rays, γ-rays interact with matter much less frequently, with a significantly longer mean free path between interactions, resulting in strong penetration capability. γ-rays with energies of several MeV can penetrate several to tens of centimeters of metal, hundreds of meters of air, and can completely pass through human bones. When γ-rays traverse matter, their intensity attenuates exponentially—higher density and atomic number materials provide better shielding.

Due to their weak direct ionization ability but strong penetration, γ-rays pose a greater external radiation hazard in practical work. Protective measures such as material shielding, time control, and distance maximization must be implemented.

8.2.2 Properties and Hazards of β-Rays

B-rays are essentially high-speed electron flows (e^- or e^+), fundamentally identical to electrons but differing in origin—β-rays are emitted from the atomic nucleus, whereas electrons originate from outer atomic orbitals.

β-rays interact with matter primarily in two ways: (1) Elastic or inelastic collisions with orbital electrons—unlike α-particles, β-rays undergo significant directional changes and energy loss per collision; (2) Interactions with atomic nuclei—energy loss is released as bremsstrahlung (X-rays), particularly noticeable when β-rays have high energy or interact with high-atomic-number materials (Wu 2017).

In the same medium, the range of β-rays is much longer than that of α-rays: Several meters to tens of meters in air; A few millimeters in concrete; Capable of fully penetrating human skin. Compared to neutrons or γ-rays, β-rays pose a less severe external radiation hazard but can still cause significant damage to skin and conjunctiva. Light materials such as plexiglass or aluminum sheets can effectively shield β-rays. However, β-rays have strong direct ionization ability. If inhaled or ingested, they can cause substantial harm to internal tissues and organs. Therefore, preventing β-emitting radioactive materials from entering the body is critical, with protective measures similar to those for α-emitting substances.

8.2.3 Properties and Hazards of Alpha Rays

Alpha rays, also known as alpha particles, are actually high-speed helium nuclei, composed of two protons and two neutrons. They are generally emitted during the decay of radionuclides with atomic numbers greater than 82, with energies typically in the range of 4 to 7 MeV.

Alpha decay can be represented as: ${}_{Z}^{A}X \rightarrow {}_{Z\text{-}2}^{A\text{-}4}Y + \alpha + E$, where ${}_{Z}^{A}X$ is the parent nucleus, ${}_{Z\text{-}2}^{A\text{-}4}Y$ is the daughter nucleus, and E is the total energy released during the decay process. The energy of alpha particles produced by alpha decay is monoenergetic, and the type of parent nucleus can be inferred by measuring its energy.

Alpha particles mainly interact with matter by directly undergoing elastic or inelastic collisions with electrons outside the atomic nucleus. Since the mass of alpha particles is much larger than that of electrons, when their speed is relatively high, their direction of motion changes very little, and their trajectory is approximately a straight line. It can be approximated that the range is equal to the path length. Alpha particles cause many atoms or molecules to ionize along their path, producing a large number of ion pairs. Their energy is quickly lost, resulting in a short range. In air, they generally only travel a few centimeters, and a regular sheet of printer paper can completely block them. Therefore, there is no need to consider the harm caused to the human body by external alpha radiation sources. However, due to their strong ionization ability, if they enter the human body through breathing or eating, they can cause significant damage to the tissues and organs at the site of entry. Therefore, it is necessary to strictly prevent alpha radiation sources from entering the human body. Generally, masks, gloves, and anti-contamination suits can effectively prevent alpha radioactive substances from contacting the skin or entering the body through the respiratory tract. Eating, drinking, and smoking should also be prohibited in radioactive workplaces to prevent alpha radioactive substances from entering the human body through the mouth and esophagus.

8.2.4 Properties and Hazards of Neutrons

Neutrons are uncharged particles, slightly heavier than protons. Neutrons confined by nuclear forces in atomic nuclei are stable, but free neutrons are unstable, with a half-life of about ten minutes. They spontaneously decay into protons, simultaneously producing an electron and an antineutrino.

The reaction cross-section of neutrons interacting with matter is also small, and its size is mainly related to the neutron energy, the type of atomic nucleus of the interacting matter (including the number of protons and neutrons), etc. The type and cross-section size of neutron interactions with matter strongly depend on the neutron energy. In some energy regions, the interaction cross-section is very small, while in others, it is much larger. In some narrow energy ranges, the cross-section may suddenly increase sharply.

Neutrons have strong penetrating power. Although their direct ionization ability is weak, their indirect ionization ability is strong, hence both external and internal radiation from neutrons pose significant risks. From a practical perspective, internal radiation from neutrons rarely occurs, with the main risk being external radiation. Materials that can effectively shield neutrons include water, paraffin, polyethylene, and composite materials containing elements with large neutron absorption cross-sections, such as boron.

8.3 Common Radiation Quantities and Units

To quantitatively describe the interaction of radiation with matter and its damage to the human body, some special physical quantities are needed. Since the object is ionizing radiation, these are often referred to as radiation quantities. The following introduces four commonly used radiation quantities in work: radioactivity, absorbed dose, equivalent dose, and effective dose. To facilitate understanding, if radioactive material is likened to an ammunition depot storing various weapons and ammunition, then radioactivity can be likened to the total quantity of various weapons and ammunition fired from the depot per unit time. The absorbed dose can be likened to the total explosive equivalent of various weapons and ammunition falling on the battlefield. The equivalent dose can be likened to the damage suffered by specific objects such as battlefield fortifications, weapons and equipment, personnel, etc., while the effective dose can be likened to the combat power damage suffered by the entire troop.

8.3.1 Radioactivity

Radioactivity (commonly represented by A) refers to the expected number of a certain type of radioactive decay or nuclear transition that occurs in a certain amount of radioactive material per unit time, which can be expressed as $A = \mathrm{d}N/\mathrm{d}t$, where $\mathrm{d}N$ is the average number of radioactive decays or nuclear transitions that occur in the radioactive material in the time interval dt.

The international unit of radioactivity is 1/s, the special unit is Becquerel, abbreviated as Bq, 1 Bq = 1/s, 1 Bq indicates that the radioactive material has undergone 1 radioactive decay or nuclear transition in 1 s. Radioactivity also has a former unit, Curie (Ci), 1 Ci = 3.7×10^{10} Bq.

The relationship between the activity A of a radiation source and the number N of atoms or atomic nuclei of the radionuclide is $A = N$. The activity A can be simply and intuitively understood as the number of radioactive atomic nuclei in an active state per unit time.

Radioactivity also follows the law of exponential decay, i.e., the activity at time t is $A(t) = A_0 e^{-\lambda t}$.

In practical work, it is often necessary to determine or differentiate the radioactivity per unit area, unit volume, or unit mass.

Surface activity (A_S) refers to the radioactivity per unit area, with the SI unit being Bq/m^2 and the commonly used unit being Bq/cm^2. It is typically used to characterize the radioactive contamination level on equipment surfaces, floors, and other areas.

The radioactivity per unit volume or unit mass of a material is referred to as radioactive activity concentration. The radioactivity per unit volume is sometimes called volumic activity (A_v), with the SI unit being Bq/m^3 and the commonly used unit being Bq/L. This is often used to measure the radioactivity of liquid effluents, gases, or aerosols generated during the operation, maintenance, decommissioning, or accident scenarios of nuclear facilities. The radioactivity per unit mass is sometimes called specific activity (A_M), with the SI unit being Bq/kg and the commonly used unit being Bq/g. This is typically used to assess the radioactivity of neutron-activated reactor structures, shielding materials, solid reference radioactive sources, and similar substances.

8.3.2 Absorbed Dose

Absorbed dose (commonly denoted as *D*) is a fundamental quantity in dosimetry, referring to the amount of energy absorbed per unit mass of a substance exposed to ionizing radiation. It is expressed as:

$$D = \frac{\mathrm{d}\varepsilon}{\mathrm{d}m} \tag{8.1}$$

Here, $d\varepsilon$ represents the mean energy imparted by ionizing radiation to a substance of mass dm. The International System of Units (SI) for absorbed dose is joules per kilogram (J/kg). The specialized unit for absorbed dose is the gray (Gy), where 1 Gy = 1 J/kg. Commonly used subunits include mGy and μGy.

Absorbed dose applies to any material, any type of radiation or particle, and any mode of exposure. The potential effects on a substance after irradiation are closely related to the absorbed dose—the amount of radiation energy absorbed per unit mass. For example, certain biological tissues may sustain damage or die once the absorbed dose reaches a specific threshold, while some metallic materials may undergo significant changes in properties such as brittleness or toughness when the absorbed dose exceeds a certain level.

Absorbed dose rate refers to the change in absorbed dose per unit time, defined as:

$$\dot{D} = \frac{\mathrm{d}D}{\mathrm{d}t} \tag{8.2}$$

The SI unit for absorbed dose rate is gray per second (Gy/s) or joules per kilogram-second (J/(kg·s)). Commonly used units include mGy/h and μGy/h.

Absorbed dose rate is often used to characterize the external radiation levels generated by sources such as neutrons and γ-rays.

8.3.3 Equivalent Dose

Equivalent dose is a weighted average of absorbed dose, where the absorbed dose refers to the mean value across an entire tissue or organ following irradiation by various types of radiation, rather than the value at a single point. Suppose radiation R delivers an average absorbed dose $D_{T,R}$ to tissue or organ T. The equivalent dose H_T received by T is the weighted sum of the average absorbed doses from all types of radiation:

$$H_T = \sum_R \omega_R D_{T,R} \tag{8.3}$$

Here, ω_R is the radiation weighting factor, which reflects the relative biological effectiveness (RBE) of radiation R in causing harm to human tissues. Since the radiation weighting factor is dimensionless, the SI unit of equivalent dose is the same as that of absorbed dose—joules per kilogram (J/kg). However, because equivalent dose represents a different concept, it is assigned a special unit called the sievert (Sv). Table 8.1 provides the radiation weighting factors for common types of radiation. As shown, photons and electrons have a weighting factor of 1, while α-particles, neutrons, and heavy charged particles have significantly higher values.

Since equivalent dose accounts for the weighted average absorbed dose across an entire tissue or organ, it provides a more comprehensive and objective measure of the potential harm caused by radiation exposure. Consequently, equivalent dose

Table 8.1 Radiation weighting factors for common types of radiation

Radiation type and energy range		Radiation weighting factor
Photons (all energies)		1
Electrons and muons (all energies, excluding Auger electrons emitted by nuclei toward DNA)		1
Neutrons	Energy <10 keV	5
	Energy 10–100 keV	10
	Energy 100 keV–2 MeV	20
	Energy 2–20 MeV	10
	Energy >20 MeV	5
Protons (excluding recoil protons, energy >2 MeV)		5
α-particles, fission fragments, heavy nuclei		20

is primarily used to assess the probability and severity of deterministic effects (also called tissue reactions) in irradiated tissues or organs.

8.3.4 Effective Dose

The risk posed by radiation exposure to biological tissues and organs depends not only on the type and energy of the radiation but also on the specific tissue or organ being irradiated. To provide a unified measure of the potential risk following radiation exposure, the effective dose is calculated by weighting the equivalent doses received by different tissues and organs according to their relative radiosensitivity.

The effective dose is used to assess the overall harm to the human body from ionizing radiation exposure. It is defined as the sum of the equivalent doses to all tissues and organs, each multiplied by their respective tissue weighting factor:

$$E = \sum_T \omega_T H_T = \sum_T \omega_T \sum_R \omega_R D_{T,R} \tag{8.4}$$

where: ω_T is the tissue weighting factor for tissue or organ T, representing the relative sensitivity of human tissues and organs to radiation. The sum of tissue weighting factors for all tissues and organs in the body equals 1. Since all weighting factors are dimensionless, the SI unit of effective dose is joules per kilogram (J/kg), with the special unit sievert (Sv). Table 8.2 provides the tissue weighting factors for certain tissues or organs. As shown, more critical and sensitive tissues and organs are assigned larger weighting factors, which aligns with intuitive understanding.

Table 8.2 ICRP recommended tissue weighting factors

Tissue or organ	Tissue weighting factor	Tissue or organ	Tissue weighting factor
Gonads	0.20 (0.08)	Esophagus	0.05 (0.04)
(Red) Bone marrow	0.12 (0.12)	Thyroid	0.05 (0.04)
Colon	0.12 (0.12)	Skin	0.01 (0.01)
Lung	0.12 (0.12)	Bone surface	0.01 (0.01)
Stomach	0.12 (0.12)	Brain	(0.01)
Bladder	0.05 (0.04)	Salivary glands	(0.01)
Breast	0.05 (0.12)	Remaining tissues/ organs	0.05 (0.12)
Liver	0.05 (0.04)	Total	1.00 (1.00)

Note The values outside the parentheses are the recommended values from ICRP Publication 60 (1991), while the values inside the parentheses are the revised recommendations from ICRP Publication 103 (2007), based on the latest research and experimental data. It is evident that the new publication significantly reduced the weighting factor for the gonads while increasing that for the breast, suggesting that previous assessments may have overestimated the risk of genetic effects while underestimating the risk of radiation exposure to the breast

8.4 Health Effects of Ionizing Radiation on the Human Body

8.4.1 The Interaction Process of Ionizing Radiation with Cells

The cell is the basic structural and functional unit of living organisms, primarily composed of the nucleus, cytoplasm, and cell membrane. The nucleus contains chromosomes, which consist of proteins and deoxyribonucleic acid (DNA), among other components. When ionizing radiation interacts with a living organism, it may induce specific physiological effects, known as the biological effects of ionizing radiation. These biological effects first manifest in the interaction process with cells and the resulting consequences on them.

At the cellular level, ionizing radiation primarily causes damage in the form of changes in the cell cycle (such as delayed or arrested division), inhibited proliferation, chromosomal aberrations, cell mutations, giant cell formation, and cell death.

Research and experiments indicate that, for the vast majority of cells, radiation sensitivity is directly proportional to their proliferative capacity and inversely proportional to their degree of differentiation. However, there are exceptions—for instance, lymphocytes and oogonia are highly differentiated yet exhibit high radiation sensitivity. Generally, continuously renewing cell populations (dividing cells) are the most sensitive to radiation, growing cell populations (reversibly post-mitotic cells) are less sensitive, and stable cell populations (post-mitotic cells) are insensitive. The most significant harm to the organism arises from radiation damage to renewing cell populations, followed by damage to growing cell populations.

DNA serves as the material basis for cell proliferation and heredity and is the key molecule responsible for biochemical and physiological changes in cells. It is the primary target of ionizing radiation and plays a crucial role in cellular radiation damage. Studies and experiments demonstrate that inducing DNA damage is the main mechanism by which ionizing radiation produces biological effects. Cells with higher DNA content are more susceptible to ionizing radiation damage than those with lower DNA content, and cells from patients with defective DNA repair mechanisms often exhibit greater sensitivity to ionizing radiation.

The consequences of ionizing radiation on DNA mainly involve alterations in DNA molecular structure or disruptions in DNA metabolism, thereby causing cellular damage.

8.4.2 *Classification of Biological Effects of Ionizing Radiation*

The biological effects induced by ionizing radiation are diverse and can be classified in different ways, as illustrated in Fig. 8.1.

Based on the subject in which harmful effects occur after exposure—i.e., the victim—radiation biological effects are divided into somatic effects and hereditary effects. Somatic effects refer to biological effects that manifest in the exposed individual themselves, such as vomiting, skin erythema, ulceration, hair loss, and cancer after irradiation. Hereditary effects refer to biological effects that appear in the offspring of the exposed individual (generally limited to two generations), such as congenital malformations, growth retardation, or developmental disorders in descendants.

Based on the timing of harmful effects after exposure, radiation biological effects are classified into early (acute) effects and late effects. Early or acute effects refer to radiation effects that appear within a short period (hours, days, or weeks after exposure), such as vomiting, skin ulceration, and hair loss. These effects typically occur only after high-dose exposure in accidents or incidents and do not appear under normal conditions. Late effects refer to radiation effects that manifest months, years, or even decades after exposure, such as leukemia, cancer, and hereditary diseases.

Based on the relationship between the probability and severity of biological effects and radiation dose, radiation biological effects are categorized as stochastic effects and deterministic effects. Stochastic effects refer to biological effects whose probability of occurrence is related to the radiation dose, but whose severity is generally independent of dose. There is no dose threshold, meaning even very low doses may induce stochastic effects. Deterministic effects refer to biological effects whose both probability and severity are related to radiation dose, increasing with higher doses.

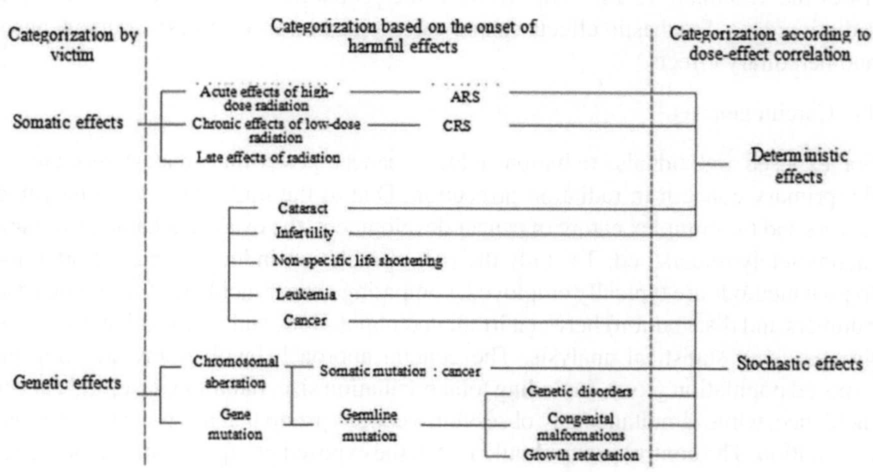

Fig. 8.1 Classification of radiation biological effects

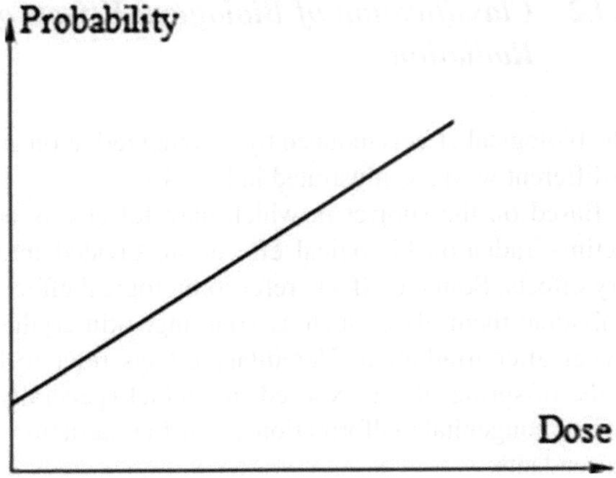

Fig. 8.2 Relationship between the probability of stochastic effects occurring and the dose received

These effects have a dose threshold, below which deterministic effects do not occur. It should be noted that different tissues and organs have different dose thresholds for deterministic effects. For example, the lens of the eye is highly sensitive, with a relatively low threshold, whereas the skin has a higher threshold for deterministic effects.

(1) Stochastic Effects

The relationship between the probability of stochastic effects and radiation dose is highly complex, particularly in the low-dose range. Researchers and institutions worldwide have proposed various models. However, for low-dose (rate) exposure under normal conditions, the linear no-threshold (LNT) hypothesis is commonly adopted from a radiation protection perspective, assuming a linear relationship between the probability of stochastic effects and radiation dose. Figure 8.2 illustrates the schematic relationship between the probability of stochastic effects and radiation dose. Stochastic effects can be categorized into two types: carcinogenesis and hereditary effects.

1) Carcinogenesis

For exposed individuals, radiation-induced cancer poses the greatest risk and is the primary concern in radiation protection. Due to the multitude of carcinogenic factors and the complex nature of cancer development, the exact mechanisms remain incompletely understood. To study the risk of radiation-induced cancer, epidemiological methods are typically employed, comparing cancer incidence (including case numbers and distribution) between irradiated populations and non-irradiated control groups using statistical analysis. The general approach involves investigating an exposed population group, including total population size, radiation dose, and cancer incidence, while simultaneously observing a control group that has not been exposed to radiation. The control group should match the exposed group in terms of race, age, sex, living environment, diet, geography, and social conditions, with radiation exposure being the sole difference. By comparing cancer incidence between the two groups, the cancer risk per unit dose can be estimated. To enhance the reliability of findings, larger population sizes are preferred, and confounding factors (e.g., age, sex, and occupation) that influence natural cancer rates must be carefully accounted for.

2) Hereditary Effects

Hereditary effects are stochastic effects that manifest in the offspring of exposed individuals, generally believed to originate from radiation damage to the testes (producing sperm cells) or ovaries (producing egg cells). Ionizing radiation may induce mutations in the testes, ovaries, or germ cells, ultimately causing harmful effects in descendants. These mutations are thought to result from DNA alterations in a single germ cell of the exposed individual, which are then inherited by the next generation. The severity of hereditary effects varies widely, ranging from early death and severe intellectual disabilities to minor skeletal abnormalities and metabolic disorders.

Animal studies confirm that radiation-induced mutations can cause harmful hereditary effects. However, many environmental factors beyond radiation can also trigger mutations and hereditary diseases. To date, no conclusive evidence has linked natural or artificial radiation exposure to hereditary defects in humans. For instance, epidemiological studies of over 70,000 children of parents exposed to atomic bombings in Japan showed no statistically significant differences compared to control groups. A 1967 cytogenetic study of more than 7,000 offspring of atomic bomb survivors found higher frequencies of somatic chromosomal abnormalities, but these differences were not statistically significant. Many studies suggest that previously estimated risk factors may have been overstated.

(2) Deterministic Effects

If the radiation dose is sufficiently high, a large number of cells may be killed, leading to clinically observable changes in certain tissues or organs. These radiation-induced biological effects, resulting from extensive cell death and subsequent tissue or organ damage, are termed deterministic effects. For clinically detectable damage to occur, a significant proportion of cells must be destroyed or killed. Therefore, deterministic effects only manifest when the radiation dose exceeds a certain threshold, and both the probability and severity of these effects depend on the dose level. Figure 8.3 illustrates the relationship between the probability of deterministic effects and radiation dose.

Deterministic effects may appear early (within hours to weeks after exposure) or late (months after exposure). Initially referred to as non-stochastic effects to contrast with stochastic effects, they were later renamed deterministic effects in ICRP Publication 60, as it became clear that once the dose reaches a certain level and a critical proportion of cells are killed, these effects inevitably occur.

Table 8.3 presents the dose thresholds for deterministic effects in some radiation-sensitive tissues and organs in adults. These values are not experimental but derived from radiotherapy studies and accident exposure data. Generally, the damage from

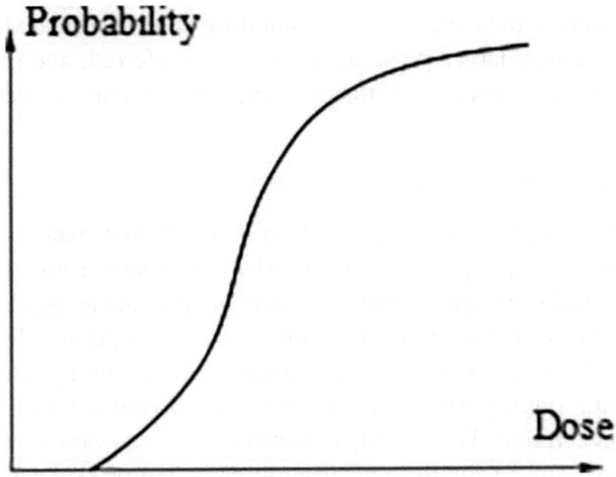

Fig. 8.3 Relationship between the probability of deterministic effects occurring and the dose received

fractionated or prolonged exposure is less severe than that from a single acute exposure. Note that the dose thresholds represent the dose at which 1% of exposed individuals develop the effect, not the dose required for all exposed individuals to experience it.

Table 8.3 Dose thresholds for deterministic effects in some radiation-sensitive tissues and organs of adults

Tissue or organ	Tissue effect	Dose threshold		
		Single or short-term exposure (total dose/Gy)	Fractionated or prolonged exposure (total dose/Gy)	Fractionated or prolonged exposure (annual dose rate/Gy·a^{-1})
Testes	Temporary sterility Permanent sterility	0.15 3.5~6.0	NA$^{(1)}$ NA	0.4 2.0
Ovaries	Sterility	2.5~6.0	6.0	>0.2
Lens of the eye	Detectable opacity Visual impairment (Cataract)	0.5~2.0 2~10	5 >8	>0.1 >0.15
Bone marrow	Hematopoietic depression Fatal aplastic anemia	0.5 1.5	NA NA	>0.4 >1.0

Note (1) NA indicates "not applicable," as the threshold depends on dose rate rather than total dose

8.4.3 Effects of Long-Term Low-Dose Radiation on Human Health

The term "low dose" lacks a clearly defined threshold but generally refers to occupational exposures below effective dose limits or environmental radiation exposure to the general public.

Characteristics of long-term low-dose radiation effects:

① The primary biological effects are late effects, with stochastic effects being the main concern for humans.
② Low incidence rates, typically requiring statistical methods for population studies.
③ Under low-dose and low-dose-rate conditions, the only potential somatic effect is cancer induction, with incidence increasing with dose.
④ Effects like lifespan shortening, lens opacity, and reduced fertility are not considered at doses below equivalent dose limits, as deterministic effects rarely occur at these levels.
⑤ Biological damage and repair processes occur simultaneously.

The radiation effects from long-term low-dose exposure are extremely subtle and nonspecific. Numerous factors influence the dose–effect relationship, and direct human data on low-dose radiation effects remain scarce. The debate continues regarding whether low-dose radiation is harmful or beneficial, with no definitive conclusion.

8.4.4 Factors Influencing Ionizing Radiation Biological Effects

The main factors affecting ionizing radiation biological effects include radiation characteristics, biological factors, and environmental conditions.

(1) Radiation Factors

1) Radiation quality: Different types and energies of radiation produce varying biological effects, with higher quality factors leading to more significant effects.
2) Dose: Generally, higher doses produce more pronounced effects.
3) Dose rate: Effects are most noticeable between 0.1 $Gy{\cdot}h^{-1}$ and 1 $Gy{\cdot}min^{-1}$. Fractionated exposures cause significantly less biological effect than single acute exposures due to repair mechanisms. For example, 2.5 Gy accumulated over 50 years causes no acute damage, while a single 2.5 Gy exposure induces acute radiation sickness.
4) Exposure area and location: 6 Gy whole-body exposure may be lethal, while the same dose to extremities may show no symptoms. Radiation sensitivity follows:

abdomen > pelvis > head > chest > limbs. Critical organ exposure requires special protection.

5) Exposure conditions: Single exposures produce more severe effects than fractionated ones. Relative biological effectiveness: external exposure: n>X>>; internal exposure P>X.

(2) Biological Factors

1) Species sensitivity: More evolutionarily advanced species show greater sensitivity (mammals > birds/fish/amphibians > single-celled organisms).
2) Individual variability: Sensitivity varies with age (embryos > fetuses > children > adults, but elderly > young adults), gender, physiological state, genetics, and nutrition. Pregnant women should avoid pelvic radiology, especially early in pregnancy.
3) Tissue/cell sensitivity: Generally correlates positively with proliferative capacity and inversely with differentiation level, though exceptions exist (e.g., lymphocytes).

8.5 Radiation Protection System and Standards

The core of the radiation protection system is the three fundamental principles: justification of practices, optimization of protection, and dose limitation. Justification is the premise, optimization is the goal, and dose limitation is the constraint.

Among these, the principles of justification and optimization are source-related and apply to all exposure situations, while the principles of dose limits and constraints are individual-related and apply only to planned exposure situations.

8.5.1 Basic Principles of Radiation Protection

(1) Justification of Practices

Any decision that alters exposure conditions must yield more benefit than harm. This means that by introducing new radiation sources, reducing existing exposures, or mitigating potential exposure risks, sufficient individual or societal benefits must be achieved to outweigh the associated harm. The considerations should extend beyond radiation hazards to include other risks, costs, and benefits, as radiation risks may only constitute a small portion of the overall impact. Therefore, justification often goes beyond radiation protection, requiring the selection of the best possible solution, which may exceed the scope of radiation protection authorities.

(2) Optimization of Protection

After accounting for economic and social factors, the likelihood of exposure, the number of exposed individuals, and individual dose levels should all be kept as low as reasonably achievable (ALARA). This implies that, in most cases, the protection level should be optimal, maximizing the net benefit. To prevent inequitable outcomes from the optimization process, dose constraints, risk constraints, or reference levels should be applied to limit individual exposure from specific sources.

Optimization must consider technological and socioeconomic developments, requiring both qualitative and quantitative judgments. The process should be systematic and carefully structured to ensure all relevant factors are considered. However, optimization does not equate to dose minimization; rather, it involves a balanced assessment of radiation risks and available protective resources. Societal impact assessments often influence final decisions on radiation protection levels, incorporating public concerns, ethical considerations, transparency, and stakeholder involvement.

(3) Application of Dose Limits and Constraints

Dose limits apply only to planned exposure situations (excluding medical exposure of patients) and not to emergency exposures. Dose constraints and reference levels, combined with optimization, are used to restrict individual doses. The primary goal is to ensure that individual doses do not exceed these levels, with the ultimate objective of reducing all doses to the lowest reasonably achievable level, considering economic and social factors.

Dose constraints are used for planned exposure (except medical exposures), while reference levels apply to other exposure situations. For the latter, optimization may initially allow doses above reference levels.

The selection of dose constraints or reference levels depends on the exposure context. Importantly, neither dose limits, constraints, nor reference levels represent a clear boundary between "danger" and "safety," nor do they indicate thresholds for changes in individual health risks.

8.5.2 Laws, Regulations, and Standards Related to Radiation Protection

Based on recommendations and publications from international organizations such as ICRP and IAEA, and in consideration of China's national conditions, relevant authorities have successively formulated and issued laws, regulations, guidelines, and technical standards.

At the national level, laws related to ionizing radiation protection include the *Occupational Disease Prevention and Control Law of the People's Republic of China* (2019), the *Radioactive Pollution Prevention and Control Law of the People's Republic of China* (2003), the *Nuclear Safety Law of the People's Republic of China* (2018), and the *Environmental Impact Assessment Law of the People's Republic of China* (2018). Regulations and administrative provisions issued by the State Council

Table 8.4 Dose limits for occupational workers and members of the public in GB 18,871

Application scope		Dose limit	
		Occupational workers	Members of the public
Effective dose		100 mSv over 5 years, averaged over the defined 5-year period; ≤20 mSv/year, with no single year exceeding 50 mSv	1 mSv/year; if the average over 5 consecutive years does not exceed 1 mSv, a single year may slightly exceed 1 mSv
Annual equivalent dose	Lens of the eye	150 mSv	15 mSv
	Skin	500 mSv, averaged over any 1 cm^2 at a nominal depth of 7 mg/cm^2	50 mSv, averaged over any 1 cm^2
	Hands and feet	500 mSv	–

and relevant departments include the *Classification of Radioactive Sources* (2005), *Classification of Radiation Devices* (2017), *Regulations on the Safety Management of Radioactive Waste* (2012), and *Regulations on the Safety and Protection of Radioisotopes and Radiation Devices* (2019).

The national standard *Basic Standards for Protection Against Ionizing Radiation and for the Safety of Radiation Sources* (GB 18,871–2002) was promulgated and implemented in 2002, replacing GB 4792–1984 and GB 8703–1988. Based on the recommendations of ICRP Publication 60 and adapted to China's national conditions, this standard serves as the fundamental framework for ionizing radiation and radiation source safety protection.

GB 18,871 specifies commonly used radiation protection limits in daily work, primarily including dose limits and surface contamination limits. Table 8.4 presents the dose limits for occupational workers and members of the public as stipulated in GB 18,871. The primary purpose of limiting effective dose is to reduce the probability of stochastic effects, while limiting equivalent dose aims to prevent deterministic effects in organs such as the lens of the eye, skin, and extremities. It is important to emphasize that the specified five-year period approved by regulatory authorities must not be retrospectively averaged beyond the original timeframe. Table 8.5 provides surface contamination limits related to occupational work.

8.5.3 Basic Methods of Radiation Protection

Radiation exposure to the human body is categorized into external exposure and internal exposure. External exposure occurs when radiation sources outside the body emit rays or particles that irradiate the body, while internal exposure happens when radionuclides enter the body and decay internally, releasing rays or particles.

Table 8.5 Surface contamination limits

Contaminated surface	α radioactive material (Bq/cm^2)	β radioactive material (Bq/cm^2)
Hands, skin, undergarments, work socks	4×10^{-2}	4×10^{-1}
Work clothes, gloves, work shoes	4×10^{-1}	4
Equipment, floor, walls	4	4×10

Correspondingly, the basic methods of radiation protection are divided into external exposure protection and internal exposure protection.

(1) General Methods for External Exposure Protection

1) Controlling Exposure Time (Time Protection)

Since the dose received is directly proportional to exposure time, longer exposure under the same conditions results in higher doses. Therefore, exposure time should be minimized while meeting operational requirements. Specific methods include: improving proficiency to enhance work efficiency, especially for complex operations, by conducting non-radioactive practice drills beforehand to increase speed and reduce exposure time; implementing shift work for high radiation levels or prolonged operations to limit individual exposure time; and avoiding unnecessary stays near radiation sources.

2) Increasing Distance from the Radiation Source (Distance Protection)

External exposure dose is inversely proportional to the square of the distance from a point source—doubling the distance reduces the dose to one-fourth. Thus, increasing distance significantly lowers exposure. Practical measures include using long-handled tools, robotic arms, or remote-control devices, and ensuring sufficient distance between radiation sources and frequently occupied areas.

3) Using Shielding Materials (Shielding Protection)

While time and distance protection are cost-effective, they have limitations, especially in confined spaces or with high-intensity sources. In such cases, shielding barriers between personnel and radiation sources are necessary. Shielding material selection depends on the radiation type: for γ-rays, materials like water, soil, concrete, iron, or lead are used; for β-rays, lightweight materials like plexiglass with additional heavy material layers are preferred.

(2) General Methods for Internal Exposure Protection

Since internal exposure results from radionuclides entering the body, the core principle is to prevent or minimize their intake. Entry pathways include inhalation, ingestion, and skin absorption, each requiring specific protective measures.

1) Measures Against Inhalation

Prevent air contamination; purify or dilute contaminated air to reduce radionuclide concentrations below permissible levels; use glove boxes or fume hoods for handling radioactive materials; and employ personal protective equipment (PPE).

2) Measures Against Ingestion

Prohibit eating, drinking, or smoking in radioactive work areas; wear gloves when handling radioactive materials and wash hands thoroughly afterward; avoid wearing work clothes in dining or living areas; and prevent contamination of drinking water sources.

3) Measures Against Skin Absorption

Cover cuts or wounds with bandages and wear gloves; avoid using organic solvents for handwashing to prevent increased skin permeability.

Discussion Questions

1) What are natural radiation sources and artificial radiation sources? What are their primary origins?
2) What types of electromagnetic radiation are emitted by electrical appliances?
3) What are the properties of γ-rays, β-rays, α-rays, and neutron radiation? In what ways are they hazardous?
4) What are the three primary effects when γ-rays interact with matter?
5) How is the radioactivity level of radioactive materials expressed? What are its units? How is surface contamination of equipment or floors by radioactive materials measured, and what units are used?
6) What are the commonly used radiation quantities, and how do they differ?
7) Briefly explain the process of ionizing radiation's interaction with cells.
8) What are somatic effects, early/acute effects, and late effects?
9 What are stochastic effects and deterministic effects? What are their key characteristics?
10) What are the biological effect characteristics of long-term, low-dose radiation exposure?
11) What factors influence the biological effects of ionizing radiation?
12) What are the three fundamental principles of radiation protection?
13) What are the general methods for protection against internal and external exposure?

References

Chen WZ, Yu L, Yu R, et al. (2019). Principles of Nuclear Reactor Engineering (Volume II). Beijing, China Atomic Energy Press.
Editorial Board of Registered Nuclear Safety Engineer Training Series. (2004). Comprehensive knowledge of nuclear safety. Beijing, China Environmental Science Press.

Chen YQ, Gong JJ, et al. (2021). Introduction to Nuclear Safety. Beijing, National Defense Industry Press.
Guo J, Zhao XF, Luo PY, et al. (2020). Atomic and Nuclear Physics. Beijing, Science Press.
Wu YC. (2017). Introduction to Nuclear Safety. Hefei: University of Science and Technology of China Press.

Chapter 9
Introduction to Nuclear Safety Culture

Nuclear safety is the lifeline of the nuclear industry's development, while nuclear safety culture is its soul. The level of nuclear safety culture serves as an indicator of the nuclear industry's safety development, impacting the future and fate of practitioners, the sustainable development of the industry, the success of the national nuclear energy "going global" strategy, and even the achievement of national overall security objectives. This chapter introduces the origin, development, definition, and fundamental elements of nuclear safety culture, as well as its role and practical implementation.

9.1 The Origin of Nuclear Safety Culture

The history of nuclear safety development can be divided into four stages: the early days of global nuclear power development, post-Three Mile Island accident, post-Chernobyl accident, and post-Fukushima accident. With the occurrence of the Fukushima accident in 2011, nuclear safety entered a new phase (International Atomic Energy Agency 2014). Reviewing the evolution of nuclear safety concepts helps us understand the historical context of nuclear safety culture and the significance of advocating for it.

9.1.1 Early Days of Nuclear Power Development

This stage was characterized by an emphasis on conservative design and equipment reliability, implementing the principle of defense-in-depth (Fan 2016). In 1942, Enrico Fermi led the construction of the world's first experimental nuclear reactor. To

L. Yu et al., *Nuclear Technology and Safety*, Nuclear Science and Technology,
https://doi.org/10.1007/978-981-95-2856-1_9

prevent uncontrolled chain fission reactions, the reactor was equipped with a neutron-absorbing control rod ready to drop into the core quickly. The issue of controlling nuclear reactions was the initial safety concern.

Notably, Fermi, as the lead engineer of the Manhattan Project, placed exceptional importance on nuclear safety. During the first criticality test of the Fermi reactor, a malfunction occurred. After repairs by the staff, it was already past noon. Fermi decided to have the team take a lunch break before restarting the experiment, easing their tension and excitement, prioritizing safety over schedule. At 3:35 PM, the reactor reached criticality, marking the first controlled chain reaction in human history.

In 1947, the U.S. Reactor Safety Committee discussed a proposal for a sealed containment structure around reactors to prevent radioactive releases during accidents. The concept of containment became a cornerstone of nuclear safety technology.

In 1955, the First Geneva Conference on the Peaceful Uses of Atomic Energy highlighted reactor safety as a key topic. Papers presented at the conference outlined fundamental safety principles for reactor design, containment, and siting, while also drawing attention to off-site radiological consequences.

By 1971, the U.S. Atomic Energy Commission established basic design criteria for light-water reactor safety systems, including hypothetical worst-case accidents. Nuclear plant safety systems were required to handle such accidents without significant radiological consequences. By this point, nuclear safety management had established three cornerstones: siting away from densely populated areas, containment structures, and design-basis accidents. The design-basis accident principle reflected deterministic safety analysis logic, without considering the probability of hypothetical or severe accidents.

By the mid-1970s, probabilistic risk assessment (PRA) matured. The U.S. Congress mandated PRA for nuclear plants, leading to the landmark Rasmussen Report (WASH-1400, "Reactor Safety Study"). This report introduced PRA to reactor safety analysis, classifying accidents by probability and modeling containment failure modes and radionuclide release scenarios. During this stage, nuclear safety management focused on the reliability of technology across design, installation, commissioning, and operation—i.e., design and procedural quality. Designs emphasized conservatism, equipment redundancy, and diversity to prevent and mitigate accidents. Procedurally, all work followed strict protocols. Key principles were established: defense-in-depth, inherent safety, fail-safe design, single-failure criterion, and safety system diversity and redundancy.

9.1.2 Post-Three Mile Island Accident

This phase was characterized by enhanced human–machine interfaces and consideration of severe accident prevention and mitigation.

The Three Mile Island nuclear power plant accident in the United States on March 28, 1979, had a significant impact on the development of nuclear safety history.

The Three Mile Island accident provided valuable lessons for the nuclear industry. It demonstrated that severe accidents could occur, often resulting from a combination of multiple equipment failures and human errors. The accident proved that the defense-in-depth concept in nuclear power plant design remained effective even under severe accident conditions, while also validating the predictions of the *Reactor Safety Study*. The accident led to two major developments: the widespread adoption of probabilistic safety assessment (PSA) technology in the nuclear field, and increased attention to beyond-design-basis accident analysis and containment behavior research.

Nuclear power professionals realized the need to focus on non-technical factors in safety, such as organization, management, procedures, personnel training, communication, public outreach, and emergency preparedness. To prevent and reduce human errors, the following measures were implemented: strengthening operator training; introducing a "safety engineer" position in addition to the operating shift to provide human redundancy during abnormal conditions; periodically using monitoring procedures to supervise core status and take appropriate measures to limit or delay core damage; improving the main control room's human–machine interface with the new "control room" system concept; consolidating essential information into the safety monitoring panel system, with separate terminals for operators and safety engineers; and incorporating severe accident prevention and mitigation research into nuclear safety regulations, standards, and plant improvements to enhance nuclear safety levels.

9.1.3 Post-Chernobyl Accident

This phase was marked by vigorous advocacy for fostering a nuclear safety culture.

On April 26, 1986, a powerful explosion occurred at Unit 4 of the Chernobyl Nuclear Power Plant in the Soviet Union, releasing large amounts of radioactive material into the environment and necessitating mass evacuations. The accident caused widespread public panic and significantly impacted the global development of nuclear power for an extended period.

The primary cause of the Chernobyl accident was severe design flaws in the reactor type, while the direct cause was inadequately planned testing procedures and operational violations by the personnel. However, the root cause traced back to the lack of a safety culture within the Soviet nuclear regulatory authorities. The design flaws of this reactor type were already known, and issues had been identified during the commissioning of similar plants, with reports submitted to the relevant authorities. Instead of addressing these concerns, the authorities disregarded them and even disabled safety protection systems during the entire test that led to the accident.

The nuclear industry conducted profound reflections and summaries of the accident, placing greater emphasis on nuclear safety management. Based on a comprehensive analysis of the causes, the concept of nuclear safety culture was developed. In 1986, the International Nuclear Safety Advisory Group (INSAG) published the *Summary Report on the Post-Accident Review Meeting on the Chernobyl Accident* (INSAG-1), highlighting the critical influence of management and human factors on safety performance (IAEA, 1986). The report introduced the concept of "safety culture" for the first time, pointing out that deficiencies in safety culture—not only during operation but also in other phases of a nuclear plant's lifecycle, including design, manufacturing, construction, and regulation—ultimately led to the Chernobyl disaster.

9.1.4 Post-Fukushima Accident

This phase is characterized by a renewed examination of nuclear safety issues and heightened emphasis on the establishment and maintenance of a nuclear safety culture system.

The Fukushima nuclear accident in Japan on March 11, 2011, shocked the world, once again drawing global attention to nuclear energy and safety issues. In the immediate aftermath of the accident, media and public scrutiny of its progression, panic over radiation hazards, and doubts about nuclear safety undoubtedly placed immense pressure on relevant international organizations, national nuclear regulatory authorities, governments, and nuclear power operators. It also prompted deeper reflection and research into the underlying challenges facing the current nuclear safety framework and policies related to nuclear energy development.

The Fukushima accident propelled China's nuclear safety culture into a new historical phase, marked most notably by the government's high-level emphasis, active advocacy, and vigorous promotion of the cultivation and advancement of nuclear safety culture (National Nuclear Safety Administration 2016). From the initial inclusion of nuclear safety culture development in national nuclear safety plans to the formulation of China's nuclear safety philosophy, and further to the release of the *Nuclear Safety Culture Policy Statement* and *Characteristics of Nuclear Safety Culture*, a top-down framework for robust nuclear safety culture development has been established.

9.2 Definition and Characteristics of Nuclear Safety Culture

"Nuclear safety culture" was proposed by the International Atomic Energy Agency (IAEA) based on lessons learned from the Chernobyl nuclear accident in the Soviet Union and the international consensus that "nuclear safety is the fundamental basis for the advancement of nuclear energy and technology utilization and is essential for world peace and development." It represents a set of shared values and behaviors that transcend national, organizational, and individual traditions to ensure nuclear safety. The introduction of nuclear safety culture has provided countries with different social systems, organizations at various levels, and employees with diverse cultural backgrounds a unified code of conduct for contributing to nuclear safety.

Different countries and organizations have varying definitions of nuclear safety culture, but all are essentially modifications of the IAEA's framework adapted to their national conditions or industry practices. The IAEA's International Nuclear Safety Advisory Group (INSAG) report INSAG-4, *Safety Culture*, provides the classic definition: "Safety culture is that assembly of characteristics and attitudes in organizations and individuals which establishes that, as an overriding priority, nuclear plant safety issues receive the attention warranted by their significance." In its subsequent Safety Report Series No. 11, *Developing Safety Culture in Nuclear Activities*, the IAEA further clarified the essence of safety culture as "a unity of values, standards, ethics, and acceptable norms." Table 9.1 presents definitions of nuclear safety culture by different organizations and countries, along with its characteristics, attributes, and principles.

The 1991 IAEA definition initially targeted nuclear power plants, positioning safety as "receiving warranted attention" (International Atomic Energy Agency 1991). Later, the U.S. NRC extended it to all nuclear-related entities (regulators, manufacturers, etc.), elevating safety to "overriding priority." WANO emphasized leadership modeling, while all agreed on "values and behaviors" as core traits.

9.3 Development of Nuclear Safety Culture

9.3.1 International Development of Nuclear Safety Culture

The term "nuclear safety culture" was first introduced as a management concept by INSAG in the 1986 IAEA safety series report *Summary Report on the Post-Accident Review Meeting on the Chernobyl Accident* (INSAG-1). For convenience, this book summarizes key events and content in the development of nuclear safety culture in Table 9.2.

In 1988, the IAEA further elaborated on nuclear safety culture in its publication *Basic Safety Principles for Nuclear Power Plants* (INSAG-3), emphasizing it as a fundamental management principle. In 1999, INSAG-3 was upgraded to *Basic Safety*

Table 9.1 Definitions of nuclear safety culture by different organizations and countries, and its characteristics, attributes, and principles

No	Definition	Characteristics/principles
1	From IAEA INSAG-4 (1991): safety culture is the assembly of characteristics and attitudes in organizations and individuals that establishes nuclear plant safety as an overriding priority	IAEA's 5 key characteristics (1994): (1) Safety is a recognized value (2) Clear accountability for safety (3) Clear safety responsibilities (4) Safety integrated into all activities (5) Safety drives learning INSAG-15 key elements (2002): (1) Safety commitment (2) Use of procedures (3) Conservative decision-making (4) Reporting culture (5) Challenging unsafe behaviors/conditions (6) Learning organization (7) Communication, clear priorities, and organization
2	From U.S. INPO (2004): *principles for a strong nuclear safety culture*: values and behaviors set by leaders and internalized at all levels, establishing nuclear safety as the overriding priority	INPO (2012 version): (1) Personal accountability (2) Questioning attitude (3) Effective safety communication (4) Leadership safety values/behaviors (5) Decision-making (6) Respectful work environment (7) Continuous learning (8) Problem identification/resolution (9) Environment for raising concerns (10) Work processes

(continued)

Table 9.1 (continued)

No	Definition	Characteristics/principles
3	From U.S. NRC (2011): *safety culture policy statement*: core values and behaviors collectively upheld by leaders and members, prioritizing safety over competing goals to protect public/environment	(1) Leadership safety values/behaviors (2) Problem identification/resolution (3) Personal accountability (4) Work processes ensuring safety (5) Continuous learning (6) Environment for safety concerns (7) Effective safety communication (8) Respectful work environment (9) Questioning attitude
4	From WANO (2006): *traits of a healthy nuclear safety culture*: organizational values/behaviors modeled by leadership and internalized by staff to prioritize nuclear safety	(1) Everyone accountable for safety (2) Leaders demonstrate safety (3) High trust within organization (4) Safety-first decisions (5) Recognize nuclear technology's uniqueness (6) Foster questioning attitude (7) Promote learning organization (8) Regular assessments/supervision
5	CGN (2013): adopted INPO's evaluation system and proposed 10 principles	INPO (2012 version, same as No. 2)
6	CNNC (2010/2015): adopted WANO's 8 principles (2006), later expanded to 10	(1)–(8) Same as WANO (9) Safety-focused communication (10) Harmonious public relations
7	From China's *nuclear safety culture policy statement* (2015): values, behaviors, and traits jointly practiced by stakeholders under the "safety-first" principle to protect public/environment	(1) Leadership safety commitment (2) Management role models (3) Employee participation/accountability (4) Learning organization (5) Comprehensive management system (6) Conducive work environment (7) Questioning/reporting/feedback mechanisms (8) Harmonious public relations

Table 9.2 Milestones in nuclear safety culture development

Year	Key events and content
1986	IAEA proposed the concept of "safety culture."
1988	IAEA emphasized the importance of nuclear safety in *Basic Safety Principles for Nuclear Power Plants*, advocating safety as the top priority and a fundamental management principle
1991	IAEA published *Safety Culture* (INSAG-4), providing an in-depth discussion of the concept, which gained widespread acceptance across industries globally
1994	IAEA developed *Guidelines for Organizational Safety Culture Assessment* (revised in 1996) for evaluating safety culture
1998	IAEA published *Developing Safety Culture in Nuclear Activities: Practical Suggestions for Progress*, outlining three typical stages of nuclear safety culture development
1999	IAEA released *Basic Safety Principles for Nuclear Power Plants* (INSAG-12), describing characteristics and principles of the first stage of safety culture, and *Management of Operational Safety in Nuclear Power Plants*, proposing safety values and goals for the second stage along with implementation methods
2001	IAEA published *Key Practical Issues in Strengthening Safety Culture*, defining goals, features, and methods for the third stage of safety culture development
2011	U.S. NRC issued the *Safety Culture Policy Statement*, elaborating on the philosophy, nature, connotation, role, and requirements of nuclear safety culture

Principles for Nuclear Power Plants (INSAG-12). In November 2006, the IAEA incorporated the main content of INSAG-12 into the *Fundamental Safety Principles* (SF-1), a nuclear safety standard jointly published by nine international organizations, including the United Nations Environment Programme and the World Health Organization. This report outlined basic safety objectives and ten related safety principles. The IAEA asserted that national governments, to fulfill their international nuclear safety obligations, should establish legal frameworks to implement management through legislation, regulation, standardization, and administrative measures, and should also be responsible for creating independent regulatory bodies.

In 1991, to enhance the practical application of nuclear safety culture, the IAEA published the specialized report *Safety Culture* (INSAG-4). INSAG-4 provided an in-depth discussion of the definition, characteristics, and essence of nuclear safety culture, aiming to foster a common understanding among relevant institutions and individuals. The report also detailed the three-tiered requirements of safety culture for decision-makers, managers, and employees, and introduced a series of questions and qualitative "indicators" to measure different levels of safety culture achievement. This gave tangible form to the seemingly abstract concept of "safety culture," offering valuable exploration and guidance for its practical implementation.

INSAG-4 laid the foundation for nuclear safety culture and remains a seminal report in the nuclear energy sector for promoting this concept. As a high-level management principle, nuclear safety culture has been advocated, implemented, and promoted globally, continuously evolving and playing a critical role in achieving excellence in nuclear power plant performance.

In 1994, the IAEA developed the *ASCOT Guidelines* for assessing nuclear safety culture. In 2008, building on these guidelines, the IAEA officially released the *SCART Guidelines* (Safety Culture Assessment Review Team Guidelines, SS-16), systematically outlining the purpose, basis, methods, and process of nuclear safety culture evaluation. These guidelines identified five key characteristics and 37 tangible manifestations of nuclear safety culture, establishing a comprehensive set of evaluation metrics.

In 1998, the IAEA published *Developing Safety Culture in Nuclear Activities* (Safety Reports Series No. 11), which described three stages of nuclear safety culture development: Stage 1, compliance with regulatory requirements; Stage 2, achieving good safety performance as an organizational goal; and Stage 3, continuous improvement of safety performance.

In 1999, the IAEA introduced the nuclear safety management system in *Management of Operational Safety in Nuclear Power Plants* (INSAG-13). This document required senior management of operating nuclear power plants to establish and implement a robust nuclear safety management system to ensure regular discussion, review, and monitoring of safety performance, as well as the continuous improvement of nuclear safety culture.

In 2001, the IAEA published *Key Practical Issues in Strengthening Safety Culture* (INSAG-15), which outlined the goals and characteristics of the third stage of nuclear safety culture development, along with methods and pathways to achieve it.

For many years, the U.S. Nuclear Regulatory Commission (NRC) has emphasized the importance of "safety first" in nuclear energy and technology for public health and safety, as reflected in two early policy statements: the *Policy Statement on Nuclear Power Plant Operations* (54FR3424) issued on January 24, 1989, and the *Freedom of Nuclear Industry Employees to Raise Safety Concerns Without Fear of Retaliation* (61FR24336) issued on May 14, 1996. After nearly a decade of effort, the NRC formally released the *Safety Culture Policy Statement* (NRC-2010-0282) in June 2011.

The 2002 Davis-Besse nuclear power plant incident drew significant attention in the North American nuclear industry. Seizing this opportunity, the Institute of Nuclear Power Operations (INPO) proposed the *Principles for a Strong Nuclear Safety Culture* in 2003. The United Station Alliance (USA) developed nuclear safety culture assessment criteria based on these principles, which, after endorsement by INPO and the Nuclear Energy Institute (NEI), have been implemented across all U.S. nuclear power plants. In 2006, the World Association of Nuclear Operators (WANO) released the *Traits of a Healthy Nuclear Safety Culture* (WANO GL2006-02) under its name.

In February 2006, the IAEA conducted its first nuclear safety culture assessment at South Africa's PBMR. In March 2008, the Japan Nuclear Technology Institute (JANTI) published the "Seven Principles of Nuclear Safety Culture" and their evaluation methods.

9.3.2 The Development of China's Nuclear Safety Culture

Since the construction of China's first nuclear power plant, the Chinese government has placed great emphasis on the research and development of safety culture. In 1991, China began introducing, studying, and promoting safety culture. Since 1993, China has conducted extensive academic discussions on safety culture, including nuclear safety culture and corporate safety culture, drawing attention from the scientific and technological community as well as government decision-makers. In recent years, safety culture has been continuously deepened and enhanced.

The National Nuclear Safety Administration (NNSA) actively advocates and vigorously promotes the development of a nuclear safety culture centered on "safety first, quality first" across all nuclear energy and technology utilization units. During nuclear safety inspection and supervision activities, the construction and evaluation of nuclear safety culture have consistently been prioritized. Since the International Atomic Energy Agency (IAEA) introduced the concept of "nuclear safety culture," relevant nuclear entities in China have conducted extensive exploration and practice guided by its fundamental principles. In recent years, China's nuclear safety regulatory authorities and nuclear power enterprises have organized numerous seminars on nuclear safety culture, thoroughly summarizing and exchanging experiences in its development.

When the Qinshan Nuclear Power Plant commenced commercial operation in 1994, it proclaimed that "safety is the life of a nuclear power plant." Since 1995, the plant has implemented nuclear safety culture education and gradually established a nuclear safety culture system. From the start of the Qinshan Phase II project in 1996, Qinshan II has firmly adhered to the guiding principle of "safety first, quality first," upholding safety culture concepts such as inherent safety, human–machine protection, hierarchical management, personnel safety, conservative decision-making, and supervised operations. Qinshan III has embraced the philosophy that "safety first and quality excellence are our solemn commitment to the public," establishing a comprehensive safety responsibility system, safety assurance system, and nuclear safety supervision system at all levels. Publications such as *The Soul of a Nuclear Safety Enterprise* have been produced to disseminate safety culture knowledge.

From its inception, the Tianwan Nuclear Power Plant set goals and requirements such as "safety first, quality first" and "everyone is a barrier to safety," striving to foster a safety culture atmosphere and actively enhance the safety culture literacy of all staff. Ensuring safety has been upheld as a solemn commitment of nuclear power employees to society. After studying and drawing on the evaluation methods of the U.S. Institute of Nuclear Power Operations (INPO), China National Nuclear Power Co., Ltd. conducted China's first nuclear safety culture assessment at Qinshan Phase I in August 2011, based on the World Association of Nuclear Operators (WANO) *Principles for a Strong Nuclear Safety Culture*. A second assessment was carried out at the Tianwan Nuclear Power Plant in July 2012.

From the very beginning, the Daya Bay Nuclear Power Plant proactively adopted safety and quality concepts from international nuclear peers, gradually establishing a safety management system and framework aligned with global standards. It actively advanced safety culture construction under the approach of "leadership demonstration, core team promotion and permeation, and full staff participation." Under the overarching goal of "safety first, quality first," the plant expanded and refined localized core safety culture principles based on its practical safety management conditions and characteristics. Among these, the concept that "nuclear safety overrides all" serves as the central tenet of Daya Bay's safety culture policy and management commitment. Conservative decision-making, the transparency principle, and the "Four Alls" guide the plant's safety culture management commitments. Meanwhile, principles such as "follow procedures," "do it right the first time," "everyone is a barrier," and "STAR (Stop, Think, Act, Review) for field operations" reflect the individual commitments of Daya Bay's safety culture. In recent years, building on advanced international management concepts and operational experience, the Daya Bay Nuclear Power Plant has actively advanced the "Two Civilizations" (material and spiritual progress) development tailored to China's national conditions. It has closely integrated ideological and political work with safety culture construction, taking into account the country's actual conditions, cultural and historical context, industry-specific characteristics, and the sensitivity of Daya Bay's proximity to Hong Kong. By combining modern management systems with operational feedback from pressurized water reactor nuclear power plants, the plant has developed a nuclear safety culture that meets its unique needs and has published *Good Practices in Safety Culture at Daya Bay Nuclear Power Plant*. On March 24, 2014, the "rational, coordinated and balanced" nuclear safety concept expounded by the President Xi elevated nuclear safety to a new level, and the development of China's nuclear safety culture entered a new stage (Jinping 2014).

With the rapid development of China's nuclear industry, the exploration and construction of nuclear safety culture have further accelerated. In January 2015, the NNSA, the National Energy Administration, and the State Administration of Science, Technology, and Industry for National Defense jointly issued the *Nuclear Safety Culture Policy Statement* for the first time. This document aims to clarify China's fundamental stance on nuclear safety culture and outline the principles for cultivating and practicing it. Building on 30 years of sound practices in nuclear safety culture development, the statement elucidates the essence of nuclear safety and nuclear safety culture, elaborates on eight characteristics of fostering and practicing a robust nuclear safety culture, and proposes initiatives for its sustained advancement.

9.4 The Status and Role of Nuclear Safety Culture

9.4.1 The Relationship Between Nuclear Safety Culture and Nuclear Safety

Nuclear safety refers to the necessary and sufficient monitoring, protection, prevention, and mitigation measures taken for nuclear facilities, nuclear activities, nuclear materials, and radioactive substances to prevent accidents caused by any technical reasons, human errors, or natural disasters, and to minimize the radiological consequences in the event of an accident, thereby protecting workers, the public, and the environment from undue radiation hazards.

Nuclear safety culture refers to the sum of values, behavioral norms, and characteristics that are agreed upon and put into practice by relevant organizations and individuals. It takes "safety first" as its fundamental principle and aims to safeguard public health and environmental safety as its ultimate goal.

Ensuring nuclear safety is the fundamental purpose of fostering nuclear safety culture, while nuclear safety culture serves as the foundation of nuclear safety, an important barrier in the "defense-in-depth" system of nuclear safety, and a summary and refinement of practical experience in nuclear safety.

To ensure the safety of nuclear power, nuclear power plants adopt a defense-in-depth approach, which includes multiple physical barriers as well as comprehensive management barriers. In addition, through the development of nuclear safety culture, the awareness and management practices of prioritizing "nuclear safety above all" are enhanced, which is also a critical component of nuclear power plant management. Ensuring nuclear safety is not only a technical issue but also a management issue, and more importantly, a cultural issue.

Nuclear safety culture serves as the ideological barrier and spiritual defense line for the safe development of nuclear power. It is the fundamental value and behavioral norm of the nuclear energy industry, as well as the professional quality and action guide for all practitioners. Nuclear safety culture emphasizes rigorous work styles, prudent attitudes, and continuous improvement. Being mindful of dangers ensures safety; reflecting on disorder brings order; contemplating decline leads to prosperity; and recognizing the risk of demise ensures survival. Nuclear safety efforts must never be relaxed at any time.

9.4.2 The Necessity of Nuclear Safety Culture

(1) Fostering Nuclear Safety Culture Is an Essential Requirement of Nuclear Safety

According to INSAG-4, safety culture refers to "the dedication and sense of responsibility of all individuals involved in activities related to the nuclear safety of nuclear power plants." A further explanation summarizes it as a key phrase: a mindset fully imbued with "safety first." This mindset implies "an inherent questioning attitude, humility and prudence, continuous improvement, and encouragement of personal responsibility and collective self-enhancement in nuclear safety matters."

Nuclear safety has five characteristics: technical complexity, suddenness of accidents, imperceptibility of impacts, difficulty in eliminating contamination consequences, and extreme sensitivity of the public. These characteristics determine that nuclear safety plays a crucial role in the development of the nuclear energy and nuclear technology utilization industry and even in national security as a whole. It can be said that nuclear safety is vital to career development, public interests, social stability, and the nation's future. Therefore, the principle of "safety first" must always be upheld. Nuclear safety culture embodies the mindset, attitude, and conduct of respecting and maintaining nuclear and radiation safety. It is a shared consensus and action guide for all practitioners in the nuclear energy and nuclear technology utilization field. Establishing the core principle of "safety first" through fostering nuclear safety culture is precisely the essential requirement of nuclear safety.

(2) Fostering Nuclear Safety Culture Is an Important Safeguard for the Development of the Nuclear Industry

Currently, the nuclear energy and nuclear technology utilization industry is developing rapidly, and the need for nuclear safety culture development is becoming increasingly urgent. In the nuclear power sector, with the rapid expansion of nuclear power, the demand for nuclear professionals is also growing. The influx of non-nuclear professionals and the outflow of operational personnel to nuclear power plants under construction have, to some extent, diluted and depleted the pool of nuclear safety experts, posing a risk of weakening nuclear safety culture. In the nuclear technology utilization sector, the lack of nuclear safety culture is severe, and awareness of radiation protection is weak. Particularly in small-scale nuclear technology utilization units, safety and responsibility awareness is poor, and radiation accidents such as loss of radioactive sources occur frequently. In areas such as nuclear equipment manufacturing and the nuclear fuel cycle, there is also insufficient cultivation of nuclear safety culture, leading to repeated violations such as unauthorized repairs and failure to follow procedures.

"Nuclear safety is the lifeline of the nuclear energy and nuclear technology utilization industry." In the course of its development, the absence or weakening of nuclear safety culture poses hidden risks to nuclear and radiation safety. Therefore, in October 2012, the State Council issued the "12th Five-Year Plan for Nuclear Safety and Radioactive Pollution Prevention and Control and the Long-Term Objectives for 2020," explicitly requiring the "establishment of a nuclear safety culture evaluation system and the conduct of nuclear safety culture assessments; strengthening the primary responsibility of nuclear energy and nuclear technology utilization enterprises and institutions for safety; vigorously fostering nuclear safety culture and enhancing the sense of responsibility among all staff, so that decision-makers, managers, and implementers at all levels can regard ensuring nuclear safety as a conscious action." Thus, fostering nuclear safety culture is an important safeguard for the development of the nuclear energy and nuclear technology utilization industry under the current circumstances.

(3) Fostering Nuclear Safety Culture Is an Effective Measure to Reduce Human Errors

The IAEA pointed out in INSAG-4: "Apart from what may be termed 'acts of God,' any problems in nuclear power plants are in some way attributable to human error. However, human ingenuity is highly effective in identifying and eliminating potential problems, which has a positive impact on safety. For this reason, individuals bear significant responsibility." Therefore, human factors play a crucial role in nuclear and radiation safety work.

On the one hand, the biggest difference between humans and mechanical systems is that "human reliability is poor." To address potential human errors, the nuclear safety quality assurance system was developed and implemented first. However, practice has shown that nuclear safety quality assurance has certain limitations—it does not account for irrational human "mistakes" or "violations," nor does it solve the problem of ensuring that people act correctly. Fostering nuclear safety culture aims to compensate for the shortcomings of nuclear safety quality assurance by cultivating a universal, recurring, and relatively stable behavioral and psychological state conducive to nuclear safety in critical nuclear safety activities, thereby reducing nuclear safety issues caused by human errors.

On the other hand, "existence determines consciousness, and consciousness in turn positively influences existence." Human ingenuity can also play an active role in identifying and eliminating potential problems. Advanced nuclear safety culture is a valuable asset created through long-term nuclear and radiation safety practices. It is a cultural form that embodies the essential characteristics of nuclear and radiation safety practices and serves as fertile ground for improving the quality of nuclear and radiation regulators and nurturing the minds of practitioners. By fostering nuclear safety culture, the positive role of humans in nuclear energy and nuclear technology utilization can be better leveraged, and the impact of human-related issues can be reduced.

Nuclear safety culture with Chinese characteristics is an advanced management culture rooted in the most representative socialist large-scale production (highly integrated nuclear power technology). It is a scientific, national, and people-oriented culture, an important component of ecological civilization construction, and a socialist advanced culture built on the guiding principles of socialist spiritual civilization.

9.5 Composition and Requirements of Nuclear Safety Culture

9.5.1 The Framework and Stages of Nuclear Safety Culture

Nuclear safety culture encompasses both intangible aspects, such as the unwavering adherence to the principle of "safety first" at the conceptual level, and its tangible manifestations. The tangible manifestations of nuclear safety culture, or the expressions of safety culture, consist of two main components. The first is the safety system established through organizational policies and management activities, and the second is the performance of individuals within this system. The success of nuclear safety culture depends on the commitment and capabilities of both these aspects regarding safety. This underscores that safety culture is both a matter of attitude and a systemic issue, involving both the organization and individuals.

IAEA's *Safety Culture* (INSAG-4) presents an overall framework for an organization's safety culture, as shown in Fig. 9.1. It involves the commitment to safety at the policy level, among managers, and by individuals. For an organization to develop an excellent safety culture, all three must demonstrate outstanding performance in addressing all safety-related matters.

Safety culture is the collective manifestation of individual behaviors, with decision-makers playing a pivotal role (Chen et al. 2021). It can be said that safety culture primarily reflects the safety philosophy of leadership, to some extent materializing their will and attitudes. The cultivation of an organization's safety culture requires leaders to lead by example. Only when leadership correctly understands

Fig. 9.1 Overall framework of safety culture

the essence and principles of safety culture can they effectively implement safety practices in daily operations—ensuring that management at all levels takes the lead in executing procedures, vigorously promotes safety awareness, and fosters a cultural atmosphere where safety is prioritized in mindset, embedded in procedures, and adhered to in execution. For every individual employee, in addition to strictly following established procedures, they are also required to act in accordance with "safety culture," which entails cultivating a questioning attitude, adopting rigorous work methods, and developing habits of open communication.

An organization at different stages of development requires corresponding organizational cultures to adapt. In its safety report *Developing Safety Culture in Nuclear Activities: Practical Suggestions for Progress*, the IAEA proposed three stages of safety culture development. In the initial stage, safety is perceived as an external requirement, with little awareness of the impact of individual behavior and attitudes on safety. Safety is largely viewed as a technical issue, where compliance with rules and regulations is considered sufficient. In the intermediate stage, an organization consciously recognizes safety as an important organizational goal, gradually paying attention to behavioral performance, spontaneously emphasizing safety performance, and beginning to explore management measures to achieve desired objectives. In the advanced stage, safety is seen as perpetually improvable, with the organization establishing a philosophy of continuous improvement applied to safety, emphasizing communication, training, management models, and enhancing efficiency and effectiveness. Cultural development is a gradual process, and a nuclear organization cannot expect to skip the first two stages and directly enter the third.

9.5.2 Requirements of Nuclear Safety Culture

In all activities of organizations at all levels and individuals, the emphasis on safety is reflected in the following aspects: First, individuals' awareness of the importance of safety; second, the knowledge and abilities acquired through training, education, and self-study; third, the commitment of senior management, who demonstrate the high priority of safety through actions and require that safety goals be recognized and accepted by individuals; fourth, the incentive mechanisms to boost enthusiasm through leadership, goal-setting, reward and punishment mechanisms, and individuals' spontaneous attitudes; fifth, effective monitoring; sixth, a clear authorization and responsibility system.

Specifically, the requirements of nuclear safety culture for organizations and individuals are reflected in the requirements for personnel at different levels within the organization, specifically including three levels: the decision-making level, the management level, and the execution level.

(1) Decision-Making Level

Whether at the government or organizational level, the policies implemented by the decision-making level create the working environment and govern everyone's behavior. The requirements for the decision-making level are as follows:

1) Publishing Safety Policies

All nuclear safety-related organizations must issue safety policy statements, making their responsibilities public and ensuring everyone understands them. This statement serves as a guide for all staff and declares the organization's operational goals and the management's public commitment to nuclear power plant safety.

2) Establishing a Scientific Management System

Safety factors must be fully considered in policy formulation, organizational structure, resource allocation, and infrastructure development. The government should establish a sound, scientific, and rational system, strict regulatory mechanisms, efficient review models, and comprehensive supervision and inspection procedures. Operating organizations must first have clear accountability in safety matters. This requires defining responsibilities in writing, establishing clear reporting channels, simplifying interfaces as much as possible, and ensuring extremely clear authority among all units involved in nuclear power plant safety. Any considerations regarding plans, schedules, or costs must not override safety, and continuous evaluation and optimization should be conducted to improve safety standards. Within organizations significantly impacting nuclear power plant safety, an independent safety management department should be established to oversee nuclear safety activities.

Additionally, all nuclear-related organizations should foster harmonious public relations. Through public communication methods such as information disclosure, public participation, and science popularization, they must ensure the public's right to know, participate, and supervise.

3) Providing Human and Material Resources

The decision-making level must ensure sufficient human and material resources for safety, particularly by having enough experienced staff, supplemented by necessary consultants or contractors. A scientific personnel management system should be established to ensure capable individuals are promoted to key positions in a timely manner. Sufficient training personnel and funding must be guaranteed. All employees must be equipped with necessary equipment, devices, and technical means when performing safety-related work. To ensure effective performance, employees' working conditions must be favorable.

(4) Continuous Self-Improvement of the Decision-Making Level

As part of safety management policy, managers at all levels should conduct regular reviews of nuclear power plant safety-related work. Review content mainly includes personnel arrangements, fostering a learning organization, operational experience feedback, and the management of design changes, plant modifications, and operating procedures.

(5) Commitment of the Decision-Making Level

The above requirements for the decision-making level must be publicly announced and widely known. These commitments demonstrate the company's stance on social responsibility and its sincere willingness regarding safety.

Top executives should personally affirm their commitment by: paying attention to regular reviews of nuclear safety-related processes; directly addressing issues significantly impacting nuclear safety and product quality; and frequently emphasizing the importance of safety and quality to employees. Notably, nuclear power plant safety should be a key agenda item in top-level meetings.

(2) Management Level

The management level of a nuclear power plant is primarily responsible for the concrete implementation of corporate safety policies and objectives. The specific requirements for their safety responsibilities are as follows:

1) Clear Division of Responsibilities

A unique and clear authorization system ensures that everyone understands their own responsibilities as well as those of their superiors and subordinates.

2) Arrangement and Management of Safety Work

Department managers must ensure that all nuclear safety-related tasks are completed to high standards. To ensure work proceeds according to regulations, managers should establish a supervision and management system that emphasizes orderly production. Work arrangements should provide employees with reasonable working hours and intensity, foster a work atmosphere of mutual respect, high trust, and teamwork, and resolve conflicts objectively and fairly.

Managers should also encourage a rigorous and questioning attitude toward safety issues; establish a mechanism ensuring employees can freely report safety concerns without fear of discrimination or retaliation; and promptly respond to and reasonably resolve potential problems and safety hazards reported by employees.

3) Personnel Qualification Review and Training

Department managers must ensure every employee is fully competent for their assigned tasks. Recruitment and appointment procedures should guarantee that staff meet satisfactory intellectual and educational standards initially; otherwise, safety could be compromised. However, specific practices must be handled carefully, and penalties should not lead to the concealment of errors.

4) Monitoring, Review, and Benchmarking

In addition to implementing quality assurance measures, department managers are responsible for enforcing a comprehensive set of monitoring or supervision measures, such as regular reviews of training programs, personnel appointment procedures, work methods, document management, and the quality assurance system. Furthermore, internal key performance indicators can be compared with external or other nuclear power plants' performance indicators to assess safety performance.

5) Commitment

Through the above measures, department managers should not only demonstrate their commitment to safety through actions but also enhance employees' safety awareness.

(3) Execution Level

The execution level primarily includes frontline management personnel and operators who are directly involved in specific tasks, particularly those related to nuclear safety. Therefore, the requirements for them are more detailed.

1) Questioning Attitude Toward Work

A questioning attitude, also called an "exploratory spirit," is a trait shared by all high performers in nuclear safety incidents. It requires employees to always ask why and never overlook any clues.

2) Rigorous Work Methods

Everyone must adopt a rigorous work approach, which includes: understanding and comprehending work procedures; following procedures strictly; remaining vigilant about unexpected situations; pausing to think when problems arise; seeking help when necessary; pursuing discipline, timeliness, and organization; working cautiously; and avoiding shortcuts.

3) Good Communication Habits

Everyone must recognize that good communication habits are crucial for safety, including: obtaining relevant information from others; providing information to others and maintaining transparency; reporting completed work results; identifying and reporting any abnormalities; accurately documenting work records, whether under normal or abnormal conditions; and proposing new measures to improve safety while valuing experience feedback.

9.6 Good Practices of Nuclear Safety Culture

9.6.1 Experience Feedback

Experience feedback involves analyzing and summarizing operational experiences and lessons from nuclear power plants to identify precursors that may compromise safety, thereby enabling necessary corrective actions before serious situations arise. This practice is a concrete manifestation of nuclear safety culture requirements and has been gradually adopted worldwide in the nuclear power sector following the Three Mile Island and Chernobyl accidents. Organizations established for this purpose, such as the Institute of Nuclear Power Operations (INPO) in the U.S. and the World Association of Nuclear Operators (WANO), have contributed significantly to theoretical innovation, practical application, and experience summarization in experience feedback.

China's nuclear safety regulatory authorities attach great importance to experience feedback and have carried out numerous effective practices. For operating nuclear power plants, the *Safety Supervision of Nuclear Power Plants* (Implementation Rules No. 2 of the Regulations on the Safety Supervision and Management of Civil Nuclear Facilities) was issued in 1988, establishing an abnormal event reporting system for nuclear power plants. In 1995, the *Reporting System for Nuclear Power Plant Operators* was formulated and released, clearly defining the criteria, procedures, content, and format for operational event reports to improve reporting efficiency and facilitate data collection on operational events. In 2004, the National Nuclear Safety Administration (NNSA) revised the *Regulations on the Operational Safety of Nuclear Power Plants*, providing clearer and more specific requirements for operators in event analysis and experience feedback, including experience evaluation, research, and international information sharing. In 2012, the NNSA issued the *Management Measures for Experience Feedback in Operating Nuclear Power Plants* (Document No. 65 [2012]), establishing the framework for China's experience feedback system, clarifying the roles and responsibilities of participating entities, and laying the foundation for the system's development.

As a management tool, experience feedback has been widely applied in nuclear power plant operations both in China and abroad, effectively enhancing operational safety. Nuclear power plants such as Daya Bay, Qinshan, and Tianwan have each established their own operational experience feedback systems.

9.6.2 Human Error Prevention Tools

In addition to experience feedback, the application of human error prevention tools is another tangible manifestation of nuclear safety culture. To prevent human errors through management measures, the nuclear power industry has developed a series of such tools based on years of operational and management experience. China's nuclear power groups have developed similar yet slightly different tools tailored to their specific characteristics. Widely used human error prevention tools include:

(1) STAR Self-Check Operation Method

The STAR (Stop, Think, Act, Review) method embodies the rigorous work approach required by nuclear safety culture for the execution level. When encountering any abnormality or safety issue, the STAR method requires:

1) Stop: Pause, calm emotions, and avoid rash actions to prevent misoperation or misjudgment. This is the most critical step of self-checking—seemingly simple yet significantly enhancing operational success and accuracy.
2) Think: Analyze the phenomenon and causes; determine the next action; assess whether external assistance is needed; identify all operation-related information; question the current environment; envision expected responses (e.g., changes in flow, pressure, water level, indicator lights, noise); plan contingency measures

for unexpected situations; and decide on actions if expected responses fail to occur.

3) Act: Take correct intervention measures and actions while visually monitoring changes in instruments or indicator lights for expected responses.
4) Review: Immediately after operation, inspect and confirm results; analyze and evaluate; verify whether actual outcomes align with expectations.

(2) Supervised Operation

Supervised operation involves one operator performing tasks while another monitors and confirms simultaneously. The steps are:

1) The operator announces the operation command verbally and completes preparatory actions (e.g., hand selecting the correct button switch, eyes observing the target instrument or indicator light).
2) The supervisor confirms the correctness of the designated equipment and completeness of preparations, then verbally authorizes the operator to proceed.
3) The operator repeats the supervisor's command, executes the operation, and reports observed information to the supervisor for joint confirmation of correctness.

Verbal commands play a crucial role, enabling the supervisor to oversee the entire process. Particularly before potential human errors emerge, the supervisor can intervene preemptively. For the operator, the supervisor acts as an additional safety barrier.

(3) Three-Way Communication

Three-way communication ensures accurate information transfer through three steps: sending, repeating, and confirming. While nuclear safety quality assurance relies on document-based management, practical experience shows oral communication must also be controlled. Hence, "three-way" communication is widely adopted in operating nuclear power plants.

(4) Procedure Adherence/Usage

Procedure usage means strictly following documents, procedures, and instructions. From a safety culture perspective, additional requirements apply to procedure users, such as classified management methods for procedure usage and rules for suspending procedures.

(5) Pre-Job Briefing

To ensure successful task completion, a pre-job briefing clarifies work-related issues, deepens understanding, and enhances preparation. Key focuses include:

1) Defining responsibilities for each role;
2) Preparing theoretical knowledge per role; if time allows, conducting public "duty statements" (oral descriptions of responsibilities and workflows) under peer supervision to ensure accuracy and completeness;

3) Jointly discussing critical steps, safety precautions, countermeasures, and evaluating the effectiveness of preventive measures and contingency plans.

Practice shows pre-job briefings are comprehensive "ritualistic" preparations covering personnel, technical, and managerial aspects to minimize human errors.

(6) Post-Job Debriefing

Post-job debriefing involves post-work summarization to identify shortcomings and implement improvements. It focuses on in-depth analysis of errors and issues, treating them as case studies for experience feedback to prevent recurrence. Both pre- and post-job briefings also cultivate good communication habits—a key nuclear safety culture requirement for the execution level.

(7) Questioning Attitude

A questioning attitude prompts operators to focus on uncertainties and potential risks, avoiding literal interpretation of ambiguous anomalies. It demands distinguishing truth from falsehood, digging deep into root causes, never overlooking clues, resisting conformity, and embracing the spirit of "making a fuss over minor issues" in nuclear safety. Specific methods:

1) Verify information source: Is it relevant to my work? Reliable and accurate? Any doubts?
2) Self-validate information: Is it reasonable? What are my expectations? Does it align with past experiences and related information?
3) Independently corroborate: Can I obtain confirmation from an independent, reliable channel?

Cultivating this attitude prevents complacency and keeps operators alert to any anomaly.

(8) Work Handover

Work handovers ensure sufficient and accurate information transfer using standardized methods. Handover locations should facilitate discussion and be sufficiently close to worksites for prompt action. Handovers occur between employees, teams, departments, and different units.

Discussion Questions

1) What is nuclear safety? What is the importance of nuclear safety culture? Please briefly describe.
2) What is the historical background of nuclear safety culture? Please briefly describe.
3) Different countries and organizations in the international community have varying definitions of nuclear safety culture. List 3–5 examples and explain their commonalities and key emphases.
4) In which year did China begin introducing, researching, and promoting safety culture?

5) Briefly describe the necessity of nuclear safety culture.
6) What are the five characteristics of nuclear safety?
7) What are the specific requirements of nuclear safety culture for the decision-making level, management level, and execution level? Please briefly describe.
8) Discuss your overall understanding of nuclear safety culture.

References

Chen YQ, Gong JJ, et al. (2021). Introduction to Nuclear Safety. Beijing, National Defense Industry Press.

Fan YM. (2016). Brief History of Nuclear Reactor Safety Evolution. Beijing, China Atomic Energy Press.

International Atomic Energy Agency. (1986). Summary Report on the Post-Accident Review Meeting on the Chernobyl Accident: (75-INSAG-1).

International Atomic Energy Agency. (1991). Safety Culture (INSAG-4).

International Atomic Energy Agency. (2014). International Basic Safety Standards for Protection against Ionizing Radiation and for the Safety of Radiation Sources. Vienna.

National Nuclear Safety Administration. (2016). Safety Requirements for Nuclear Power Plant Design, HAF102.

Xi Jinping. (2014, March 25). Speech at the Nuclear Security Summit in The Hague. People's Daily.

MIX
Papier aus verantwortungsvollen Quellen
Paper from responsible sources
FSC® C105338

If you have any concerns about our products,
you can contact us on
ProductSafety@springernature.com

In case Publisher is established outside the EU,
the EU authorized representative is:
Springer Nature Customer Service Center GmbH
Europaplatz 3, 69115 Heidelberg, Germany

Printed by Libri Plureos GmbH
in Hamburg, Germany